Cyberfeminism and Artificial Life

'A wonderful work which, for the first time ever, balances an even handed weaving together of A-Life and cognitive science alongside evolutionary biology, cultural studies and feminist theory. Sarah Kember's adroit mastery of such an astonishing range of sources is an incredible achievement – not least because she is equally at home with all of them and, importantly, has excellent powers of exposition to reach all of these audiences.'

Alison Adam, Salford University

Cyberfeminism and Artificial Life examines the construction, manipulation and redefinition of life in contemporary technoscientific culture. The book takes a critical political view of the concept of life as information, and traces it through the new biology and the discourse of genomics incorporating the changing discipline of Artificial Life and its manifestation in art, language, literature, commerce and entertainment. Using examples from cloning to computer games, and incorporating an analysis of hardware, software and 'wetware', Sarah Kember demonstrates how this relatively marginal field connects with – and connects up – global networks of information systems.

Refocusing concern on the ethics, rather than the 'nature' of life-as-it-could-be, Kember proposes that Artificial Life is in part an adaptation to the climate of opposition surrounding Artificial Intelligence. From a feminist perspective, and with a set of concerns related to the role of the body, the self and the species in the production of life-as-it-could-be, Kember points to a strategy for change that rests on a dialogue between 'nature' and 'culture', ontology and epistemology, science and the humanities.

Sarah Kember is a senior lecturer in the Department of Media and Communications at Goldsmiths College, University of London. She is the author of *Virtual Anxiety. Photography, New Technologies and Subjectivity*, 1998.

Cyberfeminism and Artificial Life

Sarah Kember

Routledge
Taylor & Francis Group

LONDON AND NEW YORK

First published 2003
by Routledge
11 New Fetter Lane, London EC4P 4EE

Simultaneously published in the USA and Canada
by Routledge
29 West 35th Street, New York, NY 10001

Routledge is an imprint of the Taylor & Francis Group

© 2003 Sarah Kember

Typeset in Galliard by
BOOK NOW Ltd
Printed and bound in Great Britain by
The Cromwell Press, Trowbridge, Wiltshire

All rights reserved. No part of this book may be reprinted or
reproduced or utilised in any form or by any electronic,
mechanical, or other means, now known or hereafter
invented, including photocopying and recording, or in any
information storage or retrieval system, without permission in
writing from the publishers.

British Library Cataloguing in Publication Data
A catalogue record for this book is available
from the British Library

Library of Congress Cataloging in Publication Data
A catalog record for this book has been requested

ISBN 0-415-24026-3 (hbk)
ISBN 0-415-24027-1 (pbk)

Contents

Preface		vii
Acknowledgements		xi

1 **Autonomy and artificiality in global networks**   1
   ALife in context, or, the 'return to Darwin'   2
   Network bioethics   7

2 **The meaning of life part 1: the new biology**   14
   Evolutionary biology – all about chickens and eggs   14
      The metaphor of life as information   16
      Biology as ideology – reductionism and determinism   19
   Molecular biology – an endgame   23
   Sociobiology and eugenics   31
   Evolutionary psychology and the 'Darwin Wars'   40

3 **Artificial Life**   53
   'Information Wants to be Alive!'   60
   The philosophy and biology of ALife   63
      ALife's autonomous agents   66
      ALife's non-vitalist vitalism   71
      Spaces of dissent   71
   The future of ALife – consolidating (digital) naturalism   78

4 **CyberLife's *Creatures***   83
   SimWorlds – ALife and computer games   85
      Stirring the primordial soup   85
      The Creation part 1: making worlds   86
      The Creation part 2: making life   88
      Cain's creation   89
   Creatures   91
      Playing the game   95
      Creatures on the Internet   98

      CyberLife – selling ALife   105
      CyberLife Research Limited – or 'real' ALife   109

5   **Network identities**   116
      Artificial agents   116
          HAL   116
          Situated and autonomous robots   118
          Bots   124
          ALife aesthetics   127
      Artificial cultures   133
      Artificial societies   138
      Artificial subjects   143

6   **The meaning of life part 2: genomics**   145
      Artificial life as wetware   145
      The species-self   150
          Confessions of a justified sinner   150
      The self as other   160
      The other species   169

7   **Evolving feminism in Alife environments**   175
      Alife-as-we-know-it   175
      Natureculture   182
      Risk   188
      Pengi and the Expressivator   193
      Alife-as-it-could-be   198
      Autopoiesis and autonomy   198
      Embodying Alife   204
      Towards situating Alife   206

8   **Beyond the science wars**   211

      Notes   217
      Bibliography   227
      Index   243

Preface

This book aims to do two things: to trace the development of identities and entities within the global information network encompassing both human and non-human environments, and to offer a pluralised cyberfeminist engagement with artificial life as both a discipline and cultural discourse. As a cultural discourse, artificial life (alife) is more than a development of artificial intelligence (AI) which seeks similar goals by means of inverse methods and philosophies. As a cultural discourse, artificial life is both a descriptor of posthuman life-as-we-know-it and a predictor of posthuman life-as-it-could-be. The posthuman is cyborgian in the sense of its enmeshment, at all levels of materiality and metaphor, with information, communication and biotechnologies and with other non-human actors. The two terms are, however, not synonymous and while they describe a similar ontology (a hybridisation of organic and inorganic forms and processes) and epistemology (a transgression of the boundaries sustaining Modern Western thought, principally those of nature and culture), they do not necessarily share a politics, history or ethics. The discourse of artificial life is informed if not contained by a discipline which developed precisely at the end of the cold war and which rejected the militarist top-down command and control and the masculinist instrumental principles of AI. The cyborg which Donna Haraway (1991a) so astutely parodied in her manifesto was the product of cold war AI. Out of this comes a new discipline based on the principles of decentralised distributed control, bottom-up self-organisation and emergence. These are at once technical principles relating to the development of embodied intelligence and, obviously, socially and historically contingent political principles governing individuals and communities. The posthuman which I wish to explore here is a product of post-cold-war ALife and it has at its heart (or soul) a fundamental anti-instrumentalism. The posthuman which the discourse of artificial life both describes and prescribes is, to a large extent, posthumanised, and as such demands a bioethics of posthumanism which is yet to be articulated. Chris Hables Gray (2002) has outlined a cyborg bill of rights (an amendment of the US Constitution) in his suggestive exploration of cyborg citizenship. But where this is based on a model of human agency, the posthumanist bioethics which is sought here is emergent in the inter- or 'intra-actions' (Barad 2000) of network (id)entities or artificial life-forms manifested in software (as computer programs), hardware (as robots) and wetware (as bio- and genetically engineered organisms). It is customary, in the short history of debates on artificial life to restrict the analysis to forms of software and hardware and to the language and discourse which generates and is generated by these forms. This book breaks with that custom by emphasising the dual constitution of artificial life as both computer science and biology, and by highlighting the increasing hegemony of biological discourse within the discipline of ALife and within technoscientific culture generally (Haraway 2000). Embodied computer programs, situated autonomous robots and transgenic organisms co-exist within the global network as kin, sharing the 'bodily fluids' (Haraway 1997) or the life-blood of information. Moreover, these entities or beings are provisional, experimental, in the process of becoming. They can only indicate, but they do indicate some key parameters of posthuman identity. They are both beings in themselves and a means of working out what may be and who may be in the future. They are the literal manifestations and the creative imaginings of artificial life which is, by definition, at the boundary between nature and culture.

To the distant gaze of cyberfeminism, artificial life seems to retreat from this precarious boundary position and engage in a form of renaturalisation by creating artificial entities in artificial worlds governed solely by Darwinian evolutionary principles. In the convergence between biology and computer science, it is the biology which speaks more loudly, and seems to speak more clearly to the feminist. From a distance, Artificial Life certainly looks like 'sociobiology in computational clothing' (Adam 1998) and Alison Adam was right to point out the dangers of a discourse which subsumes cultural into biological explanations of life. But the challenge for cyberfeminism – which, like other aspects of feminism, has had to interrogate its own relation to biological discourse through the problems of universalism and essentialism – is to recognise the plurality of positions which simultaneously undermine and strengthen not just its own case but that of its supposed adversary. There is no 'biology' any more than there is a homogenous 'feminism'. What fails immediately on the recognition of non-homogenous discourses is the possibility of resistance or opposition. Resistance and opposition are only rhetorics in the face of an enemy which is neither unified nor static. The 'science wars', based on the history of two cultures competing for a singular idea of value (Snow 1998 [1959]) and perpetuated by old fashioned institutional rivalry and insecurity (Gross and Levitt 1998 [1994]) are nothing if not futile. This is shown most clearly in the debates on evolutionary psychology which have both a direct and indirect bearing on the field of Artificial Life; direct in that as ALife takes on the synthesis of more complex life-forms it is seeking a complementary theory of mind, and indirect in that evolutionary psychology is another clear and contemporaneous example of what Segal (1999) terms 'the return to Darwin'. Debates for and against evolutionary psychology are currently locked in a bitter stalemate despite the fact that they are not clearly held within disciplinary boundaries. In fact, arguably the strongest critiques of evolutionary psychology, like those of its ancestor sociobiology, come from within the sciences (Rose and Rose 2000; Gould 2000; Rose *et al*. 1984). Where does this leave a cyberfeminism no longer safe within the confines of a strangely familiar and boyishly banal disembodied cyberspace? The conditions for dialogue are already set up within the heterogenous fields of the 'new' biology – that which is interwoven with information systems – and I argue for a strategic dialogue between cyberfeminism and artificial life which does not seek resolution or equilibrium but in which cyberfeminism takes the risk (Stengers 1997) of relinquishing some long-held conceptual ground mapped and divided by the opposition between essentialism and constructivism, ontology and epistemology, nature and culture. My point is that it is pointless not to take such risks. The choice is between making a difference and making no difference in the construction of ideas and artefacts which constitute life-as-we-know-it and life-as-it-could-be. If the network identities of artificial life are the products of both biology and computer science – if they are 'naturecultural' (Haraway 2000) – then in order to be involved in their construction and recognition, cyberfeminism itself requires a less singular, less oppositional stance; one, in short, which approaches biology in the same spirit as technology and as an opportunity rather than merely a problem. As well as facilitating the pursuit of core political goals pertaining to gender and other aspects of identity, cyberfeminism is then, in my view, uniquely positioned to make a vital contribution to the bioethics which nurtures and natures the development of posthumanism.

With regard to the organisation of the text, readers might wish to note that in order to clarify my argument with regard to the discipline and discourse of artificial life, I have distinguished them through the use of upper and lower case initial letters respectively. This is, of course, a false distinction and one which has been made only to underline that very point. Chapter 1 'Autonomy and Artificiality in Global Networks' locates artificial life within the discourses of globalisation and posthumanism by highlighting the construction of autonomous information agents – the humanised HALs or intelligent agents of the post-coldwar era. It also has an introductory role, containing brief summaries and references to successive chapters and underlining the bioethical as well as the political emphasis of the book as a whole. Chapters 2 and 3 establish the key themes within the disciplines of the new biology and Artificial Life as they currently stand. There is a review aspect to these two chapters which also seek to outline and critically assess the main facets of a contemporary discourse of alife which, as I argue in Chapter 6, extends from software and hardware to wetware and encompasses genomics. Where Chapter 4 offers a case study of the design and use of a prominent ALife computer game, drawing heavily on an extended dialogue with its original designer, Steve Grand, Chapter 5 explores in more depth the dynamics of network identities as they are described and created within alife. Chapter 7 'Evolving Feminism in Alife Environments' synthesises and explicates the critical position outlined in earlier chapters and manifested, I hope, in a degree of patience and at least the willingness to risk deferring judgement until a sense of the complexity of the field had been presented. The length of this chapter is the result of quite a long wait, whereas the final chapter briefly restates the case within the context of the science wars which both precede and succeed this text, but which in the end were not permitted to hang over it.

Acknowledgements

I am grateful to a number of people who have directly and indirectly contributed to this book. Special thanks go to Steve Grand, who was consistently patient, open-minded and generous with his time during our long conversations about cyberfeminism and artificial life. Much of my insight into the field and how best to approach it was facilitated by these conversations. Even though the arguments represented here are very much my own, I hope Steve feels that they do justice to his extraordinary work. Other Alifers who discussed their work and provided helpful comments and resources include: Paul Brown, Andy Goffey, Inman Harvey, Phil Husbands, Jane Prophet, Phoebe Sengers and Emmanual Chiva and Nick Jacobi at MASA. Friends and colleagues at Goldsmiths who offered help and encouragement include Lisa Blackman, Mariam Fraser, Celia Lury, Angela McRobbie, Dave Morley and Bill Schwarz. Thanks to Gavin Mackie for his efforts as my research assistant and for lending me his computer games (for so long). I'm grateful for the support of family and friends including Michéle Allardyce, who designed the cover for the book, and Liz Vasiliou, who did her best to keep me at my desk until it was finished. Finally, I would like to acknowledge the careful and constructive criticism provided by anonymous readers of the manuscript, and Goldsmiths College Research Committee for its generous provision of a research grant in 1999 and the Richard Hoggart Research Fellowship in 2001.

Chapter 1
Autonomy and artificiality in global networks

The convergence between biology and computer science provides the context for an emergent technoscientific culture within which the status of autonomy and artificiality are highlighted and problematised. Autonomy designates the self-organisation of actors referred to as agents, and artificiality signifies the condition of agents and environments as they co-evolve in/as global information networks. Initiated in the cybernetics of the 1940s which regarded mind and machine as analogous information processing systems, the convergence between biology and computer science has developed through systems theory, complexity theory, cyborg discourse or 'cyborgology',[1] and is now most clearly represented in a development of Artificial Intelligence (AI) known as Artificial Life (ALife). In order to set the stage for an examination of this relatively recent discipline and its cultural significance it is useful to turn briefly to science fiction – partly because of the widely acknowledged slippage between science and fiction and partly because many AI and ALife researchers were clearly informed by it.[2] Steve Grand, a prominent ALife engineer recently hailed as 'one of the 18 scientists most likely to revolutionise our lives in the coming century' (ICA 2000), was responsible for a popular computer game called *Creatures*, praised by Richard Dawkins as 'the most impressive example of artificial life I have seen' (CyberLife 1997).[3] He freely acknowledges that 'many of us grew up with Dan Dare comics, Star Wars movies and Kubrick's *2001*' (CyberLife Research 2000: 1). The star of *2001. A Space Odyssey* (1968) is, of course, the smart but wayward computer HAL 9000, and he in turn is responsible not only for contributing significantly to popular fears about the development of intelligent machines which turn out to be disastrously disobedient towards their creators, but also for contributing to a professional sense of the failure of AI as a project. In 'The Year 2001 Bug: whatever happened to HAL?', Grand (1999a) argues that, as a fictional example of artificial intelligence, HAL's homicide and subsequent demise stemmed not just from unfriendliness but also from the fact that, although smart in the technical sense, he wasn't really very bright. HAL, according to Arthur C. Clarke, 'could pass the Turing Test with ease' (Clarke 2000 [1968]: 99), but during a space odyssey in which he was programmed to assist astronauts, he was obliged (by mission control) to conceal the real purpose of the trip to Saturn from them. His major malfunction stemmed not from maliciousness but from a failure to deal with conflict – with a task which was not clearly right or wrong, black or white, binary. HAL did not understand the purpose of little white lies or what might now be termed complexity. His failure, for Grand, was the failure of the Turing Test as a measure of intelligence. The basis of the Turing Test is a concealed computer being able to pass as a human during a dialogue.[4] In 1950, Alan Turing predicted that 'within fifty years . . . the idea that machines can think will be common place and computers will routinely pass the Turing Test' (Grand 1999a: 73). They don't, and in a nutshell, ALifers such as Steve Grand argue that this is the fault of top-down as opposed to bottom-up processing. AI can build things as intelligent as a chess computer but nowhere near as intelligent as a mouse: 'A mouse will always lose at chess to a computer, but try throwing them both in a pond and see how they fare' (CyberLife Research 2000: 1). The keywords here are 'adaptive', 'robust', 'flexible' (and 'friendly') and to achieve these characteristics, the principles of AI must literally be turned on their heads. Adaptive, robust, flexible and friendly artificial intelligence is now in the process of being *grown* biologically (from the bottom up) rather than built or programmed from the top down, and as such it is beginning to acquire the status of *agency* and *autonomy*.

Another example drawn from science fiction serves to illustrate this shift towards autonomous agency. In the third of Orson Scott Card's 'Ender' trilogy (*Xenocide*, 1991), the character Jane is rather more than just a computer program (even a 'Heuristically programmed ALgorithmic computer' program like HAL). She is described as 'a being' who dwells in the 'web' or 'network' connecting computers on every world, and this web or network is 'her body, her *substance*' (Card 1992 [1991]: 67). Jane, unlike HAL, is alive, or at least capable of asking 'Am I alive . . . ?', and also unlike HAL, she is friendly. Jane is a fictional example not of artificial intelligence but of artificial life:

> And the image on the screen changed, to the face of a young woman, one that Valentine had never seen before . . .
> 'Who are you?' asked Valentine, speaking directly to the image.
> 'Maybe I'm the one who keeps all those . . . connections alive . . . Maybe I'm a new kind of organism'.
> (Card 1992 [1991]: 66)

ALife in context, or, the 'return to Darwin'[5]

One way to define ALife is as an attempt to literalise the machine/organism analogy which is prevalent within biology and technoscientific culture as a whole. The discipline was developed in the late 1980s at the end of the cold war, and its stated aims are twofold: to create viable computer simulations of biological forms and processes as a method of studying 'natural' life (the simulation of 'life-as-we-know-it') and to synthesise new forms of artificial life in both hardware (as robotics) and software (as computer programs) (Chapter 3). This is about creating 'life-as-it-could-be' (Langton 1996 [1989]: 40).[6] These two goals may be characterised as 'weak' and 'strong' ALife respectively. Synthesised artificial life-forms are not deemed to be metaphorically alive, but literally so as the definition or criteria for life are limited to self-replication, self-organisation, evolution, autonomy and emergence. Emergent life is that which is not programmed in, but which evolves spontaneously and from the bottom up through interaction with the artificial environment. First order emergence refers to 'any behaviour or property that cannot be found in a system's individual components or their additive properties', while second order emergence signifies the appearance of a behaviour which stimulates the development of adaptive behaviours (Hayles 1999a: 9). Second order emergence, then, involves the evolution of the ability to evolve and it is the goal of strong ALife. At the heart of ALife is the concept of life as information,[7] and this is derived from molecular biology's notions of the genetic code, and its fetishisation of the gene as the fundamental unit of life. Life is a property of form not matter, or as Christopher Langton (the originator of ALife) put it: 'life is a kind of behaviour, not a kind of stuff' (Langton 1996 [1989]: 53).[8] No stuff, no matter, no fleshy bodies, no experiences associated with physicality and nothing beyond the one-dimensional functionality of information processing. In her critique, Alison Adam (1998: 155) points out that there is 'no room for passion, love and emotion' in ALife worlds because passion is subsumed by sex, sex is all about reproduction and reproduction is all about competition, survival and the evolution of (genetic) information. ALife is concerned with evolving new life-forms, new species in autonomous artificial environments or worlds where the laws are prescribed entirely by biology. It is sometimes tempting to dismiss ALife as the frustrated endeavour of alien-loving scientists brought up on science fiction and disappointed by the failure of NASA to provide specimens from outer space. Artificial life is in part about the creation and investigation of alien life. But with software projects aimed at evolving artificial cultures and societies (Gessler 1994; Epstein and Axtell 1996), and with the proliferation of online virtual ecosytems, ALife might also exemplify the danger of what Adam calls 'sociobiology in computational clothing' (1998: 151).

The sociobiological basis of ALife research and the re-rooting of culture within biology appears to be naturalised and applied in the contexts of the military, medicine and the entertainment industry where, for example, games such as *Creatures* have proven to be popular (Chapter 4) and artificial life-forms known as autonomous agents are being readied for use on the Internet (Chapter 5). Pattie Maes has outlined her attempts to build agents 'that perform a practical purpose and really help people deal with the complexity of the computer world' by 'foraging' for interesting documents for a particular user on the world wide web (Dennett 1995b). These agents would watch and learn from the user and would reproduce and evolve according to their usefulness. Through the use of genetic algorithms (computer code with a simulated genome), mutations which occur in reproduction produce offspring agents which look for different kinds of documents than their 'parents'. The documents they obtain may be more or less interesting than those of others and 'if they're less interesting then that offspring won't survive'. Fitness is then determined by usefulness. Maes's work raises questions of control, ethics and evolution and has led Kevin Kelly (1994) to reflect that although 'there was probably a wider agreement that evolution was a way to do things than I thought . . . I wonder if we can get everything we want by evolution?' (Dennett 1995b).

The precursors of autonomous agents are bots: 'the first indigenous species of cyberspace' (Leonard 1997: 8). Bots are to software what robots are to hardware; algorithms rather than animated machines governed by rules of behaviour. Andrew Leonard is lyrical about the range and diversity of this new species and highlights the role of anthropomorphism in the successful mediation between digital and biological entities (22). Bots, for Leonard, 'stoke our imaginations with the promise of a universe populated by things other than ourselves, beings that can surprise us, beings that are both our servants and, possibly, our enemies' (10). Bots are impure, partial or incomplete life-forms which achieve autonomy only when they are released into their unnatural environment where they operate 'out of direct control' (21). And autonomy is 'the crucial variable', the dividing line between what counts as digital life and what does not. Autonomy is a key criterion in the development of artificial life. The distinction between bots and agents is, however, by no means absolute, especially as information and communication industries begin to realise the consumer marketability of useful, user-friendly or believable agents.

Both bots and agents are being designed to adapt to the Network ecology. The Net is regarded, in the context of alife culture, as a suitable environment in which agents can grow and evolve and agents are also produced by artists with various degrees of allegiance to the nebulous, non-homogenous interacting spheres of Artificial Intelligence and Artificial Life (for example, *TechnoSphere* by Jane Prophet and Gordon Selley 1995). Agents feature in computational models of human cultures and societies developed by anthropologists, economists and sociologists. Where these offer a method and epistemology for the study of human life-as-we-know-it, they do so within a narrative framework where life-as-we-know-it is in the process of being superseded by life-as-it-could-be. This is not an apocalyptic scenario as much as it is an evolutionary one in which the next stage in the evolution of life is digital life – and the aliens are (be)coming. Within this evolutionary scenario the concept of culture regresses from a social to a bio(techno)logical context from which it is expected to re-emerge. Within the paradigm of computational anthropology, for example, culture is viewed as a computational system and as a manifestation of the ubiquitous evolutionary process of information exchange. This then is a memetic view of culture where memes or cultural units reproduce and evolve (autonomously) in the same way as genes or biological units. Nicholas Gessler realises this view of culture in a software model he terms *Artificial Culture* (1994, 1999). This functions as a test bed for the theory of cultural evolution, and it builds on foundations which already exist in ALife. In other words, all ALife worlds have an incipient bioculture. Artificial Culture enacts a theory of culture which is evolutionary and emergent. Evolution operates through cultural variation and the emergence of behavioural patterns from individual local rules. The aim of the program is to create a population of evolving mobile autonomous agents including 'personoids' which are both embodied and situated, and a 'god' which is neither. Gessler's software program offers a reflexive view of the evolution of human life-as-we-know-it which sets out to compute but not necessarily to naturalise it. This may not be said of Dawkins's theory of memetics or recent research in evolutionary psychology – other aspects of the cultural evolutionism which is said to be indicative of a new epoch variously named as the Information Age (Castells 2000), the Biological Age (Grand 2000) or the Neo-Biological Age (Kelly 1994). In this new epoch – which due to processes of renaturalisation is not synonymous with postmodernism but rather with globalisation – it is not simply the computer, but increasingly the Net which defines the evolutionary parameters of culture and identity. The Net does this partly in so far as it is regarded as an ecosystem for emergent artificial life-forms and as an entity or intelligent life-form in itself.

Kevin Kelly's (1994) vision of a 'Network Culture' incorporates 'all circuits, all intelligence, all interdependence, all things economic and social and ecological, all communications, all democracy, all groups, all large systems' (1994: 25) in a co-evolved single organism which is analogous to an emergent hive mind. It is a decentralised distributed intelligent entity which assimilates and elides identity. Individuals are bees in the hive, neurons in the network, cogs in the wheel which is more than the sum of its parts. Derided as being emblematic of the Californian Ideology developed on the pages of *Wired* magazine, Kelly's vision of the early 1990s is nevertheless strangely echoed in a millennial issue of *New Scientist* dedicated to the role of the Internet as a 'Global Brain'. Michael Brooks (of Sussex University) examines the claim of Francis Heylighen (of the Free University of Brussels) that the global brain will grow out of attempts to manage the store and flow of information on the Internet (Brooks 2000). Here, web links function as synapses which build and grow from the bottom up with use and diminish and die with lack of use, as in the model of neural networks. Moreover, Grand, following Kelly, has offered a 'how-to' guide to the creation of life incorporating cybernetic building blocks, bio-informatic networks and the process of emergence. The non-vitalist vitalism of emergence is this biological age's answer to the physics of entropy – and the job of postmodern science and culture would appear to be done.[9] The paradigms of nature – concealing and revealing life itself – have survived (if not unchanged) the search and destroy missions of post-structuralist epistemologies and are newly deployed in and through the artefacts of information and communication. These artefacts weld together 'engineered technology and unrestrained nature' (Kelly 1994: 471) producing a bioculture which is at once more and less than the sum of its parts, but identical to none. Bioculture is not the biological culture of the petri dish any more than it is the forms and processes of everyday (human) life. Bioculture is the culture of analogous (organic and inorganic) information systems, self-organised within what has been referred to as the network society. The network society is constituted in part by an investment in technocultural forms of autonomy and agency which relies on a dialectic and not a division between the Net and the self. There is no clear opposition between 'global networks of instrumentality' and 'the anxious search for meaning and spirituality' (Castells 2000: 22). The search for meaning through identity – and perhaps even the search for spirituality – occurs neither outside nor inside the Net but in a dialectic articulated in part through the reproduction (symbolic and material) of agency and autonomy. The transfer of agency and autonomy to the (id)entities of the Network – to Maes's little helpers or Gessler's personoids (Chapter 5) – although apparently anti-humanist, is, in one sense, a process of externalisation which enables agency and autonomy to be renegotiated and reclaimed within the identities of the self. The posthuman self thus engages with the forms and concepts of posthumanism.

The posthuman is an epistemology and ontology of the self in the post-cold-war Information Age and one which necessarily engages with historical constructions of humanism. The universality and subsequent disembodiment inherent within liberal humanism has been critiqued in feminist, postcolonialist and postmodern theories which share a concern with the erasure of difference (Hayles 1999a). Hayles argues that although the loss of a concept 'so deeply entwined with projects of domination and oppression' is not to be regretted, it might still be necessary to reconsider the role of specific characteristics of the humanist subject – such as agency and choice – in a contemporary context. Posthumanism represents, for her, an opportunity to 'keep disembodiment from being rewritten, once again, into prevailing concepts of subjectivity' (5). I explore this opportunity partly by entering into a dialogue with ALife engineers,[10] principally Steve Grand, whose work on computer software and robotics might be summarised as an attempt to humanise HAL and is, for me, most affective in this regard. Grand's representation of a primarily liberal humanism is complicated by his investment in simulating autopoiesis rather than autonomy – in embodying and situating his creations within their environment – and is therefore a potent resource for debating the increasingly symbiotic relation between humans and machines. I also explore the opportunity to re-embody post(liberal)humanism by examining (mainly in Chapter 7) engineering projects at the margins of AI/ALife which have already entered into a dialogue with contemporary cultural theory or 'the humanities'. These projects are concerned with the generation of a new kind of agent technology which is based on a practical and theoretical critique of liberal humanist concepts such as autonomy and agency. These novel agents do not so much evolve as co-evolve in the dynamic interplay between observer and object, and they are more a facet of communication – the desire for alife – than of computation – alife itself. The concept of dialogue is derived from the work of Mikhail Bakhtin, and is developed strategically as a possible means to prevent the encounter between cyberfeminism and artificial life being reduced to a continuation of the science wars. The science wars, in this context, would hinge on the biologisation of computer science which is, I maintain, indicative of the increasing biologisation of contemporary technoscientific culture.

Arguably because the new biological hegemony subsumes what was thought to belong to the realms of culture and society (notably technology), because it reproduces a naturalised culture and more importantly adapts to forms of de-naturalisation and evolves (as evolutionary psychology has adapted to critiques of sociobiology and as ALife evolves from AI) then feminism needs to offer more than the familiar critique. Feminists – and it is here that cyberfeminists, with their expertise in negotiating the boundary between the body and technology, may take a leading role – can no longer rely on the oppositional discourses of nature versus culture because those categories, though not 'imploded', collapsed or elided are not discrete (Franklin *et al.* 2000). Rather cyberfeminism, while observing the exchanges between nature and culture in specific contexts has the opportunity of intervening in processes of renaturalisation by learning, through dialogue, new languages and/or new skills with which to engineer embodied forms of subjectivity (which come to matter) and perhaps a posthumanism with a difference. This dialogue entails risk, and what cyberfeminism risks in a dialogue with ALife is not complicity but the complacency of a secure, well-rehearsed oppositional stance. Cyberfeminism risks its anti-biologism (and residual technophobia) by entering a more dynamic relationship based on contest and consent, on centrifugal and centripetal forces. These forces, it is important to recognise, are at play internally and are therefore made available to what Haraway terms 'diffraction' (1997). The internal critique of autonomy in ALife is symptomatic of a broader contest over gender, power and knowledge played out through the reinscription or renegotiation of top-down masculinist AI versus bottom-up feminised ALife programming; physics versus biology; hard versus soft epistemologies; Cartesian versus anti-Cartesian philosophy; form versus matter; embodiment versus disembodiment; creation versus evolution; holism versus reductionism and so on. These oppositions are culturally productive and have, not least through the proliferation of metaphors, more than local significance. Primarily they correlate with the end of the cold war, the displacement of the arms race by ideologies of greater co-operation, decentralisation and globalisation. The global bioculture is that which is evolving, emergent, self-producing and informational. It constitutes the open, distributed world of an era in which the individual and species-self is becoming other, becoming artificial/alien life.

Network bioethics

The connection between artificial life (Chapter 3), evolutionary psychology (Chapter 2) and genomics (Chapter 6) has, after the first decade of research, been made explicit (Bedau 2001) as researchers seek greater scientific rather than science fictional credibility under the increasing arc of biology. The simultaneously expanded and restricted agenda for ALife in the new millennium is organised around three principle questions: 'How does life arise from the nonliving? What are the potentials and limits of living systems? How is life related to mind, machines and culture?' (Bedau *et al.* 2001: 263). These questions extend biology into the social sphere and return ALife to the ethical problems held largely in suspense since Langton's initial programme. The problems which are highlighted almost exactly reproduce the humanist bioethics articulated within genomic discourse: the sanctity of the biosphere; the sanctity of human life; responsibility towards new forms of artificial life and the risk entailed in exploring the possibilities of artificial life. The practice of engineering life explicitly includes natural or artificial systems, and the biosphere or ecosystem explicitly incorporates 'indispensable' elements such as the Internet (374). The sanctity of human life is underscored not undermined and ALife 'like the theory of evolution' is set to have 'major social consequences' including 'our future increasing dependence on artificial life systems' (375). This agenda, unlikely as it is to be unanimous, nevertheless clearly indicates the connectedness of ALife with other biological discourses and practices. As part of a wider attempt to manufacture and manipulate life and to produce biological models of minds, machines and cultures, ALife cannot be contained and reduced by a rationalist assessment of its strong claim to synthesise life-as-it-could-be. Life-as-it-could-be is the science fact and fiction of the present neo-biological age characterised by the convergence between natural and artificial systems. The hype is integral to the practical, commercial application of these systems, as evidenced by the range of academic and corporate research involving the development of autonomous agents (Chapters 4 and 5). These populate computer games, war simulations, the toy and other industries, virtual ecosystems, computational models of culture and society, and are evolving on the Internet. The question of whether or not they are really alive depends of course on shifting and ever contestable definitions of life, and is irrelevant considering the evident drive to humanise HAL; to create artificial life in software, hardware and wetware. Alife is the enaction, through biotechnology, of a creationist and post-humanist fantasy. More than a bad case of anthropomorphism, alife exemplifies a global metaphysics centred on autonomy, artificiality and network systems which are more than the sum of their parts. The configuration of human and machine here is not merely instrumental but imaginatively instrumental – it (arguably) works, but more by (biological) association than by logic. Complex systems such as minds, machines and cultures are no longer deemed to be under control: programmable, analysable, reducible to their component parts. Rather, they are self-organising and emergent, and it is these designations of distributed agency and potentiality which – much more than the master/slave rhetoric of AI – constitute the zeitgeist.

Where ALife (often referred to as nouvelle AI and broadly associated though not synonymous with connectionism/neural networks)[11] challenges the methods, and to an extent, the epistemology of AI with its emphasis on bottom-up processing, embodiment and situatedness, the chief legacy of AI in ALife centres on the computability of complex systems. Based on his assessment of the supra-rational capabilities of computer technology, Paul Cilliers (1998) argues that complexity can be modelled computationally through neural networks or connectionism. Complex systems are, for him, impossible to analyse in terms of their component parts and are characterised as being open (situated), far from equilibrium, temporal and constituting a large number of elements which have no knowledge of the system as a whole but which interact in a dynamic, nonlinear or looped fashion (Cilliers 1998: 3). Cilliers offers a rather abstract, post-structuralist perspective on complexity which concentrates on the semantic emptiness of the individual sign or element and on the richness of the system as a whole: 'when we look at the behaviour of a complex system as a whole, our focus shifts from the individual element in the system to the complex *structure* of the system' (4). Meaning in complex systems is not objective (language is not a nomenclature) but relational (signs signify through a system of differences) and identity is conferred through a revised concept of autonomy as self-organisation; an environmentally embedded, unfinishable biological process of becoming. Cilliers draws the analogy between post-structuralist semiotics and neural networks as complex information processing systems (18), and by doing so he abstracts each from its context, missing the crucial distinction between processes of denaturalisation and those of renaturalisation.

Hubert Dreyfus's critique of classical AI and aspects of nouvelle AI is based on his rejection of the information processing theory of mind and matter. Focusing on the opposition of knowledge and 'know-how', information and meaning, he rejects both computational attempts to capture complex systems. AI's rationalist, rule-based approach to knowledge can model only that of the beginner since expertise is a function not of knowledge per se but of context sensitive know-how. Drawing on the phenomenology and sociology of knowledge (on Heidegger, Merleau-Ponty and Bourdieu) which is dependent on the bodily experience and cultural savoir faire of the knower, Dreyfus characterises know-how in a way which is compatible with, though exceeded by Donna Haraway's formulation of situated knowledge. There is, he maintains, no abstract context-free knowledge either in the social or physical realm, and no viable distinction between facts about, and skills within the world. One is reminded, of course, of Roland Barthes's brilliant phenomenology of photography in which he rejects post-structuralist analysis in order to 'make myself the measure of photographic "knowledge"' and to ask 'what does my body know of Photography?' (Barthes 1980: 9). Dreyfus, like Adam (1998), is critical of Douglas Lenat's attempt to formalise know-how or overcome the commonsense knowledge problem of AI by building a vast database ontology of objects, individuals, space, time, institutions and social situations (Dreyfus 1999: xix). According to Dreyfus, the Cyc database has relatively modest but still unrealisable goals since it is based on the 'dubious rationalist assumption' (xxii) that humans store and retrieve information from a vast library of commonsense knowledge which is ultimately context-free:

> No one in AI believes anymore that by 2001 we will have an artificial intelligence like HAL. Lenat would be satisfied if the Cyc data base could

understand books and articles, for example, if it could answer questions about their content and gain knowledge from them. In fact, it is a hard problem even to make a data base that can understand simple sentences in ordinary English, since such understanding requires vast background knowledge.

(Dreyfus 1999: xix)

Neural networks may undermine the rationalist assumption 'that one must have abstracted a theory of a domain in order to behave intelligently in that domain' (Dreyfus 1999: xxxiii) but the problem of computing know-how nevertheless resurfaces in the gap between what counts as commonsense for computers and what counts as commonsense for humans. In an early military application, the army tried to train a neural network to recognise tanks in the forest. They used photographs taken over a period of time and discovered that the network had conflated tanks with shadows, giving somewhat mixed and confused results. The network clearly lacked an appropriate commonsense understanding of the world (xxxvi). Phenomenologically then, it is unlikely that computers could ever model the complexity of the human mind:

One might still hope that networks different from our brains will make exciting new generalisations and add to our intelligence. After all, detecting shadows is just as legitimate as detecting tanks. In general, though, a device that could not learn our generalisations and project our practices to new situations would just be labelled stupid.

(Dreyfus 1999: xxxvii)

Rosalind Picard's (2000) hope for the future of a humanised HAL rests on the possibility of imparting emotional know-how to machines. Unlike Lenat, she maintains that emotion is an integral aspect of intelligence and that effective computing is a facet of 'affective' computing: 'because *emotional computing* tends to connote computers with an undesirable reduction in rationality, we prefer the term *affective computing* to denote computing that relates to, arises from, or deliberately influences emotions' (Picard 2000: 281). Although HAL could recognise and simulate affect – 'I'm afraid, Dave' – he ultimately lacked emotional intelligence and could not effectively cope with his internal contradictions. While asserting that 'lack of emotions may be a key reason why artificial intelligence has failed' (302), Picard wonders whether people are ready to face the ethical challenges posed by truly affective computers.

Francisco Varela (1999) explores the ethical dimension of the distinction between know-how and know-what using phenomenology and the wisdom traditions of the East in preference to the rationalist philosophies of Kant, Habermas and Rawls.[12] Rawls, after Kant, articulates an ethics based on the rational autonomous subject. Kant 'begins with the idea that moral principles are the object of rational choice' (Rawls 1999: 221) and Rawl's theory of justice holds that 'a moral person is a subject with ends he has chosen, and his fundamental preference is for conditions that enable him to frame a mode of life that expresses his nature as a free and equal rational being as fully as circumstances permit' (491). Moreover, the unity of the person 'is manifest in the coherence of his plan' (491). Varela (1999) seeks to move away from ethics as reason and to focus on actions which do not spring from judgement and reasoning but from an 'immediate coping' with a given situation. He gives two examples which he argues are more than merely reflexes and illustrate a pervasive mode of ethical behaviour which does not rely on a central I performing deliberate and willed action:

> You are walking down the sidewalk thinking about what you need to say in an upcoming meeting and you hear the noise of an accident. You immediately see if you can help. You are in your office. The conversation is lively and a topic comes up that embarrasses your secretary. You immediately perceive that embarrassment and turn the conversation away from the topic with a humorous remark.
>
> (Varela 1999: 5)

Both Varela and Dreyfus attempt to describe the intuitiveness and immediacy of (ethical) human behaviour which may be non-computable and impossible to formalise but which is, nevertheless, being described within biotechnological models which seek, somehow, to capture it. For Varela, ethical know-how rather than know-what is based on coping rather than judgement, response rather than reason, and ultimately on the 'empty' rather than autonomous self. This empty social/spiritual self is to an extent modelled on Varela's biological formulation of autonomy as autopoiesis where the autopoietic organism's actions upon and within its world constitute cognition. Varela's ethical subject is cognitive in this autopoietic fashion – not autonomous in the sense of being detached and rational, not in possession of a central I, but having a shared or coupled agency and a capacity for situated action. Action or 'enaction' is central to knowledge-formation, and though not computable through the informationalist methods of classical AI or cognitive science, it might, according to Varela, be captured in bottom-up emergent ALife systems which are embodied and situated. Ethical expertise stems not from rules or reasoning but from the skilled behaviour which people engage in on a daily basis (working, moving, talking, eating), and so an ethical expert is 'nothing more or less than a full participant in a community' (Varela 1999: 24). Here, Varela turns to Confucianism in order to elaborate an ethics based on extension, attention and intelligent awareness rather than rational judgement (27). What is central to the ethical know-how and responsiveness of Varela's cognitive subject is its constitution in and by networks: biological (neural networks) and social (language or representational networks). Agency within networks is distributed, and autonomy is at best partial or shared between the cognitive system coupled with its environment – including that of other cognitive systems (56). The psychologically and socially necessary illusion of a 'personal "I"' can be construed as an ongoing interpretative *narrative* of some aspects of the parallel activities in our daily life' (61). At best exposed in modern Western science, the fundamentally selfless rather than nihilistic self is nurtured in belief systems of the East as the basis of ethical expertise (63). Ethical expertise is figured when 'the street-fighter mentality of watchful self-interest slips away gradually to be replaced by an interest in others' (66). It is a factor of a developed sense of relatedness and connectedness resulting in spontaneous gestures or 'actions that embody and express the realisation of the emptiness of self in a nondual manifestation of subject and object' (69). Ethical expertise is intuitively or affectively responsive, not rule-based even though it emerges from beginner level rules (72). So how, asks Varela, can 'decentred, responsive, compassionate concern be fostered and embodied' if not by norms and rationalistic injunctions? It must be developed, he suggests, through disciplines which facilitate the relinquishment of ego-centric habits (73). Without conflating this with self-improvement schemes which serve only to reinforce the ego, he insists that 'we simply cannot overlook the need for some form of sustained, disciplined practice or practique de transformation de suject, to use Foucault's apt term' (75). Some form of technology of the self based on realising decentralisation is, for Varela, both necessary and facilitated within social networks of distributed agency which are not always already responsive or compassionate. He offers what he refers to as 'strong measures for the troubled times we have at hand, and the even more troubled ones we are likely to have' (75).

Within global networks characterised by autonomy (self-organisation) and artificiality (the emergent convergence between nature and culture, biology and technology), ethical know-how is located at the level of the distributed agent, the self-transforming posthumanist subject. A response to such networks, I will argue, exceeds the politics of resistance, is necessarily and strategically dialogic and might best be characterised as bioethical. ALife is an intrinsically interconnected discipline within which dialogical strategies and bioethics are being articulated, and it is constituted in and by processes of globalisation. The perspective on ALife offered here has been reinforced by the events of 11 September 2001 which, at the time of writing, seem likely to end the end of the cold war period (1989–2001?). The illusion of individual autonomy – of safe and self-determined lives – was dramatically shattered here in the at times horrific and at times moving images of lives lost and recalled on thousands of flyers and on pages of portraits. Surrounding such very human representations is a more mythical archetypal structure and iconography of good versus evil. Where artificiality was central to the spectacle of biotechnological death and destruction and is central to the outbreak of 'bioterror', the after-effects have concentrated on the problem of a global terror network which intrinsically resists conventional military and political acts of resistance. The key question from the Western alliance perspective concerns the means of combating a threatening and seemingly organic, self-organising, robust, regenerative network of terrorist 'cells'. The more marginalised question concerns the essentially ethical issue of perspective itself, and the risks of strategic dialogue. The dialogic method in this project emerged from (remains limited by) a more standard feminist oppositional standpoint and was facilitated by email, the networker's network. The dialogue is partial, minimal, unfinished and merely indicative. It involves relatively few individuals but is supported by a shift in attitude, a suspension of (at least some) cynicism, an openness in relation to a diverse range of materials and practices. It produces an account of the field which few specialists would recognise but does not seek to offer a definitive or empirical study (see Emmeche 1994; Levy 1992; Risan 1996; Helmreich 1998a). This is an embodied encounter which became an embodied relation to representations (including popular and technical, written and verbal) of alife in a neo-biological age. What follows is an attempt to situate the discipline of ALife within a wider technoscientific culture which includes, crucially, genomics, evolutionary psychology, memetics and organic metaphors of agency, technology and social processes. Without losing touch with its roots in feminist epistemologies, this project seeks to contribute to important debates in bioethics and to offer both feminist and ALife researchers some insight into the potential of dialogue.

Chapter 2

The meaning of life part I
The new biology

The new biology is foundational to the discipline and discourse of artificial life, and its redefinition of life as information makes alife projects possible. Although founded in a molecular biology which divests life of both meaning and vitality, alife perversely if not ironically seeks to restore meaning and vitality in its bio-engineering practices. In other words, it resurrects the dead body of life from information and is, in this sense, a spiritual discourse thoroughly characteristic of the post-cold-war era of globalisation. The seeds of alife's spirituality are sewn in the internally contested ideology of the new biology which may, for analytic purposes, be considered across the nominal divisions between evolutionary biology, molecular biology, sociobiology and evolutionary psychology. This latter and most recent manifestation is also arguably the most problematic given its disingenuous claims to liberalism. An incipient aspect of alife, it is most urgently in need of a response from cyberfeminists and others invested in the opportunities afforded by a hegemonic yet far from homogenous field.

Evolutionary biology – all about chickens and eggs

What came first, the chicken or the egg? For most of us, this is the ultimate rhetorical question. We perceive only its polarised form and infinite circular structure. Substantively though, and in its reference to the origins of life, the chicken or egg question remains meaningful to those engaged in the life sciences – evolutionary biologists in particular. Indeed, the field of evolutionary biology may be said to be divided along the chicken/egg axis.

The debate within biology is of course not only with the origin and evolution, but also with the meaning of life, and this chapter will trace the shift in the meaning of life which has taken place within evolutionary, molecular and sociobiology (including its new incarnation, evolutionary psychology).[1] This shift may be characterised (loosely) as one which prioritises the egg over the chicken – or in its current translation the gene over the organism. In contemporary biology (and particularly in molecular biology) the gene is regarded as the fundamental unit of life; that which gives rise to the organism and which (to a greater or lesser extent) determines its characteristics and behaviour. Molecular biology is gene-centred or, arguably, genetically deterministic, ascribing autonomy and agency to genes rather than to organisms or species. This represents a quite radical reaction to the organism/species-centred focus of population, developmental and other forms of biology (Rose 1997).

The turning point (from chicken to egg, from organism to gene) coincided with the discovery of DNA by James Watson and Francis Crick in the 1950s. This discovery was dependent on the contribution of Rosalind Franklin who, both knowingly and at times unknowingly, provided the Cambridge scientists with X-ray diffraction pictures which gave them the necessary clue to the structure of the double-helix:

> The pictures were the technology Watson and Crick required, for they immediately provided the clue to the now famous double-helix structure of DNA, and to the fact that its component *nucleotides* (sub-units) – adenine, guanine, cytosine and thymine – could fit together only within particular configurations which pointed unmistakably to how chromosome duplication and copying could occur.
>
> (Rose 1997: 118)

Watson and Crick realised that 'if the two strands of DNA were to unwind, each could provide the template on which its matching strand could be copied, without error' (Rose 1997: 118). So genes could now be considered to be constructed of DNA. A further 'vital complexity' lay in the relationship between DNA and the manufacture of proteins which comprise the tissues and organs of the body. DNA (deoxyribosenucleic acid) exists in the cell nucleus and comprises the chromosomes. Another form of nucleic acid called RNA (ribosenucleic acid) is present in the nucleus and in the cell cytoplasm. DNA is double-stranded whereas RNA is single-stranded. Proteins are synthesised in the cytoplasm and the mechanism involves the partial unwinding of the DNA double-helix which is copied by a single strand of RNA which then moves from the nucleus to the cytoplasm where it enables the synthesis of specific protein chains. What is important is that this process of synthesis is one-way only. Francis Crick's 'Central Dogma' of molecular biology entailed 'a one-way flow of information' (Rose 1997: 120). In 1957 he established the principle 'that once "information" has passed into the protein *it cannot get out again*' (Crick in Keller 1992: 161). Evelyn Fox Keller outlines what she refers to as the hyperbolic elaboration of Crick's central dogma. She relates Jacques Monod's argument that:

> What molecular biology has done . . . is to prove beyond any doubt . . . the complete independence of the genetic information from events occuring outside or even inside the cell – to prove by the very structure of the genetic code and the way it is transcribed that no information from outside, of any kind, can ever penetrate the inheritable genetic message.
>
> (Keller 1992: 161)

Keller (1992) argues that the structure of the genetic code does not prove the one-way flow of information and that the transcription process is more complex than originally supposed, leading to the theoretical possibility of reverse translation and weakening over all of the central dogma. Both Keller and Rose point to the epistemological power of the information theory metaphor which includes the 'transcription' of DNA to RNA, and the 'translation' of RNA to protein. The information theory metaphor constructs DNA as the master-molecule and its coinage, according to Rose (1997: 120), may have been at least as influential on the future direction of biology as the discovery of DNA itself.

The metaphor of life as information

Lily Kay (2000) offers a critique of the literalisation (the transformation from analogy to ontology) of information in molecular biology. Though driven by global capital, the human genome projects in the UK, US and Japan 'are perceived as a mission of "reading" and "editing", and though "problematic"', this view of the genome as an information system or linguistic text written in DNA code 'has been guiding theories and practices of molecular biologists since the 1950s' (Kay 2000: xv). Information theory and cybernetics co-evolved during the post-war and emergent cold war period when the militarisation of science was strongly established. Derived from weapons research,[2] the central notion of Norbert Wiener's cybernetics 'was that problems of control and communication engineering were inseparable . . . and centred on the fundamental notion of the message' (Kay 2000: 84). This was complemented by Claude Shannon's work on coding and decoding signals before and during the Second World War. The premise of information theory is the cleavage of information from semantics or meaning: 'The fundamental problem of communication is that of reproducing at one point either exactly or approximately a message selected at another point. Frequently the messages have *meaning* . . . These semantic aspects of communication are irrelevant to the engineering problem' (Shannon in Kay 2000: 96). In other words, communication involves the transmission of the message as information, not meaning, from source to destination, transmitter to receiver, with minimal noise or interference. This mathematical scheme was designed for communications between machines, but encouraged projection and speculation: 'Provided the components were properly interpreted, communication between muscle and brain and chromosomes and cells could be, in principle, analogised to machine communications' (97). Kay argues that information theory entered biology partly through the study of automata or cybernetic human/machine (cyborg) systems, and that the problematic centres on the transfer of information to a context in which – unlike telecommunications – semantics matter (100). She details the attempt by Henry Quastler to apply information theory proper to biology (in 1949). Quastler reworked biochemical concepts in informational terms, applying them to the study of genes and chromosomes. He met with only modest success due to lack of an experimental agenda and the ease with which his research premises became outdated (Kay 2000: 25). Other similar efforts also failed, but the result of this failure was not the obsolescence of information theory in biology, but rather its transformation into information discourse: 'the system of representation – *information, messages, texts, codes, cybernetic systems, programs, instructions, alphabets, words* – that first emerged in the late 1940s' (26). At this point, information becomes less of a technical term, more of a productive, constitutive if contested metaphor in biology which is intimately interwoven with religious connotations of the Word and the Book of Life. Molecular biologists since the 1950s transact a 'kind of divine biopower' and this biopower entails mastery of the genomic Book of Life 'first through secular exegesis and subsequently through secular (re)creation' (36). As the Word, the DNA sequence 'has brought molecular biologists as close to the act of creation as could be experienced, invoking supernatural, Faustian powers' (37). Or, as Watson, in his mandate for the human genome project and the quest for genetic salvation put it: 'If we don't play God, who will?' (Watson in Kay 2000: 37). The problem outlined by Kay is that, given the plastic and contingent relation between genes, structures and functions, the Book of Life cannot be read or edited unambiguously (Kay 2000: 326). Information as the ontological unit of life has promised rather more than it seems able to deliver (xvi).

For Rose (1997), the metaphor of life as information has been consolidated by the contemporaneous development of molecular biology and computer technology and by influential popular science writers – notably Richard Dawkins: 'Crick may have originated the metaphor, but it has taken Dawkins to draw it to its logical conclusion' (in Rose 1997: 121). Rose cites a passage from Dawkin's *The Blind Watchmaker* in which he looks out of his window at a willow tree in seed:

> It is raining DNA . . . It is raining instructions out there; it's raining tree-growing, fluff-spreading algorithms. This is not a metaphor, it is the plain truth. It couldn't be any plainer if it were raining floppy discs.
> (Dawkins in Rose 1997: 121)

Rose's reaction is indicative of the heated (at times vitriolic and somewhat personal) disputes between Darwinists of different hues:

> You might ignore the trivial fact, irritating to a biochemist like myself but airily dismissed in the paragraph containing this extract by the grand theorist, that seeds contain a great deal more than DNA: there are proteins and polysaccharides and a multitude of other small molecules without which DNA would be inert. But you cannot ignore the blunt statement that 'this is not a metaphor', for this is precisely and at best what it is. It certainly isn't 'the plain truth' . . . It is a manifesto.
> (Rose 1997: 121)

Here, Rose exposes a connection between language and knowledge that Dawkins himself is reluctant to acknowledge. Dawkins's most influential book is *The Selfish Gene* (1989 [1976]), and in it he states that his purpose 'is to examine the biology of selfishness and altruism' (Dawkins 1989 [1976]: 1) while his argument 'is that we, and all other animals, are machines created by our genes' (2). Genes are inherently 'selfish' (autonomous and determinant), and they produce selfish behaviour (behaviour designed to perpetuate the individual's genes). In the preface to the 1989 edition of his book, Dawkins argues that the 'selfish gene theory is Darwin's theory, expressed in a way that Darwin did not choose . . . It is infact a logical outgrowth of orthodox neo-Darwinism, but expressed as a novel image' (viii). Instead of focusing on the individual organism, the selfish gene theory takes a 'gene's-eye view of nature', and this, for Dawkins 'is a different way of seeing, not a different theory' (viii). While acknowledging that 'a change of vision can . . . usher in a whole climate of thinking, in which many exciting and testable theories are born, and unimagined facts laid bare' (ix) he (rather disingenuously) disclaims 'any such status for my own modest contributions' (ix). Dawkins retreats from the epistemological and ideological force of his own narrative, not just in deference to the grand narratives of Darwinian evolution on which it depends (itself, of course dependent on the Truth of Nature) but because he does not want the responsibility of a social Darwinist politics:

> I am not advocating a morality based on evolution. I am saying how things have evolved. I am not saying how we humans morally ought to behave. I stress this, because I know I am in danger of being misunderstood by those people, all too numerous, who cannot distinguish a statement of belief in what is the case from an advocacy of what ought to be the case.
> (Dawkins 1989 [1976]: 3)

So, a 'statement of belief in what is the case' is, for Dawkins, completely neutral. According to N. Katherine Hayles (1994), Dawkins expresses a view of language which is very much in keeping with that of most mainstream practising scientists – 'what I call the giftwrap model of language' (Hayles 1994: 115). This model 'sees language as a wrapper that one puts around an idea to present it to someone else. I wrap an idea in language, hand it to you, you unwrap it and take out the idea. It does not really matter what the language is – whatever the wrapping, the idea is conveyed intact' (115). But, Hayles continues, ideas are surely not independent of language or 'giftwrap'. Rather, 'the idea is present only in and through the giftwrap' (119). Dawkins's anthropomorphic metaphor of the gene is performative, constructing rather than unwrapping ideas about agency and subjectivity. His story 'is about displaced agency, about a subjectivity that has the illusion of control while the real locus of control lies with another agent who inhabits the subject and uses him for its own ends' (120). Hayles integrates Dawkins's narrative into the grand cultural narratives of the time, and principally to the attack on individualism and the notion of coherent and autonomous subjects which came from post-structuralist thinking. She cites the influence of

Michel Foucault, who regarded subjectivity as an effect of discourse, and control as a facet of systems of power and knowledge rather than of individuals (1979, 1981, 1987, 1997). In this context, she regards Dawkins's theory as one which both acknowledges and contests the Foucauldian model of cultural inscription:

> Like Foucault's archaeologies, Dawkins' text posits a shift in the locus of control away from the individual, with a consequent erosion of the autonomy of a human being. Unlike Foucault, however, Dawkins' rhetoric recuperates the basic formation of autonomous individualism by attributing it to the gene.
> (Hayles 1994: 120)

Dawkins's text is therefore regarded as 'an adaptive strategy in a particular intellectual climate' (120) and one which successfully ensures the survival of individualism. There is a direct analogy here between biology and the discourse of alife which it informs. Within alife, autonomous agency is attributed not just to genes but to entities whose ontological status rests on the assertion that life is a facet of information and that this is more than just a metaphor. Through biology, alife inherits the creationist response and Faustian counter-response to the imagined exegesis of the Book of Life. It also shares an adaptive strategy which reinstates liberal humanism through the back door of post-structuralist and post-cold-war philosophy.

Biology as ideology – reductionism and determinism

In attempting to resolve the tension within the selfish gene theory between gene and individual body as the fundamental agent of life, Dawkins deploys the terms 'replicator' and 'vehicle'. Replicators are 'the fundamental units of natural selection, the basic things that survive or fail to survive' (Dawkins 1989 [1976]: 254). DNA molecules are replicators and they inhabit 'survival machines' or vehicles. Individual bodies are vehicles, not replicators: they 'don't replicate themselves; they work to propagate their replicators' (254). So, ultimately, 'the prime mover of all life, is the replicator' (264). Individual bodies as we know them are merely mutations, accidents of evolution – mere survival machines. They 'did not have to exist. The only kind of entity that has to exist in order for life to arise, anywhere in the universe, is the immortal replicator' (266). Evelyn Fox Keller distinguishes this view of the organism as a chemical machine from the machine metaphors of the eighteenth and nineteenth centuries, where the machine was 'capable only of executing the purposes of its maker' (Keller 1992: 113). The twentieth century organism is 'a cybernetic machine par excellence: absolutely autonomous, capable of constructing itself, maintaining itself, and reproducing itself' (113). The entire system is self-enclosed and impervious to external influence. Its only purpose is its own survival and reproduction, or more specifically, 'the survival and reproduction of the DNA that is said to program and to "dictate" its operation' (114). In Dawkins's terms, we are 'survival machines' or 'lumbering robots' housing self-preserving and inherently selfish genes. In response to such a marked and resonant image, Keller poses the basic but fundamental question of where exactly that image comes from. What motivates it? To what extent can we allow it to remain, in its own terms, a purely technical or descriptive image?

> To what extent can our contemporary scientific description of animate forms, culminating in the description of man as a chemical machine, be said to be strictly technical, and to what extent does it actually encode particular conceptions of man – conceptions that derive not so much from a technical domain as from a social, political, and even psychological domain?
> (Keller 1992: 114)

For her, the incorporation of social values into the substance of scientific theory is 'unwitting' – an effect of language which subverts 'objective, value-free description' (Keller 1992: 127). For Richard Lewontin (1993), allied with Steven Rose in the anti-deterministic, anti-reductionist, anti-Dawkins camp, biology simply *is* ideology and there are only different ideological positions to adopt. Both locate the contemporary emphasis on atomic individualism, autonomy and competition clearly within the Darwinist world-view, while Lewontin is more concerned to underline the social and political economy of Darwinism. Darwin's theory of natural selection was influenced by Thomas Malthus, an economist concerned with population growth and the scarcity of socio-economic resources. Lewontin suggests that 'what Darwin did was take early nineteenth-century *political* economy and expand it to include all of *natural* economy' (Lewontin 1993: 10).³ Not only did his theory of evolution by natural selection bear 'an uncanny resemblance to the political economic theory of early capitalism' (10) but his theory of sexual selection reflected the gender relations established in a Victorian patriarchal society where 'the chief force is the competition among males to be more appealing to discriminating females' (10). Lewontin maintains that the rise of industrial capitalism produced a new view of society 'in which the individual is primary and independent, a kind of autonomous social atom that can move from place to place and role to role' (11). At this point society is regarded as being the effect rather than the cause of individual characteristics. Individuals make society, and the driving force of the economy is consumerism. Crucially, this newly atomised view of society is matched by a newly reductionist view of nature:

> Now it is believed that the whole is to be understood *only* by taking it to pieces, that the individual bits and pieces, the atoms, molecules, cells, and genes, are the causes of the properties of the whole objects and must be separately studied if we are to understand complex nature
> (Lewontin 1993: 12)

Another feature of the transformation in scientific views is, according to Lewontin, 'the clear distinction between causes and effects' (12). So, for Darwin, organisms were the effects of the environment: 'they were the passive objects and the external world was the active subject' (12). Organisms find the world as it is and either adapt to it or die. Molecular biology replaces external cause with internal cause and regards the organism as the effect of its genes:

> The world outside us poses certain problems, which we do not create but only experience as objects. The problems are to find a mate, to find food, to win out in competition over others, to acquire a large part of the world resources as our own, and if we have the right kind of genes we will be able to solve the problems and leave more offspring. So in this view, it is really our genes that are propagating themselves through us.
>
> (Lewontin 1993: 13)

Biochemist Steven Rose concurs with (geneticist) Richard Lewontin over the ideological basis of biology, and is equally critical of reductionism and simple cause and effect determinism. He attributes these limitations within biology to the influence of physics (pre-eminent in the hierarchy of science) and the quest for universal laws expressed in simple and/or mathematical terms. One consequence of the primacy of physics 'has been the power of technological metaphor in biology, whereby living systems become analogised to machines' (Rose 1997: 19) – hearts as pumps, brains as computers and so on. The reverse influence, the biologisation of technology has come along only recently where parallel computer processing or neural networking is based on 'analogies with the organisation of the brain' (19). Rose usefully breaks down the meaning of reductionism, or rather he breaks down the criticisms of the meaning of reductionism. To some, he suggests, 'it is an unqualified boo-word' which represents 'a way of emptying life of its manifold rich meanings, of turning individual personal experience into chemistry and physics, mere mechanisms' (73). For others, the critique is more systematic and politically coherent. These others include Rose himself and Lewontin who see 'modern science as the inheritor of nineteenth-century mechanical materialism, itself tightly linked ideologically to a particular phase of the development of industrial capitalism' (Rose 1997: 73). He also outlines the feminist critique of reductionism which relates it to limited, rationalist masculine ways of thinking. These typically reject 'the validity of subjective experience' in favour of a notion of objectivity (74). Above all though, reductionism is a methodology which involves isolating particular phenomena in order to study and understand them. It is a way of limiting variables or eliminating extraneous information. In this sense it is a means to an end, and echoing the tenets of structuralist semiotics (where the gene is the unit of life, so the sign is the unit of language), this is one of the ways in which Dawkins defends his own practice: 'the properties of complex wholes can be explained in terms of the units of which those complex wholes are composed' (Dawkins 1986: 73). Dawkins goes on to distinguish what he calls 'step-by-step' reductionism from the 'precipice' version:

> When a computer goes wrong, a repair engineer who is a step-by-step reductionist (they always are in practice), would try to isolate the major part of the computer in which the fault lies before narrowing it down by successive steps, until reaching a unit that can be replaced economically, say a faulty chip or a component-board. The precipice reductionist (who does not really exist) lays an oscilloscope lead on every one of the billion connections in the computer, and dies of old age before find the fault. The anti-reductionist engineer (who also does not really exist) blames the 'whole' computer and throws the whole computer away.
>
> (Dawkins 1986: 75)

Precipice reductionists are, according to Dawkins, the straw men created by anti-reductionists like Steven Rose, and step-by-step reductionism is the only viable scientific methodology, the alternative to which is 'religious mysticism' or a misguided holism. In rejecting the concomitant accusation of cause and effect determinism, Dawkins inadvertently raises an interesting question as to the fitness of Darwinian evolutionary theory as a means of explaining complex (or in his terms 'distorted') societies. Generally speaking, bodies are machines for propagating genes, brains are the body's on-board computers and behaviour is the output of those on-board computers. But, Dawkins asks, does this general statement tell us anything about human social behaviour? It is quite possible, he answers, that it does not: 'It could be that, although the human brain exists in the first place as part of a gene-preserving machine, the conditions under which it now lives have become so distorted that it is no longer helpful to interpret the detailed facts of human social behaviour in Darwinian terms' (Dawkins 1986: 66). This might seem to reinforce Rose's point that reductionism 'often seems to work, at least for relatively simple systems' (Rose 1997: 78). The problem is of course that living systems (not to mention societies) are not simple, not uniform and involve many interacting variables. For Rose, the constraints of a reductionist methodology might be appropriate to the study of, for example, chemistry because '(so far as is known) the chemical world is the same everywhere', but they function as straitjackets in the living world where 'the exception is nearly always the rule' (79).

The internal debate within biology is, as Hayles indicates, situated within wider cultural narratives of subjectivity, power and knowledge where the competing claims of modernity and postmodernity are articulated with respect to complex (information) systems. Such claims are not resolved either within or between the differing Darwinian approaches. Both Rose and Lewontin insist on a notion of individual agency which, although opposed to Dawkins's brand of individualism is still at odds with a broad post-structuralist notion of subjectivity. In opposition to Dawkins's notion of the passive organism (determined by its genes), they posit the active organism which exists in a dynamic relationship to the environment and which is capable of acting upon it. Transposed into human terms 'this means that we have the ability to construct our own futures, albeit in circumstances not of our own choosing' (Rose 1997: 309). This active/passive dichotomy in biology resonates with the culturalist/structuralist dichotomy in cultural studies which has given rise to polarised views of the active or passive cultural consumer.[4] Whereas the structuralist position and its passive subject may be regarded as anti-humanist, the culturalist position (formulated as a reaction against structuralism) with its active subject is essentially humanist (Hall 1981). Similarly, in terms of biological debate, it would seem that Rose and Lewontin are keen to rescue the organism from its 'lumbering robot' status conferred by the 'ultra-Darwinist' Dawkins, and to attribute to it what we might call a revised humanism or limited agency.[5] Organisms and environments share a degree of agency and Rose is clear that 'the idea of a stable, unchanging environment, affected only by human and technological intervention, is a romantic fallacy' (Rose 1997: 307). Rose characterises this concept of the organism not as autonomous but as autopoietic.[6] Autopoietic or self-producing organisms are 'active players' in their environment; neither determining nor determined by it (245). In Rose's view, the history of life (or the story of evolution) is 'one of organisms, not of mere molecules' and in terms of origins he is in no doubt that 'chickens . . . came before eggs' (270). The internal debate within biology regarding the ontological and epistemological status of agency is enacted in alife through the relation between autonomy and autopoiesis. A novel but influential development in theoretical biology, autopoeisis (Maturana and Varela 1980 [1971]) is in part a rejection of the reductionist definition of life as information and of the endgame played out within molecular biology.

Molecular biology – an endgame

One of the primary concerns in critical cultural studies of the life sciences is 'the installation of DNA as the sovereign agent of life' (Doyle 1997: 6). Sharing Evelyn Fox Keller's view of the performative power of language, Richard Doyle looks at how the rhetoric of molecular biology has 'ordered' and defined the body by arranging it around a molecule. His reading of the rhetoric of molecular biology goes beyond a hermeneutics of heredity and life to an attempt to mark out the vectors 'that formed the shape of our bodies today – those bodies whose illness, intelligence, and sexual preference is "ordered" through the gene' (5). For Doyle, 'the body that fits, and is fitted to, molecular biology' is a postvital body (8). The postvital body is, for him, a matter of contemporary fact which incorporates the cyborg, the virtual body, the fetishised foetus and other figures which 'act out the technoscientific construction of the body as a site of genetic remote control' (8). The postvital body is a lifeless construct within the symbolic and material framework of molecular biology. It is no more than the effect of a molecule: 'an extension or supplement to the real, timeless, deathless bit of immanence known as DNA' (8). Doyle's project does not merely consist of tracing the effects of the statements of molecular biology. Rather, he is concerned with what has been 'unsaid' – what in Christopher Bollas's (1987) psychoanalytic terms might be described as 'thinking the unthought known'. As if a genealogy of the life sciences would prove to be too teleological (inscribing 'a temporality of before and after, a logic of either/or that is not prima facie applicable to historical change': Doyle 1997: 9), Doyle chooses a rhizomatic approach to the connections and particularly to the gaps of discourse. There is, he suggests 'no once-and-for-all branching of discourse or history; there are murmers and shouts and scents of possibility at every place and moment' (9). Nevertheless, it is to Foucault that he turns in order to say 'the great unsaid of the life sciences' which is that not only has life ceased to exist but it never existed at all before the nineteenth century. What this means, more precisely, is that before the nineteenth century the conditions of possibility, the conceptual matrix framing biology as a science of life had yet to be articulated (10). Before then, Foucault argues, 'all that existed was living beings, which were viewed through a grid of knowledge constituted by natural history' (Foucault 1997: 127). The shift from natural history to biology incorporated the transition from the study of living beings to the study of life itself, where life is not only interiorised but comes to occupy a 'sovereign vanishing point within the organism' (in Doyle 1997: 10). It becomes invisible, a secret at once outside and deep inside the bodies of individual organisms which had been studied and classified on the basis of superficial or visible criteria (Foucault 1997: 273). Doyle argues that this reorganisation of living beings, the objects of the life sciences, facilitated the development of both vitalism and molecular biology:

> Despite this apparent opposition, both vitalism, the idea that life exceeds known physiochemical laws, and molecular biology, the science that has claimed the reduction of life to those same physiochemical laws, relied on an unseen unity that traversed all the differences and discontinuities of living beings, 'life'.
>
> (Doyle 1997: 11)

No such unity was available to the natural sciences, based as they were on visual comparisons between organisms as radically different as, for example, vertebrates and invertebrates. But Foucault argued that the construction of life as an invisible unity made biology possible. Ultimately, it 'ordered' and organised molecular biology around a single quest – to discover the 'secret' of life. And if biology annexed the body by regarding 'beings as the mere epiphenomena of life, a secret force beyond being' (Doyle 1997: 13) molecular biology has elided the body now that the secret of life is out – and all there is is information. Where the body once had this deep unity, the postvital body is, for Doyle, a memorial. It is flat, depthless and transparent with 'nothing behind or beyond it' (13). The discovery of DNA as the secret of life, and particularly the map of the human genome may be said to mark the end of the story of life. Biology is faced with the realisation 'that there is nothing more to say' (20). There is no-thing at the heart of the organism which thus becomes a virtual object upon which a new story of information (as life) is based: 'In molecular biology, the end of the grand narrative of life, the "death" of life is overcome through a new story of information, in which a sequence of "bits" is strung together or animated into a coherent whole through the discourse of "that is all there is", a story of coding without mediation or bodies' (22). With the advent of the human genome project, secrets give way to sequences – of nucleic acids which are really 'all there is' (22). This may be said to be something of a paradigm shift in biology. Life is now displaced from the body and dispersed 'through the narratives and networks that make up the interpretations of genetic databases' (24). But what of the status of this information? For Doyle it is a new sublime, matching the unrepresentable vision of life in the nineteenth century with 'the story of resolution told in higher and higher resolution' – that is 'the continual story that there is nothing more to say' (20). Doyle's sense that the genetic language of life as information is exhausted in molecular biology is Beckettian and evokes a form of endgame: 'Finished, it's finished, nearly finished, it must be nearly finished' (Beckett: 1964 [1958]: 12). The ontology of information, like that of the Word, succumbs to a form of nihilism.

Evelyn Fox Keller (1992) is concerned with the psychology of molecular biology's endgame. She explores that perennial motif underlying scientific creativity – 'namely, the urge to fathom the secrets of nature, and the collateral hope that, in fathoming the secrets of nature, we will fathom the ultimate secrets (and hence gain control) of our own mortality' (Keller 1992: 39). Secrets have a function which may be expressed in psychological, social and economic terms. They articulate a boundary: 'an interior not visible to outsiders, the demarcation of a separate domain, a sphere of autonomous power' (40). What is more, life has traditionally been seen as a secret kept by women from men. The concept of life which Keller employs is one of creation, or more properly reproduction: 'By virtue of their ability to bear children, it is women who have been perceived as holding the secret of life' (40). And, due to the historical identification of women with nature 'it is a short step from the secrets of women to the secrets of nature' (40). Keller argues that throughout most cultural traditions the secrets of the female body of nature have been threatening and/or alluring to men 'simply by virtue of the fact that they articulate a boundary that excludes them' (40). So, she suggests, Western science is precisely culture's method for undoing nature's secrets and it is based on a highly gendered, patently sexual set of metaphors: 'The ferreting out of nature's secrets, understood as the illumination of a female interior, or the tearing of Nature's veil, may be seen as expressing one of the most unembarrassedly stereotypical impulses of the scientific project' (Keller 1992: 41).

The story of scientific enlightenment is based on the inversion of epistemological binaries: surface and interior, invisible and visible, light and dark, masculine and feminine. It is a story which constantly needs to be retold, or 'a drama in need of constant reenactment at ever-receding recesses of nature's secrets' (41). In molecular biology, the drama is heightened as Watson and Crick's stated aim was a 'calculated assault on the secret of life' (Keller 1992: 42). Like Steven Rose, Evelyn Fox Keller attributes this unusual and unfashionable hubris in biological research to the influence of physics. Rather than giving biology new skills, physics gave it a whole new attitude and conviction that life's secrets could be found. It enabled biology to claim 'that in the decoding of the mechanism of genetic replication, life's secret *had* been found' (43). In his introduction to the 1999 edition of *The Double Helix*, Steve Jones reframes James Watson's hubristic tale as 'as much an account of the sociology of science as of science itself' (Jones 1999: 1). Chiefly, he reflects on the conflict between science and feminism which is played out through the characterisation and narrativisation of Rosalind Franklin: 'Watson's discussion (somewhat redeemed by a curiously embarrassed postscript) of the role of Rosalind Franklin in the work ("The thought could not be avoided that the best home for a feminist was in another person's lab") is particularly offensive to the modern reader' (Jones 1999: 2). Rank sexism does not, he maintains, detract from Watson and Crick's cleverness and their right to 'feigned modesty' at having discovered the structure of DNA: 'It has not escaped our notice that the specific pairing we have postulated immediately suggests a possible copying mechanism for the genetic material' (in Jones 1999: 3). Between 1953 and 1967 Crick contributed to the quest to crack the code and reveal the secret of life or how, exactly, DNA 'makes' proteins. The 'so-called code' consists of sixty-four codons (units consisting of three bases) 'specifying the assembly of twenty amino-acids into myriads of exquisitely complex proteins' (Kay 2000: 3). Watson's subsequent work on RNA produced his central dogma 'that DNA makes RNA makes protein' although 'he had not then known what "dogma" actually meant' and was 'confused' in his assertion of the one-way, linear flow of information: 'the flow of information may be reversed' (Jones 1999: 4). What appeared, in 1953, to be a clear and legible code, now appears to be somewhat 'baroque' in Jones's terms. Far from being a simple set of instructions, DNA may actually have a highly convoluted structure (4). Jones points out that the current realisation that 'the working genes of higher organisms make up only a small proportion of their DNA' came as something of a shock to the founders of molecular biology, and he seems to substantiate the idea that genetic language is not only ambiguous (Kay), but also largely empty, redundant, even partially extinct and far from revelatory:

> Often genes themselves are interrupted by strings of bases that code for nothing. The whole sequence, discontinuous though it may be, is read off into RNA and – with a perversity alien to physics – edited to cut out redundant sections. Even worse, much of the DNA consists of repeats of the same sequence. A series of letters is followed by its mirror image, and then back to the original, thousands of times. Scattered among all this are the corpses of genes that expired long ago, and can be recognised as such only by their similarity to others that still function. The image of genetic material has changed. No longer is DNA a simple set of instructions. Instead, it is a desert of rigidity and waste mitigated by decay.
>
> (Jones 1999: 4)

More Beckett than baroque then, the death of genetic language informs the beginning of the end (the end of the beginning) of the story of life as told by and against molecular biology at the start of the twenty-first century.

Through her association between secrets of life and secrets of death, and for 'historical and psychological reasons', Keller goes on to juxtapose the story of the discovery of DNA with the development of the atomic bomb. One of the historical events which heightened the authority of physics, the development of the A bomb was contemporaneous with the emergence of molecular biology. What is more, the making of the bomb 'was perhaps the biggest and best kept secret that science ever harbored' (Keller 1992: 43). From the secrecy and interiority of the Manhattan Project came '"Oppenheimer's baby" – a baby with a father, but no mother' (44). Brian Easlea (1983) argued that the metaphor of pregnancy and birth prevailed during the production and testing of the atomic and then the hydrogen bomb. Maternal procreativity was effectively co-opted, but then of course the bomb was dropped and the secret of life became the secret of death: 'I am become Death, the shatterer of worlds' (Oppenheimer in Keller 1992: 45).

The story of DNA may also be said to be one of masculine autonomous creation. Keller cites an article by Mary Jacobus,[7] in which 'the author finds *The Double Helix* notable not for a simple but for a complex elision of both the real and the symbolic woman' (Keller 1992: 51). Keller adds to Jacobus's examples (where women are referred to as 'popsies') by suggesting that

> the story of the double helix is first and foremost the story of the displacement and replacement of the secret of life by a molecule. Gone in this representation of life are all the complex undeciphered cellular dynamics that maintain the cell as a living entity; 'Life Itself' has finally dissolved into the simple mechanics of a self-replicating molecule.
> (Keller 1992: 51)

The discovery of DNA, Keller argues, has not produced death but has given rise 'to a world that has been effectively devivified'. The base pairing of the double helix is then, not life-threatening, but 'lifeless' (Keller 1992: 52). The production of lifeless and life-destroying forces converge through the displacement of flesh and blood – of bodies, and the exclusive reign of the omnipotence fantasies out of which both are produced may well signify 'a kind of ultimate psychosis' (55).

In *The Code of Codes*, Kevles and Hood (1992) offer a range of social, historical and technical essays on the holy grail of molecular biology – the Human Genome Project. Genomics is integral to the masculinist creation of alife at the end of the twentieth century and Kevles and Hood suggest that both legitimate fears and 'science fiction fantasies' regarding the genome might be diminished if not abolished by being closely 'tied to the present and prospective realities of the science and its technological capacities' (1992: viii). For them, fantasy is not a facet of science but of its misrecognition. Accordingly, they present an essay by the Nobel scientist Walter Gilbert, who co-developed one of the major techniques for DNA sequencing and is 'an eloquent advocate for the genome project' (Kevles and Hood 1992: 386). In 'A Vision of the Grail', Gilbert paints the simplest picture of the human genome sequencing project as 'an attempt to define all of the genes that make up a human being' (Gilbert 1992: 83). Still in the name of simplicity, he goes on to suggest that knowledge of the human genome is indeed the basis of all knowledge about the body and most fundamental 'secret' of life. Being able to elucidate 'our' DNA sequence is a 'historic step forward in all knowledge', and although it will be necessary to return to the sequence in order to gain more and more knowledge over time, 'there is still no more basic or more fundamental information that could be available' (83).

Gilbert outlines three phases of the project. The first phase is referred to as physical mapping and involves breaking down the two meter long strip of DNA 'into ordered smaller fragments' (Gilbert 1992: 85). The second phase involves determining the sequence of all the base pairs of nucleic acids in all the chromosomes, and the third phase – 'understanding all the genes' – will be the problem of biology throughout the next century' (85). In Gilbert's view, molecular biology is now the driving force of all biological research and 'all sorts of questions are now being studied by finding a gene and seeing what it does to the organism, or by deducing a feature about the pattern of inheritance' (93). By referring to 'all sorts of questions', Gilbert is keen to suggest a degree of open-endedness at least in the social dimensions of the project, but his belief in genetic determinism is clear. There are limitations to the project in his view; it can tell us what specifies the human organism (what makes us human or different from other animals) but it cannot tell us how we differ from one another. This is because 'molecular biologists generally view the species as a single entity, sharply defined by a set of genes' (84). Genes cause or determine the organism and it is this cause and effect relationship – or predictability – which defines the usefulness of the project, notably in medicine and the treatment of disease. The possession of a genetic map and the DNA sequence of a human being will, according to Gilbert, transform medicine and not simply by identifying the genes which cause rare and specific genetic diseases. He claims that there will be genetic causes for common diseases such as heart disease, cancer or high blood pressure: 'Along with many other common afflictions, these will turn out to have multiple genetic origins in populations, as will such mental conditions as schizophrenia, manic-depressive illness, and susceptibility to Alzheimer's disease' (94). The presence of 'multiple' genetic origins does not seem to trouble Gilbert's linear cause and effect determinism and the likely presence of multiple social, psychological and economic factors in disease is simply not considered. It is not that he is ignorant of social and ethical issues, rather that he regards molecular biology and its ultimate manifestation as being separate and prior to them – a realm of pure science which is itself asocial and apolitical but which can have social and political 'effects'. This is, of course, typical of mainstream science and its claims to neutrality and therefore truth – what Haraway refers to as disembodiment. An 'effect' of being able to recognise 'defective' genes in the embryo will be to exacerbate the abortion controversy 'with which society is wrestling at the moment' (Gilbert 1992: 95). Nevertheless, Gilbert's own views are narrowly disguised in his statements of fact. The ability to isolate these same defective genes 'will mean a constant improvement and extension of prenatal diagnosis, which will lead to the elimination of much human misery' (95). So-called 'positive eugenics' is more likely to enter the back door of molecular biology than 'negative eugenics' due in part to the lessons learned from history. Thus, Gilbert declares that 'racism is a danger' which may be alleviated by extending the project of DNA sequencing from one individual to 'different individuals from around the world' and so reflecting 'an amalgam of the underlying human structure' and our 'common humanity' (96). Gilbert here refers to the Human Genome Diversity Project set up as a result of criticisms of normalisation inherent within the Human Genome Project. Finally, Gilbert (like Dawkins) struggles and falters over that small but persistent question of human agency. While we are (necessarily within his own paradigm) 'dictated by our genetic information' and will have to come to terms with this new view of ourselves, we are nevertheless 'not slaves of that information' (96). Like Dawkins, Gilbert creates a category of 'shallow genetic determinism' to which he does not subscribe because it is quite clearly 'unwise and untrue' (96). By creating this repository of faulty thinking about the extent of genetic influence, Gilbert frees himself and his discipline from a kind of totalitarianism or at least totalism which does not allow the possibility of free will (and therefore responsibility). This move ensures that where power is (re)located to the genes, responsibility still lies with the individual.

Geneticist Richard Lewontin is unequivocal about the political implications of the Human Genome Project. He outlines both the political economy of the project and its relation to sociobiology and eugenics. He suggests that genes are now held to be responsible not simply for medically defined diseases but for social ills such as alcoholism, criminality, drug addiction and mental disorders. We are led to believe that if we can find the genes that underlie these problems, then all will be well in our society. The human genome sequencing project is, for him, the current manifestation of this belief in the importance of our inheritance as a social as well as medical determinant. The project is, in the second instance very big business: 'a multibillion-dollar program of American and European biologists that is meant to take the place of space programs as the current great consumer of public money in the interest of conquering nature' (Lewontin 1993: 46). If the first impetus for the project lies in the ideology of simple unitary causes, then the second 'is a rather crass one' which revolves around power and money: 'The participation in and the control of a multibillion-dollar, 30- or 50-year research project that will involve the everyday work of thousands of technicians . . .is an extraordinarily appealing prospect for an ambitious biologist. Great careers will be made. Nobel Prizes will be given' (51).

The project will clearly profit biotechnology companies with which professional scientists are often associated, and it will consume 'vast quantities of chemical and mechanical commodities' (Lewontin 1993: 52). There are commercial machines which manufacture DNA and there are machines that automatically sequence DNA (Haraway 1997). These require a variety of chemicals which are sold at 'an immense profit by the companies that manufacture the machines' (Lewontin 1993: 52). Kevles (1992) points out that the instigators of the project, including Walter Gilbert and James Watson, regarded it as a 'Big Science' endeavour comparable to the conquest of space but with more certain economic rewards: 'It did no good to get a man "a third or a quarter of the way to Mars . . . However, a quarter or a third . . . of the total human genome sequence . . . could already provide a most valuable yield of applications"' (in Kevles 1992: 22). Moreover, industrialists argued that the project would be 'essential to national prowess in world biotechnology, especially if the United States expected to remain competitive with the Japanese' (26). Kevles outlines the growth of Japanese and European interest during the 1980s, culminating in the formation in 1988 of an international organisation – the Human Genome Organisation (HUGO) – described as 'a U.N. for the human genome' (28). Indeed, so big is the science and business of the project that one of the most compelling realities surrounding it 'was the consequences of remaining out of the human genome sweepstakes' (29). Such a compelling reality effectively quashed concerns within the scientific community that human genome sequencing was, in a sense, premature – 'somewhat as if one were "to list the millions of letters in an encyclopedia without having the power to interpret them, ignoring practically all vocabulary and syntax"' (29).

Lewontin (1993) also sets out the project in three phases where the first two involve physical mapping and the sequencing of nucleotides in DNA. It is after the second phase, when the genome project as such has ended, that the 'fun' begins 'for biological sense will have to be made, if possible, of the mind-numbing sequence of three billion A's, T's, C's and G's' (1993: 62). Lewontin asks what exactly this is going to tell us about health, disease, happiness, misery and the meaning of life. One of the real difficulties in making sense, or deriving causal information from DNA messages is hermeneutical as 'the same "words" have different meanings in different contexts and multiple functions in a given context, as in any complex language' (66). In other words, the units of life, like the units of language, are polysemic and the cell has to decide how to read a given genetic message. Unfortunately, Lewontin points out, we don't know how it does that. If polysemy represents one problem, polymorphism represents another. Every human genome is different, and 'the final catalogue of "the" human DNA sequence will be a mosaic of some hypothetical average person corresponding to no one' (68).

Polysemy and polymorphism undermine the legitimacy of the Human Genome Project by disrupting the relationship between genetic causes and medical or behavioural effects. Lewontin argues that we do not know all of the functions of the different nucleotides in a gene, or how the context in which a nucleotide appears will affect DNA interpretation. Moreover 'because there is no single, standard, "normal" DNA sequence that we all share, observed sequence differences between sick and well people cannot, in themselves, reveal the genetic cause of a disorder' (1993: 69).

Lewontin concedes the limited benefits of gene therapy where mutated genes are replaced or removed, while outlining some of the objections to germ-line therapy where such alterations would be inherited by future generations in no position to make a choice of their own. He points out that genetic disease is by no means the major cause of death in developed countries where heart disease, cancer and strokes account for 70 per cent and 60 million people suffer from cardiovascular disease. This however does not prevent the 'rage for genes' and even the editor of *Science* magazine from having visions of genes for everything from disease to drug abuse. The appeal of simple causes for complex phenomena can, as Lewontin makes clear, be rather dangerous:

> What we had previously imagined to be messy moral, political, and economic issues turn out, after all, to be simply a matter of an occasional nucleotide substitution. While the notion that the war on drugs will be won by genetic engineering belongs to cloud cuckoo land, it is a manifestation of a serious ideology that is continuous with the eugenics of an earlier time.
> (Lewontin 1993: 72)

As Kay, Jones and Doyle indicate, the hopes and fears expressed about molecular biology's endgame in the early 1990s seem by the late 1990s to be misplaced and to an extent replaced by a characteristically post-structuralist approach to genetic language. If this more nihilistic approach is the opposite inverse of conventional naturalism, an almost apocalyptic lament for the absence/end of presence, then this too may be said to be misplaced and replaced at the turn of the new millennium which coincides with the completion of the first draft of the human genome (chapter 6). At this point, the image of genetic material undergoes further changes and the spectre of eugenics is transmuted through the ideology of biotechnological consumerism just as the status of the individual human self is transmuted through (the) species.

Sociobiology and eugenics

In his essay on the history and politics of the human genome, Daniel Kevles (1992) argues that the scientific search for the 'Holy Grail' of biology dates back to the rediscovery, in 1900, of Gregor Mendel's laws of inheritance. These were based on peas, but it soon became apparent that the transmission of dominant and recessive factors (what became known as genes) took place in other organisms as well. The application of Mendel's laws to human organisms was motivated, according to Kevles, by the association between heredity and eugenics. He defines eugenics as 'the cluster of ideas and activities that aimed at improving the quality of the human race through manipulation of its biological heredity' (Kevles 1992: 4). The aim of improving the species through breeding may, as he suggests, go back to Plato, but the elaboration of eugenics as a science is attributed to Darwin's cousin, Francis Galton, in the late nineteenth century. Galton sought to eliminate 'undesirables' from the human species and to increase the number of 'desirables'. His ideas took hold after the turn of the century and mostly in the UK, US and Germany among white middle-class professionals concerned about increasing signs of social degeneration in urban industrial societies. Urban incidents of poverty, crime and disease became attributed to 'bad blood' (Kevles 1992: 5). Essentially, eugenics was an early form of sociobiology where scientists and policy makers sought biological explanations and resolutions to social problems ranging from physical and mental illness to criminality and alcoholism. For example, eugenicist Charles Davenport (1911) maintained that if there was a high incidence of a given characteristic in a family, that characteristic must be biologically inheritable. He argued that 'patterns of inheritance were evident in insanity, epilepsy, alcoholism, "pauperism", and criminality' and was chiefly concerned with the characteristics of different races which he held to be biologically distinct (Kevles 1992: 7). Inter-racial breeding or 'race-crossing' was therefore thought to be 'biologically and socially deleterious' (7). His work influenced studies of 'feeblemindedness' which was thought to be inherited and 'linked to lower-income and minority groups' (7). Kevles suggests that his early eugenics often neglected polygenic complexities (the dependence of characteristics on many genes) in favour of single-gene explanations, and ignored cultural, economic and environmental influences (Kevles 1992: 8).

The eugenics programme took two forms: one referred to as 'positive' eugenics and focused on increasing the population of desirable or 'superior' people; the other known as 'negative' eugenics concerned with eliminating undesirable or 'inferior' people from the species. Elimination might be achieved by discouraging reproduction, or preventing certain groups from entering the population through immigration (Kevles 1992: 9). The positive eugenics programme was never really pursued even though 'eugenic claims did figure in the advent of family-allowance policies in Britain and Germany during the 1930s' and were implied in the 'Fitter Family' competitions in the US during the 1920s. Entrants in this competition had to take an IQ test and the Wasserman test for syphilis (10). Negative eugenics was more clearly manifested in the sterilisation laws in the US and, notoriously, in Nazi Germany. The activities of Josef Mengele and the Nazi death camps produced a backlash against eugenics in the US and the UK, although a kind of 'reform eugenics', free of racial and class bias survived in the vision of British scientists such as J.B.S. Haldane (Kevles 1992: 11).

The development of molecular biology, and particularly Watson and Crick's determination in 1953 of what genes actually are, served to weaken the eugenics movement. Kevles presents it as a case of more knowledge equals less understanding and argues that most human geneticists would agree with Lionel Penrose's (1967) statement that 'our knowledge of human genes and their action is still so slight that it is presumptuous and foolish to lay down positive principles for human breeding' (Kevles 1992: 16). While it might be tempting to suggest that the human genome project has put an end to such modesty, Kevles claims (at least in 1992) that the search for the biological grail has become totally emancipated from eugenics, and that the study of human heredity is purely a means to understand, diagnose and treat disease (18).

In so far as human heredity is in fact manifestly regarded – by some – as a means to understand, perhaps diagnose and possibly even treat social characteristics and behaviours including, for example, homosexuality, then how can it be free from eugenics? Does human genetics not inherit eugenics – if not biologically then historically? What many critics and commentators are concerned with is the short step that exists between supposedly identifying genetic causes and effects and deciding to act upon them. If there is a gay gene which causes homosexuality, what exactly are we going to do with it? In a society structured by inequality and by no means free from prejudice, such knowledge may indeed be a dangerous thing. Eugenicist sentiments are organised around the frequent references to Francis Galton in Darwin's *The Descent of Man* (1871). Darwin compares the effects of natural selection on 'savages' and on 'civilised nations', observing the differential treatment afforded to 'the weak in body or mind' (Darwin 1901 [1871]: 205). Through the provision of asylums, poor laws and medical care, only in so-called civilised societies are the weak encouraged to propagate their kind, and 'no one who has attended to the breeding of domestic animals will doubt that this must be highly injurious to the race of man' (206). An incidental result of the necessary sympathy instinct, the one consistent check on this (atavistic) process is 'that the weak and inferior members of society do not marry so freely as the sound; and this check might be indefinitely increased by the weak in body or mind refraining from marriage, though this is more to be hoped for than expected' (206). Again, Darwin borrows Galton's insight that a major hindrance to increasing the number of men of 'superior' class in civilised society is 'the fact that the very poor and reckless, who are often degraded by vice, almost invariably marry early, whilst the careful and frugal, who are generally otherwise virtuous, marry late in life, so that they may be able to support themselves and their children in comfort' (212). Countering this, there is at least a higher rate of mortality among the 'intemperate', and the effects of urban overcrowding among the poor in general. Darwin constructs historically contingent regulative and static hierarchies of race, gender and class through the story of natural selection which – applied to the animal kingdom in general – celebrates plasticity, variation and diversity. A fundamentalist and a pluralist Darwinism therefore becomes ideologically available. In his discussion of myth and science, François Jacob looks at how the over-extension of a single theory can supplant a heuristic with a sterile belief and he cites Freudianism, Marxism and Darwinism: 'A theory as powerful as Darwin's could hardly escape misuse' (Jacob 1993: 22). As if a theory which, according to Darwin, is at times 'highly speculative' and possibly 'erroneous' (1901 [1871]: 926) were actually innocent, Jacob argues that 'evolution by natural selection was immediately used in support of various doctrines and, since there are no moral values in natural processes, it could just as well be painted in pink or in black and be claimed to uphold any thesis' (Jacob 1993: 23). Thus, socialist, capitalist, colonialist and racist ideologies have co-opted Darwinism as has, more recently, 'scientism' in the form of sociobiology (24). Scientism is 'the belief that the methods and insights of the natural sciences will account for all aspects of human activity'. It underlines the terminology of sociobiology and some of sociobiology's 'unwarranted suppositions and extrapolations from animal to human behaviour' (24). For Jacob, modern biology has little to say about human behaviour (64) but rather more to say about human diversity and the dialogue between the possible and the actual. If that diversity and possibility (of human nature as afforded by natural selection) is ideologically undermined and undervalued – by those who seek to establish conformity – then the more imaginative scenario is afforded by biology itself:

> Diversity is one of the great rules in the biological game. All along generations, the genes that constitute the inheritance of the species unite and dissociate to produce those ever fleeting and ever different combinations: the individuals. And this endless combinatorial system which generates diversity and makes each of us unique cannot be overestimated. It gives the species all its wealth, all its versatility, all its possibilities.
>
> (Jacob 1993: 66)

Like Darwin himself, Jacob has an almost science-fictional vision of the future of (alien) life under natural selection. Mutation, variation and selection may not (just) perfect the species but generate new ones – necessarily at the expense of the old. Here, more than anywhere, size matters: 'We can so far take a prophetic glance into futurity as to foretell that it will be the common and widely-spread species, belonging to the larger and dominant groups, which will ultimately prevail and procreate new and dominant species' (Darwin 1985 [1859]: 458). The discourse of alife which interlinks with that of biology is precisely about the generation by one dominant species of a new dominant species of artificial or alien life-forms. This generative praxis relies on the plasticity and potentiality of life which lies alongside more normative and regulatory tendencies at the heart of Darwinism. A Darwinism of very different hues is available not just to alife but to those disciplines, like feminism, which engage with alife. A dialogic engagement between feminism and alife must therefore amount to more than a response to the threat of sociobiology in a computerised context.

Sociobiology is that discipline which claims to identify biological and specifically genetic causes of human social behaviour. It may stop short of advocating public policy (unlike evolutionary psychology) and is primarily a naturalised view of society. But the naturalisation of social phenomena is itself a deeply political act (it is the politics of de-politicisation) which justifies the status quo and absolves us of the (respons)ability to act. Sociobiology is no less political than eugenics by virtue of being non-interventional – the basic ideology (of genetic determinism) is the same. Lewontin describes sociobiology as 'the ruling justifying theory for the permanence of society as we know it' (1993: 89). It combines both evolutionary and molecular biology and is often translated in coffee-table books, magazines, newspapers and, of course, popular science books. For Lewontin, 'sociobiology is the latest and most mystified attempt to convince people that human life is pretty much what it has to be and perhaps even ought to be' (1993: 89). Lewontin identifies three steps in the construction of sociobiological theory. The first involves describing what human nature is like and identifying the key features which are common to all people everywhere and throughout time. The second step is to argue that those universal features or characteristics are genetic, and the third step is to explain and justify why those genes and not others have come to define human nature. Natural selection is given the principal role in the third and final step and accounts for the inevitability not only of individuals but also of society. Through the course of natural selection, the kinds of genes – and therefore the kinds of societies – we have are regarded as being the only ones we could have. They are the inevitable outcome of natural law; the struggle for existence and the survival of the fittest. And if '3 billion years of evolution have made us what we are' then what is the point in trying to effect change through one hundred days of revolution' (Lewontin 1993: 90) or a few decades of feminism?

Among the universal human characteristics identified within sociobiology are; warfare, male sexual dominance, love of private property and hatred of strangers (Lewontin 1993: 91). These are formed against an environmental background of scarcity, competition and survival of the fittest. For Lewontin, step one of the sociobiological argument already betrays an 'obvious ideological commitment to modern entrepreneurial competitive hierarchical society' and to a deeper ideology of individualism (93) This individualism is reinforced by the assertion of genes for entrepreneurship, male dominance, aggression and so on. The evidence that these characteristics are genetic is, to say the least, circular: 'Often, it is simply asserted that because they are universal they must be genetic' (94). There is, Lewontin suggests, no evidence for the hereditability of traits such as aggression or dominance. Genetics judges things to be heritable if they are shared by close relatives (it is a study of similarity and difference between relatives), but 'similarity between relatives arises not only for biological reasons but for cultural reasons as well, since members of the same family share the same environment' (96). In fact, claims for the hereditability of certain characteristics are often based on the simple observation that parents and children resemble each other in some way. Clearly, the observation that parents and children are similar 'is not evidence of their biological similarity' (96).

The theory of natural selection seems to offer a coherent account of the existence of some characteristics; aggressive ancestors would have left more offspring because they are likely to have eliminated the competition. Similarly, the 'more entrepreneurial would have appropriated more resources in short supply and starved out the whimps' (98). These are plausible stories which can account for the reproductive superiority of some types. Sociobiologists have found it more difficult to account for other suppposed universals, notably altruism. Altruistic behaviour does not obviously produce a reproductive advantage, but one way in which it has been explained is through the theory of kin selection. Altruistic individuals may promote their genes indirectly through supporting their relatives and helping to ensure their reproductive success. This kind of behaviour has been observed in birds (where non-reproductive 'helpers at the nest' help to raise more than the ordinary number of offspring) and is thought to be applicable to humans. General altruism (towards all members of the species) is explained through the theory of reciprocal altruism which is essentially the idea that if I scratch your back, you will scratch mine.

Between them, the accounts of direct advantage, kin selection and reciprocal altruism effectively cover all eventualities of natural selection. They may be plausible accounts, but, Lewontin suggests, at the end of the day that is all they are: 'At the very minimum, we might ask whether there is any evidence that such selective processes are going on at present, but in fact no one has ever measured in any human population the actual reproductive advantage or disadvantage of any human behaviour' (100). In fact, the whole process of sociobiological thinking might be characterised as follows: make a general observation based on superficial evidence, link it to genes without any evidence at all and then tell a plausible (ish) story (101).

One of the founding documents of sociobiological theory, and the most influential is E.O. Wilson's *Sociobiology. The New Synthesis* (1975). Wilson asserts that the (Darwinian) theory of natural selection precedes all philosophical and epistemological questions. The most fundamental philosophical questions, such as 'who am I?' can be answered by recourse to biology which provides the answer that we are merely gene carriers: 'In a Darwinist sense the organism does not live for itself. Its primary function is not even to reproduce other organisms; it reproduces genes, and it serves as their temporary carrier' (Wilson 1980 [1975]: 3). Sociobiology solves the riddle of the chicken and egg by indicating that 'the organism is only DNA's way of making more DNA' (3). The 'morality' of the gene lies in its ability to determine the right mixture of selfishness and (instrumental) altruism in the organism. The right balance of love and hate, fear and aggression is important not to the general happiness of individuals but to the maximisation of gene transmission. Wilson defines sociobiology as 'the systematic study of the biological basis of all social behaviour' (4). It applies in the first instance to animal societies (population structure, castes, communication, physiology and related social adaptations), and then to 'social behaviour of early man and the adaptive features of organisation in the more primitive contemporary societies' (4). Wilson distinguishes sociobiology from sociology which he regards as being structuralist and nongenetic. Nevertheless, he calls for a complete biologisation of the social sciences, or their incorporation into what he calls the 'Modern Synthesis' in which 'each phenomenon is weighted for its adaptive significance and then related to the basic principles of population genetics' (4). Wilson is looking for a universal or general theory of sociobiology where the principal goal 'should be an ability to predict features of social organisation from a knowledge of [these] population parameters combined with information on the behavioural constraints imposed by the genetic constitution of the species' (5).

It is perhaps this predictive feature of sociobiology which has led subsequent theorists, notably Richard Dawkins, to apply it to modern societies. But Wilson (1980 [1975]) does suggest that there might be severe limitations here. In the process of evolving into 'a very peculiar species' (271) – upright, naked and with big heads and therefore big brains we may well be defying, in some sense, natural law: 'The mental hypertrophy has distorted even the most basic primate social qualities into nearly unrecognisable forms' (272). Human societies have evolved to levels of 'extreme complexity' because individuals have 'the intelligence and flexibility to play roles of virtually any degree of specification, and to switch them as the occasion demands' (278). But in Wilson's own terms this does not represent a problem for the legitimacy of the sociobiological project as much as a problem for individuals themselves. Our innate intelligence and flexibility have led to the 'acute inner problem of identity' which characterises complex contemporary societies (278). The identity crises of this very peculiar species are exacerbated within the discourse of genomics encompassing transgenesis and cloning. Meanwhile, sociobiology's doctrinal progeny – evolutionary psychology – fights a rearguard action to shore up the shaky foundations of human nature.

The sociobiological model of culture is, for Wilson, one of indirect genetic influence. He suggests that although culture is not determined by genes, race is, and different cultures can be defined by the behavioural differences between races (274). Wilson reminds us that there are two stages of human evolution: the first, from primates to 'man-apes' took some 10 million years; the second, or cultural evolution began a mere hundred thousand years ago. And even if this second stage is 'mostly phenotypic in nature' it is still built on the 'genetic potential that had accumulated over the previous millions of years' (291). So genes are, at least in this sense, inescapable. For Dawkins, in what is arguably another founding text in sociobiology – *The Selfish Gene* – cultural transmission is analogous to genetic transmission 'in that, although basically conservative, it can give rise to a form of evolution' (Dawkins 1989 [1976]: 189). Dawkins performs a sleight of hand by simultaneously deposing and reinstating the gene as an agent of cultural evolution. He recognises that a basic genetic determinism does not account for the complexities of modern culture and society and states that 'as an enthusiastic Darwinian, I have been dissatisfied with explanations that my fellow-enthusiasts have offered for human behaviour' (191). But genes re-enter his account through the back door as analogical units of replication which he calls 'memes'. Seeking 'a monosyllable that sounds a bit like "gene"', Dawkins shortens the Greek 'mimeme' (meaning imitate) to arrive at his new cultural replicator, examples of which include 'tunes, ideas, catch-phrases, clothes fashions, ways of making pots or of building arches' (192). Memes pass from brain to brain, just as genes pass from body to body; and by a similar process of imitation. Memes, like genes, are subject to the laws of natural selection. A meme, such as the idea of God, may acquire a high survival value because of its 'great psychological appeal' (193), but in general memes, like genes, must display qualities of 'longevity, fecundity, and copying-fidelity' (194). Another common feature of genes and memes is agency: 'Just as we have found it convenient to think of genes as active agents, working purposefully for their own survival, perhaps it might be convenient to think of memes in the same way' (196). In other words, the designation of agency to memes may be a convenience and a metaphor but it may well be one with a rather high 'survival' value – 'we have already seen what a fruitful metaphor it is in the case of genes' (196). Quite. What is more, because memes, like genes, must compete, they may also be said to be selfish. Competition in this case is enforced by conditions of scarcity (of time not space) in the environment of the human brain: 'The human brain, and the body that it controls, cannot do more than one or a few things at once. If a meme is to dominate the attention of a human brain, it must do so at the expense of "rival" memes' (197). Dawkins's theory of memetics has been developed as a general theory of mind and culture by philosophers and psychologists such as Daniel Dennett (1995a) and Susan Blackmore (1999). Dawkins himself denies retreating from memetics, arguing that he developed it initially as an example of 'Universal Darwinism' (the idea that the real unit of natural selection is not specifically the gene but 'any kind of *replicator*') and as a means to cut the selfish gene down to size: 'I would have been content, then, if the meme had done its work of simply persuading my readers that the gene was only a special case: that its role in the play of Universal Darwinism could be filled by any entity in the universe answering to the definition of Replicator' (Dawkins 1999: xvi). In her memetic challenge to human autonomy and agency, even Dawkins fears that Blackmore may be too ambitious (Dawkins 1999: xvi). In effect though, she merely extrapolates from Dawkins's gene-meme determinism, developing memetic agency and autonomy as an increasingly separate and potentially conflicting process (Blackmore 1999: 30). For her, crucially, cultural replication is not synonymous with biological replication, and memetic agency is not encompassed by the agency of genes. In this way she marks a distinction between her theory of memetics and the associated (in fact in her account conflated) fields of sociobiology and evolutionary psychology. Evolutionary psychology, in her account, 'is based on the idea that the human mind evolved to solve the problems of a hunter-gatherer way of life in the Pleistocene age' (36). So human behaviours and beliefs are all adaptations to a prehistorical environment. For example, 'sexual jealousy and love for our children, the way we acquire grammar or adjust our food intake to deal with nutritional deficits, our avoidance of snakes and our ability to maintain friendships are all seen as adaptations to a lifestyle of hunting and gathering' (36). Where evolutionary psychology may, according to Blackmore, usefully underpin memetics – by offering a theory of mind and culture based on natural selection – it is limited by its underlying biologism and rejection of a second replicator: 'In other words, the world of ideas, technology and toys, philosophy and science are all to be explained as the products of biology – of evolution by the natural selection of genes' (115). For Blackmore, the essence of memetic replication is imitation, and it is a form of memetic rather than genetic determinism which can account for elements of modern behaviour such as the separation of sex and reproduction. This separation can be accounted for as a 'mistake' only in terms of genetic advantage, but may be said to work with a concept of memetic advantage where mate selection favours the 'best imitators or the best users and spreaders of memes' (130). Memetic agency may therefore be at odds with genetic agency, and indeed, with any sense of advantage for the species. Memes like genes are necessarily selfish replicators and 'blind' or without foresight – the ability to 'think' long term. Once again, however, Dawkins retracts from the consequences of genetic and memetic determinism by suggesting that our unique capacity for conscious foresight 'could save us from the worst selfish excesses of the blind replicators' (1989 [1976]: 200). Dawkins uses the *metaphor* of genetic and memetic agency as a loop-hole in the law of his 'ultra'-Darwinism. It allows him and his disciples to escape to safer ideological territory. Genetic agency and its cultural analogue is not a metaphor that he, at least, wants to take responsibility for: 'We have the power to defy the selfish genes of our birth and, if necessary, the selfish memes of our indoctrination' (200). Free will, it would seem, simultaneously counters and legitimises determinism. Metaphors of genetic and memetic agency and the ideological loop-hole which Dawkins constructs within them permeate the creation of artificial life forms in artificial life worlds which are, to this extent, biologically determined.

Concepts of human agency articulated in the philosophical realms of social and cultural theory are invariably at odds with biological determinism. Dawkins faces his critics (rather than merely appropriating their concepts) in a defensive essay entitled 'Sociobiology: The New Storm in a Teacup' (1986). The sturdy defence of Darwinism which Dawkins puts up here clearly betrays the ideological and methodological bias of sociobiology, and ultimately its limitations. It becomes clear that the questions that sociobiology asks of human society are predetermined by Darwinian answers. In answering accusations of genetic determinism made by Steven Rose and Richard Lewontin (Rose *et al.* 1984), Dawkins states that: 'Anybody wishing to offer a Darwinian explanation for some biological phenomenon has to postulate genetic variation in that phenomenon, otherwise natural selection could not have led to its evolution' (Dawkins 1986: 77). This then is why sociobiologists 'postulate' genes for 'x' or 'y'. Whatever 'x' is, 'we have to postulate genes for x if we want to discuss the possibility of the Darwinian evolution of x' (77). So, Darwinism is the underlying Truth of social and biological phenomena. But whose Darwinism? While arguing that his genetic determinism is not 'inevitable' or immune from environmental influence, Dawkins and other sociobiologists (arguably including evolutionary psychologists) fail to incorporate a model of mutual influence. Their Darwinism does not seem to provide them with this, whereas other Darwinisms do (Rose 1997; Rose and Rose 2000; Lewontin 1993, 2000; Gould 2000).

Sociobiological theory depends on a linear cause and effect determinism – genes are the cause and societies the effect. The claims of sociobiology are weakened to the same extent as the loosening relationship between genetic causes and behavioural effects, and this loosening relationship is most clearly characteristic of complex modern societies. Here, even Dawkins acknowledges 'that it is downright naïve to look at the social life around us and try to interpret the actions of individuals directly in terms of survival value or gene preservation' (1986: 66). Dawkins then has to invent a non-deterministic or indirect genetic determinism, or rather, he has to retreat to more strictly biological origins whose evolutionary trajectory stops short of modern societies, multiple causation and complexity. Dawkins puts this rather well himself:

> Our bodies, then, are machines for propagating the genes that made them; our brains are the on-board computers of our bodies; and our behaviour is the output of our on-board computers. This is all very well as a general statement, but does it tell us anything, in particular, about human social behaviour? It is quite possible that it does not. It could be that, although the human brain exists in the first place as part of a gene-preserving machine, the conditions under which it now lives have become so distorted that it is no longer helpful to interpret the detailed facts of human social behaviour in Darwinian terms.
>
> (Dawkins 1986: 66)

Recent debates in evolutionary psychology take up, in effect, where sociobiology left off, rendering cause and effect genetic determinism more complex and seeking to restore the principles of natural selection within the context of modern human behaviour. Where sociobiology is foundational to alife as-we-know-it, evolutionary psychology looks likely to become a feature of alife-as-it-could-be in as much as the discipline is in search of increasing biological authority.

Evolutionary psychology and the 'Darwin Wars'

Andrew Brown casts the Darwin Wars as 'nastier than most' scientific disputes (1999: ix). For him, the main reason for this is that they are about the nature and importance of human beings (more specifically, 'mankind') – an issue which has the status of a scientific (as opposed to Judaeo-Christian) belief. Darwinian beliefs may be malleable and have been used 'to justify anarchy, fascism, liberal capitalism, and almost anything in between', but they have the force of a moral imperative. All justifications, says Brown (1999: ix), agree that 'the study of our evolution can help us discover how we ought to live'. Being central to philosophy, ethics and a sense of personal salvation, the subject of Darwinism 'is so important, getting it right is rewarded with great fame and large amounts of money' (x). Clearly, the scale of the prize is not unconnected to the scale of the dispute. Lewontin (1993) and Rose (1997) argue that organisms result from the interaction between genes and environment, and their work discusses the relationship between biology and ideology. Dawkins (1976) and Dennett (1995a) argue that organisms are no more than vehicles or 'survival machines' for their selfish genes, and that this may well have unpleasant implications but is merely a statement of fact. There may be no room for ideology in Dawkins's biology, but he is at the centre of a vitriolic battle over significant ideological territory. Brown narrates the Darwin Wars since the early 1970s in terms of allegiances and oppositions to Dawkins and his apparent ultra-Darwinist genetic determinism established in *The Selfish Gene*. Dawkins is regarded as the key representative or populariser of sociobiological arguments set out by E.O. Wilson in 1975, despite his clear but less popularised disclaimers (Dawkins 1986). In the preface to the 1989 edition of his most influential book, Dawkins states that he is not recommending a morality based on evolution, but rather, simply describing the process of evolution. He insists that he is not advocating a biological code of behaviour (3). This revision of his original thesis precisely predicts the way in which the Dawkinsean discourse of the 1990s – evolutionary psychology – revises (and restores) the Darwinian discourse of the 1970s – sociobiology. The argument at the turn of the century appears to be over what Darwin really said but is, in fact, about what Darwin really meant, and therefore what Darwinian ideology is. Where Dawkinsean scientists remain allergic to the concept of ideology, Dawkins's opponents do not and their arguments insist on the materiality of his 'novel image' or metaphor (Rose 1997; Rose and Rose 2000; Lewontin 1993; Gould 2000). Stephen Jay Gould and Richard Lewontin led the critique of sociobiology and Brown points out the irony that the development of evolutionary psychology from sociobiology was, in ideological terms 'a triumph for Gould and Lewontin, who have seen almost all of their original objections incorporated' (1999: 147). In so far as their incorporation serves to refine and purify sociobiology, Gould remains critical of and reviled within the field of evolutionary psychology, whose exponents then claim that 'the central problem is that Gould's own exposition of evolutionary biology is so radically and extravagantly at variance with both the actual consensus state of the field and the plain meaning of the primary literature that there is no easy way to communicate the magnitude of the discrepancy in a way that could be believed by those who have not experienced the evidence for themselves' (Tooby and Cosmides in Brown 1999: 151).

In their seminal text, Barkow, Cosmides and Tooby (1992) define evolutionary psychology as simply that which is informed by evolutionary biology 'in the expectation' that the architecture of the human mind will be elucidated by the process which led to its formation (Barkow *et al.* 1992: 3). Evolutionary psychology also provides a bridge between biology and sociology (3) and is an instance of the 'conceptual integration' which the authors (after E.O. Wilson) perceive as being for the good of the human sciences. The human sciences must, they argue, become compatible both with the natural sciences and with each other – must constitute an epistemological unity in order to be viable. This then, subtextually, is a quest for a theory of everything which relies on a notion of universal Darwinism. The central premise of evolutionary psychology betrays what Rose and Rose (2000: 8) refer to as its 'revisionist' nature. There is, the authors argue, a universal human nature 'but this universality exists primarily at the level of evolved psychological mechanisms, not of expressed cultural behaviours' (Barkow *et al.* 1992: 5). Cultural variation or relativism therefore poses no threat to biological universalism but rather provides data which illuminates the structure of the psychological mechanisms which gave rise to the possibility of variation (5). The second premise is that evolved psychological mechanisms are adaptations constructed by natural selection during the Pleistocene. For most, if not all evolutionary psychologists 'the evolved structure of the human mind is adapted to the way of life of the Pleistocene hunter-gatherers, and not necessarily to our modern circumstances' (5). This Dawkinsean disclaimer raises such obvious questions of timing and relevance as to seem somewhat disingenuous. The historical and empirical basis of evolutionary psychology is weak (little is known from archaeological and palaeontological records about the precise details of the Pleistocene way of life) and its claims rest on the plausibility of its retrospective stories about the likely origins of observable behaviour. Part of this plausibility is provided by the relative lengths of evolutionary history (measured in millions of years) and modern history (a hundred thousand) (5). These relative time spans 'are important because they establish which set of environments and conditions defined the adaptive problems the mind was shaped to cope with' – given, as Darwin constantly maintains, that evolution is a very slow process. There is still, however, the questionable reduction of (observable) behaviour to processes of evolution and adaptation. Despite, or because of the concessions within evolutionary psychology to critiques of sociobiology, it is hard not to see it as an extended hypothesis generated from a belief in social Darwinism. Evolutionary psychology resonates with the historical contingencies, the ideologies of class and gender (if not race) in *The Descent of Man*. In a supposedly classless, post-feminist (if not yet antiracist) present, the re-emergence of Darwin's (genetically legitimised) theory of natural selection extending to sexual selection is, to say the least, a strange coincidence. This is a reactionary discourse whose emergence coincides with marked biotechnological challenges to the category of human nature. On the principles of sexual selection, Darwin writes: 'That the males of all mammals eagerly pursue the females is notorious to every one' (1901 [1871]: 341). On the other hand 'the female . . . with the rarest exceptions, is less eager than the male' (342). Though 'comparatively passive' she 'exerts some choice', and though coy she demands to be courted and charmed (350), leading to the development of secondary sexual characteristics – like big horns and colourful feathers – in males.

On sexual selection in humans, Darwin writes that not only is man stronger and more hairy, but also 'more courageous, pugnacious and energetic than woman' and with 'a more inventive genius' (847). Darwin is aware of the contentiousness (at least scientific) of his assertion that men and women are mentally or psychologically distinct: 'I am aware that some writers doubt whether there is any such inherent difference; but this is at least probable from the analogy of the lower animals which present other secondary sexual characters' (857). Women of all races are then more tender and less selfish, displaying these qualities towards children 'in an eminent degree' and extending them towards others (857). Conversely, since men are in competition with each other, they are more ambitious, more selfish. Women share with the 'lower races' greater powers of intuition but weaker intellects while (white, European) man attains to 'a higher eminence, in whatever he takes up, than can woman – whether requiring deep thought, reason, or imagination, or merely the use of the senses and hands' (858). Since there are many more eminent men than women in music, art, philosophy and science, Darwin – after Galton – infers that 'the average of mental power in man must be above that of woman' (858). The determined if not heroic contest and struggle between men for choosy women is what leads, through sexual selection and natural selection, to their ultimate superiority (860). If there is a moral here, it is shallowly repressed. In *Divided Labours. An Evolutionary View of Women at Work* (1998), Kingsley Browne's sexual morality and politics is even less repressed. Referring to Darwin's psychological characterisation of the sexes, he argues that both the glass ceiling and the gender gap in earnings should be reviewed, and the causes perceived as being not structural and social but sexual. These phenomena 'may reflect evolved differences between the sexes, which would require the merits of 'correction' to be debated rather than assumed' (Browne 1998: 4). If men are more competitive, more willing to take risks (originally in the quest for reproductive advantage) then 'would it be unfair if a disproportionate number of men achieved the highest positions in the hierarchy?' (4). His thesis is that work inequalities are the result of the operation of evolved biological differences operating in the modern labour market (5). Biology is not presented as the exclusive cause and social attitudes 'also play a role'. Browne makes no argument against attempts to reduce economic disparity and offers no policy statement. However, he does state that 'policy makers should take humans as they are – rather than as they would like them to be – when crafting their policies', suggesting at least some adherence to the status quo (6). More regressive thinking is betrayed in his comments on the relationship between women's work and domestic roles. The bio-logic presented here is that women do less well at work than men because they are less aggressive, less competitive and put less time and effort in. This in turn is because of their commitment to family life and to children (no mention is made of single and/or childless women) which is largely determined by our evolutionary heritage. What is more, 'the notion that women will in large numbers completely overlook their children in favour of careers is no more realistic than the notion that one can have a completely satisfying high-powered career and be fully involved with the lives of one's children' (47). Browne does not engage with feminism but rather seeks to turn back the clock, to precede and to preclude feminism through a rearticulation of Darwinian sexual selection which allows for but subsumes social differences.

Evolutionary psychology is cognitivist, informational and so 'an account of the evolution of the mind is an account of how and why the information-processing organisation of the nervous system came to have the functional properties that it does' (Barkow *et al.* 1992: 8). The stress on function and adaptation in evolutionary biology is contested by Stephen Jay Gould (2000), who characterises evolutionary psychology as ultra-Darwinist and ultra-Darwinism as 'a conviction that natural selection regulates everything of any importance in evolution, and that adaptation emerges as a universal result and ultimate test of selection's ubiquity (2000: 86). He reminds us that Darwin himself questioned the ubiquity of natural selection. Indeed, in the preface to the second edition of *The Descent of Man* Darwin writes:

> I may take this opportunity of remarking that my critics frequently assume that I attribute all changes of corporeal structure and mental power exclusively to the natural selection of such variations as are often called spontaneous; whereas even in the first edition of the 'Origin of Species', I distinctly stated that great weight must be attributed to the inherited effects of use and disuse, with respect both to the body and mind.
>
> (Darwin 1874)

Ultra-Darwinists are then both more and less Darwinian than Darwin, who also discusses the effects of 'the direct and prolonged action of changed conditions of life', reversions of structure, and correlation – or the effect of changes in one part of the organism on others (Darwin 1901 [1871]). For Gould, where Darwin was a pluralist, evolutionary psychologists are fundamentalists and it is 'an odd time to be a fundamentalist about adaptation and natural selection – when each major subdiscipline of evolutionary biology has been discovering other mechanisms as adjuncts to selection's centrality' (Gould 2000: 89). In developmental biology, for example, recent work has focused on the high degree of 'conservation' or similarity between developmental pathways in separately evolving and distinct organisms. This suggests a stability at odds with potentiality, and that in evolutionary explanations, historical constraints might be equal to the short term benefits of adaptation (2000: 90). In the context of original and subsequent evolutionary pluralism, the fundamentalism of evolutionary psychology appears reactionary (90). For Gould, the malevolent manifesto is not Dawkins (1976) but Dennett (1995a): 'Dennett explains the strict adaptationist view well enough, but he defends a blinkered picture of evolution in assuming that all important phenomena can be explained thereby' (Gould 2000: 91). Gould runs through the three main claims in evolutionary psychology: modularity (the idea that human behaviour and mental functioning is divided into relatively discrete areas or organs); universality (the focus on apparently universal aspects of behaviour and psychology) and adaptation (where assumptions differ from those of sociobiologists in that not all universal behaviours are thought to be adaptive to modern humans and may indeed be destructive) (99). Because universals are not necessarily adaptive now, but arose as adaptations in an earlier time, 'the task of evolutionary psychology', says Gould, 'then turns into a speculative search for reasons why a behaviour that may harm us now must once have originated for adaptive purposes' (100). For example, Robert Wright's argument that the sweet tooth was designed for an environment in which fruit existed but chocolate didn't 'ranks as pure guesswork in the cocktail party mode; Wright presents no neurological evidence of a brain module for sweetness and no palaeontological data about ancestral feeding' (100). Claims for a prehistoric 'environment of evolutionary adaptation' (EEA) cannot be tested but only subjected to speculation. At least, Gould (2000) argues, claims about modern societies can be tested by analysing the impact of a given psychological feature on reproductive success: 'Indeed, the disproof of many key sociobiological speculations about current utility pushed evolutionary psychology to the revised tactics of searching for an EEA instead' (100). Tools and bones deposited on the African savannahs do not offer enough evidence to substantiate the validity of adaptive stories and say little about 'relations of kinship, social structures and sizes of groups, different activities of males and females' and so on (101). Ultimately, for Gould, it is the adaptationist premise which weakens evolutionary psychology, and the 'failure to recognise that even the strictest operation of pure natural selection builds organisms full of non-adaptive parts and behaviours' (103). Gould refers to non-adaptive by-products of evolution as 'spandrels' and where these may be co-opted for secondary use, it is a mistake to argue that the secondary uses explain their existence (104). In the context of the posthuman present of artificial life engineering, evolutionary psychology's observations regarding maladapted modern man returns us to a mythical time, a prehistoric environment in which men were men, women were women and, equally importantly, humans were definitely human.

The authors of *The Adapted Mind* (Barkow *et al.* 1992) aim not for a dialogue but a synthesis of disciplines, a conceptual integration or unified scientific epistemology which will counter the fallacies of what they term the Standard Social Science Model (SSSM) with an Integrated Causal Model (ICM). The ICM holds that the mind is a modular information processing system and that these modules or mechanisms are adaptations 'produced by natural selection over evolutionary time in ancestral environments' (1992: 24). Some of these mechanisms are functionally specialised to deal with adaptive problems such as mate selection and language acquisition, and they are (therefore) content-specific. Content-specific functional information processing mechanisms 'generate some of the particular content of human culture, including certain behaviours, artifacts, and linguistically transmitted representations' (24). Barkow, Cosmides and Tooby acknowledge creating the caricature of an SSSM against which to proffer their preferred model (31) but nevertheless claim to offer a more balanced, more synthetic approach to the understanding of human nature and culture by reintroducing biology and evolution to the social sciences. So, culture and behaviour are not (exclusively) the products of learning and socialisation, and variation does exist but is constrained. Culture is 'the manufactured product of evolved psychological mechanisms situated in individuals living in groups', and behaviour, like culture 'is complexly variable, but not because the human mind is a social product' (24). Variability is generated by the functional mechanisms or '"programs" that use and process information from the world, including information that is provided intentionally and unintentionally by other human beings' (24). The SSSM, in comparison, tends to be culturally deterministic and based on a 'faulty analysis' of the nature–nurture problem 'stemming from a failure to appreciate the role that the evolutionary process plays in organising the relationship between our species-universal genetic endowment, our evolved developmental processes, and the recurring features of developmental environments' (33). This statement betrays a certain inconsistency in the claim to effect a more balanced approach. It is still the evolutionary process which organises the relationship between genes, development and the environment. So when Barkow *et al.* (1992) argue against the division between genetically and environmentally determined traits (33) they are, in effect, arguing for the subsumption of environmental by genetically determined traits. Tom Shakespeare and Mark Erickson make a similar argument that evolutionary psychologists effect a synthesis (between natural and social science, biologism and environmentalism) 'only by collapsing much of the social world into an ultra-Darwinian model in which biological imperatives predominate' (2000: 190). My argument regarding the particular manifestations of a new subsuming biological hegemony will be developed throughout the analysis of artificial life in context, and the nature–nurture problem will be further explored and developed in the context of feminist epistemology, ontology and methodology in technoscience studies. With Shakespeare, Erickson, Rose and others I maintain that evolutionary psychology – as an example of the new biology – demands a response from those who are characterised as adversaries, and this response, crucially should move beyond the two cultures (Shakespeare and Erickson 2000: 190) or the science wars (Barkow *et al.* 1992: 36). Where Shakespeare and Erickson explore a 'biosocial' model of disability, I will explore a biosocial model of gender. Like them, I remain sceptical of the rise of/return to Darwinism which is commensurate with the decline of religion and politics and which offers 'simplistic answers to the question of origins and causes' (2000: 191). Like Barkow, Cosmides and Tooby, I remain mindful, respectful of the historical and political imperatives of anti-biologism; the need to reject biological explanations of race, gender, class, sexuality and disability which facilitate oppression and discrimination (Barkow *et al.* 1992: 34). It is a mistake, though, to conflate social or environmental causes (of human nature/culture) with freedom and to conflate biological or evolutionary explanations with determinism. 'Biophobia' may even be costly (36). The celebration of human diversity in evolutionary psychology, and its attribution of variation and diversity to inherited universal psychological mechanisms may sound hollow and unconvincing, but the relationship between biological and sociological, natural and cultural explanations of mind and matter are certainly up for review. Evolutionary psychologists may try to close the gap between genotype and phenotype – 'between the inherited basis of a trait and its observable expression' – but they are right to indicate its current significance and the potential compatibility of biological and social explanations (45).

The rise of or return to Darwinism was widespread in the 1990s and evolutionism has penetrated computer science (evolutionary psychology is compatible with artificial intelligence and its cognitive emphasis on programmed computational or information processing machines), the academy, industry, management, law and the media (Brown 1999: x). An apparently random example goes some way toward illustrating this point. In the year 2000, MIT published a book by Randy Thornhill and Craig Palmer entitled *A Natural History of Rape. Biological Bases of Sexual Coercion*. The authors sought and achieved a great deal of media exposure so that they could 'change the way people think about rape – from a social to an evolutionary context – and therefore affect a better method of rape prevention' (Dano 2000: 2). Their campaign is detailed on Thornhill's website, which contains letters for and against the book by students and colleagues at the University of New Mexico, as well as the outline of a course entitled Darwinism Applied. The book opens with a criticism of established feminist and sociological theories of rape on the grounds that they are scientifically illiterate and purely ideological. In a quite Dawkinsean tautological move which sets out to demonstrate Scientific Truth having already defined it, the authors state that existing theories of rape are 'uninformed about the most powerful scientific theory concerning living things: the theory of evolution by Darwinian selection' (Thornhill and Palmer 2000: xi). They go so far as to suggest that non-evolutionary approaches may have increased rather than prevented incidences of rape by teaching men and women that the causes are social rather than sexual. Some of the negative publicity around the book has seized on the more obviously controversial arguments, for example, that boys should complete courses in evolutionary biology before being granted a driving licence and girls should be 'made aware of the costs associated with attractiveness' and advised to dress conservatively in order to attract 'good investors' rather than trouble (180). We have, of course, heard it all before and may be tempted to dismiss it on this basis, but the references to current evolutionary as well as sociological theory are present and the authors are unfortunately right to claim that there is a renewed climate of interest at least in the theoretical basis of their work. This is why I would suggest that the best response to it may not be to simply reassert feminist and sociological arguments about rape (although this may inevitably be part of the response) but to explore the wider social and scientific environment which makes a return to sociobiology through evolutionary theories of rape possible.

Darwin@LSE is a research programme on evolutionary theory with a leading interest in evolutionary psychology and the links between Darwinism and social policy. The programme has produced a series of books under the heading *Darwinism Today* which claim to introduce 'the Darwinian ideas that are setting today's intellectual agenda' (Browne 1998: vii). The series editors, Helena Cronin and Oliver Curry, seek to popularise evolutionary theory and maintain that the Darwin@LSE programme 'is having an enormous impact' (Browne 1998: vii). The website promotes *Demos Quarterly*, published by what Segal terms 'the Blair government's favourite think-tank' – Demos (Segal 1999: 80). In a special issue (edited by Curry and Cronin 1996), John Ashworth (former Director of the LSE) hails Darwinism as 'An "ism" for our times' (Ashworth 1996: 3) which displaces Marxist philosophy at the (supposed) centre of the social sciences. Ashworth locates evolutionary theory between right-wing individualism and the 'communitarian' left, gently promoting the 'tentative' and apparently middle-ground suggestion that 'evolutionary theories might work where both individualistic and group (or class) based explanations of behaviour have proved unsatisfactory' (3). Success, in his account, seems to rest on the 'sensitive' repackaging of the old sociobiological conflict between altruistic behaviour and the doctrine of the selfish gene. Herbert Spencer at the turn of the twentieth century, and E.O. Wilson in the 1970s, lacking a sufficiently sophisticated genetics, were too 'assertive' where the new Darwinists can afford to adopt a more conciliatory tone which 'might now lead to something other than a dialogue of the deaf' (3).[8] The olive branch offered by evolutionary biologists to social scientists takes the form of a non-deterministic morally corrected Darwinism in which genetic programming does not preclude social conditioning, and the way things are is not at all synonymous with the way things should be. The big stick not waved by evolutionary biologists at social scientists takes the form of memetics (in which society is an effect of biology and culture's 'memes' are analogical with nature's genes) and risk management (where the cost of correcting or containing antisocial impulses might well outweigh the cost of indulging them). Proffering the branch instead of the stick, Ashworth (1996) hopes that a fruitful dialogue will take place.

Some exponents of evolutionary psychology, notably Richard Wright (1996) in his exploration of 'the dissent of woman', prefer didacticism to dialogue and so contribute to the perpetuation of the science wars. Wright asserts that 'history has not been kind to ideologies that rested on patently false beliefs about human nature' (1996: 18). He aligns feminism with the decline of communism on account of the false beliefs feminists have about gender and Marx had about the inheritance of acquired characteristics: 'the falseness of the doctrine is increasingly evident and its adherents can admit as much only at some risk, if not of imprisonment, then of an extremely chilly reception from fellow feminists' (18). Wright's argument proceeds with fictional caricatures of 'difference feminists', 'radical feminists', 'liberal feminists' and 'assorted others' homogenised and contained through their supposed ignorance of modern Darwinism: 'none is interested in the well-grounded study of human nature' (18). The Darwinian theory of natural selection is not so much 'grounded' (empirically) as normalised within the 'science' of evolutionary psychology which establishes innate but not immutable differences between the minds of men and women. Mutability or the effect of culture is, however, limited and so, therefore, are the realistic goals of feminism: 'many of the differences between men and women are more stubborn than most feminists would like, and complicate the quest for – even the definition of – social equality between the sexes' (18). This tendency to recuperate culture within biology might lead to the suspicion that evolutionary psychology subjects sociobiology to a notion of complexity which ultimately strengthens and reinforces it. Wright rehearses the natural history of rape (in which the cost to women is ultimately genetic) in order to argue for more effective laws and policies of protection which recognise 'female vulnerabilities' (20). He revisits the madonna–whore dichotomy as an effect of genetics rather than patriarchy (men don't marry promiscuous women because their children may not be carrying the right genes) and co-opts essentialist arguments within 'radical' feminism. Men are 'by nature oppressive, possessive, flesh-obsessed pigs' but they are not beyond cultural improvement 'thanks to the fact that love, compassion, guilt, remorse and the conscience are evolved parts of the mind, just like lust and jealous rage' (22). Culture then, is an effect of nature and by definition constrained and limited. Ultimately, it does a 'lacklustre job' of improving men. What evolutionary psychology contributes to social policy is not (quite) a sanctioning of the 'natural order' which underlined critiques of sociobiology, but a heightened awareness of the costs and benefits of 'alternative norms' and of where (on whom) those costs and benefits fall. It is no coincidence that feminism is simply not needed in the pursuit of answers to these rhetorical and implicitly gendered questions, since evolutionary psychology absorbs many of its concerns and feminists, according to Wright (1996), should 'know their enemy' as their friend.

Anne Fausto-Sterling (2000) responds to Wright's attack by indicating the long history of the dispute between Darwinism and feminism which began with Antoinette Brown Blackwell's publication of *The Sexes throughout Nature* (1875). Blackwell takes Darwin to task over his assertion of male superiority, arguing that 'for every special character males evolved, females evolved complementary ones' (Fausto-Sterling 2000: 174). Fausto-Sterling also outlines recent feminist primatology research which counters Darwin's theory of male dominance and greater 'eagerness' (182). She points out that while evolutionary psychology reproduces Darwin's sexism, it retreats from racism. While she offers no explanation for this, it is surely no coincidence, as mentioned earlier, that what Gould (2000) describes as fundamentalist, reactionary theory should be commensurate with an allegedly classless, post-feminist but significantly not non-racist society. Antiracism, it would seem, is evidently not sufficiently well established to warrant such a reaction. Fausto-Sterling, like Gould, Rose and Lewontin, regards evolutionary psychology as primarily a hypothesis, a 'thought-experiment' producing flat 'cardboard' representations of men and women, and very much in need of more detailed and rigorous research. She cites 'Latour and Strum's Nine Questions' which aimed at specifying hypotheses about human evolution. These include questions about the specific units of evolution (genes, individuals, family, species) in question, the specific prehistorical time frame, method and political implications but are also significant, for Fausto-Sterling, for constituting a 'model collaboration' between an anthropologist of science and a primatologist. Scientifically sound theories about the evolution of human behaviour and the relevance of evolutionary thinking to modern societies could emerge, she argues, from collaborations between evolutionists, behavioural biologists and social scientists – especially those 'who have been so bitterly attacked in the current science wars' (Fausto-Sterling 2000: 179).

Lynne Segal (1999) argues that the goal of the current 'return to Darwin' is not only the 'conceptual containment of potentially unlimited shifts in gender beliefs and practices' but also the 'return to the allegedly more rigorous authority of the biological sciences of much that has recently been understood as cultural' (Segal 1999: 78). The aim is to silence or 'defeat' feminists and other theorists who perpetrate 'the idea that gender is essentially a construct, that male and female nature are inherently more or less identical' (Wright 1996: 18). This aim is certainly clear in Wright's combative evolutionary theory and Segal's counter-argument constitutes a sceptical response to any invitation to dialogue within a discipline which has been characterised as fundamentalist (Gould 2000). Nevertheless, her exploration of the 'enemies within' both feminism and biology (which, in a properly situated reading includes Darwin himself) opens up spaces of contestation and 'epistemic diversity' (Segal 1999: 111) which I would argue are possible spaces for change. The heterogeneity of biological discourse destabilises both evolutionary psychology and artificial life where the goal might be said to be the conceptual containment of the diversity of life itself. Life has no cultural meaning or dimension in alife and is epistemologically and ontologically contained within the notion of information processing and replication. ALife, like evolutionary psychology, ascribes to the memetic theory of culture established by Richard Dawkins in his influential treatise on the selfish gene. The materiality of Dawkins's metaphors have been noted in cultural theory (Hayles 1994; Keller 1992) which nevertheless, and with few exceptions, remains dismissive of them, missing an opportunity to contest the meaning of agency and autonomy within biological and social discourse. This currently remains an internal debate, fought principally through the concept of autopoiesis. Autopoietic organisms have limited agency and are not determined by internal or external environments. They are more than the sum of their genes. Steven Rose uses the concept of autopoiesis to oppose Dawkins's construction of 'lumbering robots' or gene/meme machines (Rose 1997: 245). Margaret Boden (2000) uses the concept of autopoiesis to highlight a similar division within the field of ALife where the contest is over not only the terms agency and autonomy but also situatedness and embodiment. Autopoietic artificial life-forms have simulated physical environments and bodies (rather than just genetic algorithms and an evolutionary trajectory). Some have a degree of complexity which – at least according to their creators – might approximate intelligence and anticipate the emergence of artificial cultures and societies (Cliff and Grand 1999). Since such creatures already populate computer games (such as *Creatures*) and are coming soon on the Internet (as autonomous agents), the exact 'nature' of their existence – hyped or otherwise – ought perhaps to be of wider concern. The world described by evolutionary psychology and the 'worlds' prescribed within artificial life run solely on Darwinian principles which *seem* to resolve complexities of mind and matter and to dissolve pluralism into fundamentalism, heterogeneity into homogeneity, questions into answers which constitute a social scientific theory of everything. The quest for a theory of everything, the *raison d'être* of the physical sciences (from Newton to chaos to superstrings) transfers to the life sciences through molecular biology, where endgames such as the discovery of DNA and sequencing of the human genome (unravelling the secrets of life itself) are staged. Drawing their inspiration from molecular biology, evolutionary psychology and ALife have taken the stage with the 'eternal principle' of natural selection. For Dorothy Nelkin, the theory of everything is the religion of non-theistic scientific cultures and within the new life sciences 'the gene appears as a kind of sacred "soul"' (Nelkin 2000: 15). The quest for a theory of everything follows 'a religious mindset that sees the world in terms of cosmic principles, ulimate purpose and design' (16) and both evolutionary psychology and artificial life have their 'missionaries' (19). As a religion, evolutionary psychology seeks to guide not only moral behaviour but also social policy in areas such as law, welfare, education and equal opportunities. Offering seemingly simple and universal answers to complex social problems guarantees media attention and wide publicity. Evolutionary theories shore up the credibility of science by returning to 'old seeming certainties' reflected in a widespread 'resurgence of fundamentalist religions, Islamic, Jewish and Christian, with their enthusiasm for militancy and their beliefs in creationism' (Rose and Rose 2000: 3). For Nelkin, evolutionism is 'especially convenient at a time when governments faced with cost constraints, are seeking to dismantle the welfare state' (Nelkin 2000: 21). Why, asks Nelkin, 'support job training, welfare or childcare programmes when those targeted are biologically incapable of benefiting from the effort?' (21). Whether or not Nelkin is right in asserting that evolutionary psychology, just like sociobiology simply naturalises the status quo, its uniquely direct challenge to the social sciences 'demands a reply' (Rose and Rose 2000: 8) and perhaps even some alternative explanatory perspectives: 'It's an old academic adage, but nonetheless true, that bad theory can never be driven out solely by criticism. A better alternative has to be offered' (9).

Cyberfeminist cultural analysis must engage with current trends in science and technology with a critical historical awareness of how they are naturalised culturally and with a strategic investment in dialogue rather than dismissal. When the current trend is sociobiological this is a difficult thing to do. Few if any genetic and evolutionary 'determinists' really believe that because human behaviour is 'ultimately' natural, it is therefore inevitable, unaltered and unalterable through socialisation and education. The problem, as Haraway (2000) points out, is not necessarily or entirely the science but the way in which it is represented by scientists and non-scientists with a primary interest in obtaining funding, status, sales or publicity. Thornhill and Palmer (2000) use evolutionary theory in place of (in the same place as) social theory to argue that current rape law works more in the interest of husbands, fathers and other dominant investors in genetic capital than in the interest of women victims and should be changed accordingly. Addressing the social environment of actual and potential offenders amounts to a biological intervention for them, and it is this tendency to subsume rather than ignore the social which marks the new biological hegemony. The task for feminism is to address what motivates this, and to explore the representational, epistemological, ethical and political dimensions without, as Haraway claims that she appeared to do in *Primate Visions* 'a kind of hands-on-your-hips negative critique where you are just standing there shaking your finger, going "this is a racist, sexist, colonialist enterprise"' (Haraway 2000: 56).

The new biology incorporates and informs artificial life discourse with a marked set of internal contradictions relating to the autonomous agency of chickens and eggs, organisms and genes and to the efficacy of reductionist and deterministic accounts of life. Alife re-engineers these contradictions in the form of autonomous and autopoietic entities which contribute to the formation of posthuman identity and invite a dialogic response.

Chapter 3
Artificial Life

> If what she had suggested was true, then Jane was more than a program. She was a being who dwelt in the web of philotic rays, who stored her memories in the computers of every world. If she was right, then the philotic web – the network of crisscrossing philotic rays that connected ansible to ansible on every world – was her body, her *substance*. And the philotic links continued working with never a breakdown because she willed it so.
> 'So now I ask . . .,' said Jane . . . 'Am I alive at all?'
> (Orson Scott Card, *Xenocide* 1991: 67)

Andrew Wiggin (aka Ender) is the hero of Orson Scott Card's science fiction trilogy. In the first novel, *Ender's Game* (1977), he saves the world from an intelligent race of insect-like creatures by destroying them in a simulated inter-Galactic battle game which turns out to be real, and not a game at all. In the third novel, *Xenocide* (1991) Ender saves the creatures (and what is left of the human race) from destruction with the help of a cosmic synthetic or artificial life form called Jane.

That there is a reciprocity of ideas – a dialogue rather than a sense of determinism – between science and science fiction is widely accepted and frequently demonstrated with reference to William Gibson's *Neuromancer* (1986) which coined the term 'cyberspace' and influenced the development of virtual reality (Woolley 1992). Orson Scott Card's trilogy also strikes me as being remarkably prescient as it speaks to developments in computer science which were debated during the 1990s and into the twenty-first century. The Ender trilogy explores, among other things, the transition from computer simulation to computer synthesis. Put another way, it looks at the transition from computer models of life to computer manifestations of life. The simulated battle in *Ender's Game* has real effects, and the technology he uses is not unlike the virtual reality technologies developed and used, most notoriously in the Gulf War. In *Xenocide*, Jane is an artificial life-form who exists autonomously in what might be called a computer web or matrix. She is sufficiently human-like to ponder the status of her own being, and in this sense is still a little in advance of recent developments in science which is only now contemplating the synthesis of consciousness.

That science 'fact' and fiction share not only ideas but fantasies, hopes and fears is also commonly accepted at least within critical cultural perspectives (Penley *et al.* 1990; Penley 1997; Haraway 1991). As Stefan Helmreich points out, fictional heroics may well have inspired scientists in their youth and desires to save the world may well be manifest in current scientific attempts to create alternative worlds in computer if not in outer space (Helmreich 1998a).[1] Perhaps the clearest manifestation of the blurring of science and science fiction is in popular science writing of the kind which dramatises the subject in order to increase its appeal. This kind of writing is likely to display the fantasies which drive both science and fiction but which science alone seeks to conceal. It often expresses a hubris which must otherwise be suppressed and is an illuminating indication of science's embodiment. Richard Dawkins is a key exponent, and this is how he describes his initial reactions to Biomorph – a computer program he wrote in order to simulate the evolution of natural forms:

> When I wrote the program, I never thought that it would evolve anything more than a variety of tree-like shapes . . . Nothing in my biologist's intuition, nothing in my 20 years' experience of programming computers, and nothing in my wildest dreams, prepared me for what actually emerged on the screen . . . I distinctly heard the triumphal opening chords of *Also Sprach Zarathustra* (the '2001 theme') in my mind. I couldn't eat, and that night 'my' insects swarmed behind my eyelids as I tried to sleep.
>
> (Dawkins 1991: 60)

Here, Dawkins is at once a located and embodied individual – a biologist with twenty years' experience capable of appetite-suppressing excitement – enthralled by the generative power of the computer, and an omniscient God-like figure capable of creating his own creatures ('my' insects) to the accompaniment of Richard Strauss. In one paragraph of his writerly text, Dawkins captures something of the essence of artificial life research; its ambition and its internal contradiction. At its most ambitious, the discipline of Artificial Life is creating, or rather evolving, life-forms in software and hardware which are said to be truly alive. The contradiction at the heart of this hubris is not so much the redefinition of life in terms of information (although this is certainly regarded by many observers and critics of ALife as being controversial),[2] as the identification of the creator. Artificial Life projects, including Dawkins's own Biomorph, are modelled on Darwinian theories of evolution incorporating natural selection and survival of the fittest in a competitive environment. They are necessarily anti-creationist and Dawkins's book is explicitly so. But ALife is driven by the desire to create, so what happens to the creationist urge? It seems to me that it is (often unsuccessfully) sublimated and that the desires of the ALife creationist gods are realised through, and seemingly by, the generative powers of computers which transcend number-crunching and appear to evolve life. It is possible to surmise that ALife research fits a pattern of sublimation and projection which has, historically, characterised the engagement between masculinity and technology (Easlea 1983; Keller 1992; Braidotti 1996; Stone 1996). Dawkins's account of Biomorph is followed by an appendix to the original edition of *The Blind Watchmaker* in which he considers computer programs and 'the evolution of evolvability' (Dawkins 1991). This also appeared in an early collection of essays on Artificial Life (Langton 1989).

The original Biomorph program represents Dawkins's attempt to model the development, reproduction and evolution (by cumulative selection) of natural forms. He begins with the principle that recursive branching is a good model or metaphor for the development of plant and animal embryos. What he is referring to is the two-way branching pattern of cell division which occurs in all embryo development. This is represented on the computer by a simplified analogue or drawing rule which begins with a single vertical line. The line then branches in two. Each of the sub-branches also divides and so on recursively. The branching pattern is recursive because it is applied locally all over the growing 'tree'. In order to model evolution as well as development, Dawkins then has the problem of representing genes which influence development and can be passed on from generation to generation. He proceeds with a manageable number of nine genes, each of which is represented by a different number or value in the computer. By altering or mutating specific genes by the value of plus or minus one (genes responsible for the angle or length of branching, for example) Dawkins produces a variety of forms or development and simulates the evolutionary process through the production of 'offspring' or 'children'. Since each new tree-like form is only one mutational step away from the original, it is regarded as progeny and this produces the model of reproduction. In this model of reproduction 'there is no sex; reproduction is asexual' (Dawkins 1991: 55). What is more, 'I think of the biomorphs as female . . . because asexual animals like greenfly are nearly always basically female in form' (55). In this model of reproduction, Dawkins is also careful to embody neo- (or what Rose calls 'ultra'-) Darwinian genetic determinism in which genes influence development but not vice versa – 'that would be tantamount to "Lamarckism"' (56). By repeating the process of reproduction by mutating successive generations of forms, Dawkins produces 'genetic' variety and the possibility of Darwinian selection whereby the criteria for selection are not the genes themselves but the 'bodies' or forms the genes produce (phenotypes). Yet instead of being able to model natural selection through survival of the fittest, Dawkins settles for 'artificial selection' by promoting the survival of recognisable forms. His agent of selection is his own eye, and what he is looking for are shapes which look more and more like animals and less and less like trees. In other words, he is selecting 'evidence' of evolution or creating an evolutionary pattern which he then uses as evidence of evolution. With this sleight of hand, Dawkins captures the paradox of creation which is at the heart of the ALife project: the God-like act of creating life is 'stolen' or appropriated by man and then credited to the computer. It is, I dare to suggest, almost as if 'man' has 'inherited' the lessons of his ancestors and learnt that playing God is dangerous and leads to disaster. As in the story of genomics, that of ALife is fundamentally informed by religious mythology – creation and the fall – particularly as it is manifested in the narratives of *Faust* and *Frankenstein*. Mephistopheles and the monster clearly haunt ALife hubris. The displacement and reappropriation of creationist power and status follows the same dynamic as that of humanism, reclaimed as it is by the generation of autonomous agents. The key to this dynamic is the concept of emergence. Dawkins disavows or displaces the creation of first insect-like forms and then higher order animals and objects. He expresses 'incredulity' at the 'evolving resemblance' to insects of the forms 'I began to breed, generation after generation, from whichever child looked most like an insect' (59). Later, he 'encountered' all sorts of things from bats and scorpions to a lunar lander on 'my wanderings through the backwaters of Biomorph Land' (60). Despite being unable, at that time,[3] to simulate natural selection by the introduction of death in a competitive environment, Dawkins insists that it is the computer and not him which successfully simulates the evolutionary process:

> There is a popular cliché . . . which says that you cannot get out of computers any more than you put in. Other versions are that computers only do exactly what you tell them to, and that therefore computers are never creative . . . I programmed EVOLUTION into the computer, but I did not plan 'my' insects, nor the scorpion, nor the spitfire, nor the lunar lander. I had not the slightest inkling that they would emerge, which is why 'emerge' is the right word.
>
> (Dawkins 1991: 64)

Emergence, as this chapter will explore, is one of the key concepts through which ALife researchers argue for the generative power of computers and secure a form of digital naturalism in the face of the evident constructivism of 'artificial' life. For Dawkins (1991), the property of emergence confers on computers not any kind of 'mystical' power (65), but the power of evolution – the power to evolve life. The power of computers to evolve life is something which Dawkins, in his appendix to *The Blind Watchmaker*, clearly thinks is evolving. Having added variations in segmentation and symmetry to the original Biomorph program, he goes on to make stronger and rather grandiose claims about the parallel between the evolution of evolutionary programming and the process of evolution itself. The 'invention' of segmentation 'has opened the flood-gates of evolutionary potential in the land of computer biomorphs' (331). Moreover, 'my conjecture is that something like that happened in the origin of vertebrates . . . The invention of segmentation was a watershed event in evolution' (331). The claims that Dawkins makes for his biomorphs do, however, fall a little short of the claims he makes for the Biomorph program. If evolution in Biomorph is real evolution, the life in the biomorph life-forms is not real life. Unlike future generations of artificial life forms, biomorphs are said to be (merely) lifelike, but not alive.

The distinction between life and not-life, or what constitutes the nature of life is the key, if largely deferred question in both biology and philosophy (Bedau 1996). ALife's biology is the new biology of genetically-more-than-environmentally (cognitive more than behavioural) programmed organisms in tension with autonomous, self-producing autopoietic entities. ALife's biology also functions to tip the epistemological scales towards its scientificity, objectivism and retention of the transcendental disembodied subject of modernity. Richard Dawkins witnesses his creatures emerge and evolve. Lars Risan (1996) traces the transition from constructivism to naturalism in ALife via the role of the detached, Boylean witness. He starts with a description of genetic algorithms (GAs) which are evolutionary computer programs designed to construct artificial populations and artificial worlds 'containing legitimate objects of scientific enquiry' (Risan 1996: 36). Genetic algorithms, first described by John Holland in the 1960s, are a successful means to solve computational problems through the reproduction, mutation, selection and evolution of 'populations' of possible solutions represented as 'chromosomes' or 'organisms' (Mitchell and Forrest 1997: 268). A simple form, as described by Mitchell and Forrest, would work as follows:

1 Start with a randomly generated population of chromosomes (e.g. candidate solutions to a problem)
2 Calculate the fitness of each chromosome in the population
3 Apply selection and genetic operators (crossover and mutation) to the population to create a new population
4 Go to step 2

(Mitchell and Forrest 1997: 268)

One or more highly fit 'chromosomes' or solutions should emerge once the process has been repeated – iterated – over a number of time steps or 'generations' (268). A list of the current applications of GAs to scientific and engineering problems might include optimisation, machine and robot learning, economic and immune system models, ecological and population models, models of social systems and the interaction between evolution and learning (the major question in evolutionary psychology) (269). The key to the success of GAs is emergence or the chaotic and unpredictable appearance of unthought-of and undesigned solutions to computational problems (Risan 1996: 37). The possible solutions to problems emerge spontaneously from the bottom up through the interaction between component parts or units of the system. They are a result of the dynamics of the system itself which proves to be more than the sum of its parts, and of a process which Risan (1996) claims is central to the way in which ALife combines reductionist molecular biology and a more holistic autopoietic biology (37). When the result of running a computer program is not predicted, not thought-of, then it may be objectivised and analysed; 'studied *scientifically* as a technoscientific "nature"' (38). ALife programming therefore opens up new worlds, new objects to explore, effectively transmuting the constructivist into the naturalist subject or witness. ALife's witness enacts the Cartesian separation of subject and object, culture and nature which is otherwise and elsewhere consciously refused in its contradictory epistemology. Risan indicates that this separation is based on the construction of a particular kind of technology or experimental apparatus which provides the necessary distance between scientist and his object(s). He retells the story, the 'origin myth' of technoscientific objectivity, told by Shapin and Schaffer (1985), Latour (1993) and subsequently Haraway (1997) about the experimental philosopher Robert Boyle and his air pump. Robert Boyle (1627–1691) resolved experimentally the philosophical debate of the time concerning the possible existence of a vacuum, or space without matter. He did so by inviting disinterested members of the establishment to observe the air pump experiment and thereby combining the machine with the trustworthy witness (Risan 1996: 17). This dual technology of machine and witness is constitutive of modern scientific authority and the license to discuss Nature independently of the State, Society and Subject (18). The scientific authority of artificial life is compounded by the generative ability of simulation, synthetic and visualisation machines and by the multiplicity of potential witnesses. Risan describes the effect of this in graphics-based conference presentations in which the emphasis is deflected from engineering to scientific discovery and the audience is invited to identify with the researcher as distanced witnesses of significant findings (1996: 91). ALife both unsettles and secures the boundaries between nature and culture, subject and object in a contradictory epistemology which centres on the idea of the witness(es) pulled back from the brink of scientific obscurity. That obscurity is always threatened in the science fictional and constructivist dimension of ALife engineering and is countered, increasingly, by the authority of biology.

From the opposite perspective, the epistemological modernity of 'artificial' life, founded on the separation between subject and object is problematised by the evidently constructivist premise of the project, and further problematised by a strong and productive current of anti-Cartesian philosophy which extends from problems of subjectivism to those of embodiment. The rejection of Cartesian philosophy is at this point conflated with that of the science of classical Artificial Intelligence with its cognitivist or mentalist emphasis. Following Dreyfus (1991) and Heidegger (1961 [1927]), Risan (1996: 6) highlights the stream of artificial life consciousness concerned with being-in-the-world. The philosophy of Artificial Life is not abstract but embedded in the production of what Agre refers to as 'discursive technologies' (Agre 1997). These discursive technologies concretise ways of thinking about life-as-we-know-it and life-as-it-could-be. They are literalised thought-experiments (Bedau 1996; Dennett 1997), explorations at the boundaries of science and science fiction which necessarily destabilise ontological and epistemological categories such as nature, culture and life itself and therefore constitute a form of postmodern technoscience. Artificial Life deals with, or rather deals in virtual reality through the construction of biological simulations which may be said to multiply, curtail and supplant the possibilities of the natural world (Baudrillard 1983). For Risan, this aspect of ALife is amplified through the role of creative engineering or the production of simulations at the intersection between art and science, and through the willing transfer of agency and autonomy from humans to machines (1996: 48). ALife is, to an extent, premised on the failure of top-down command and control programming in classical AI, and the tendency toward anthropomorphism is, as Risan points out, only the most apologetic, ultimately recuperative manifestation of a subjectification of technology. The engineering of autonomous agents marks the decentralisation, distribution and dehumanisation of the subject whose status is at least partially restored by the evident enchantment of witnesses who proclaim the wonder not of their hands, but of their eyes. The ALife scientist, in this account, may be characterised (perhaps caricatured) as an engineer masquerading as a scientist, a creationist masquerading as an evolutionist, a constructivist masquerading as a naturalist. The caricature is, of course, undermined by the radical differences and non-homogeneity of the field. Risan, for example, argues that the Artificial Life practised in the Department of Cognitive and Computer Science at the University of Sussex (the site of his anthropological fieldwork) tends to embrace its creative engineering (45). In what is clearly too neat a technological and epistemological division, the claims of constructivism and naturalism, postmodernity and modernity are assigned to simulation and synthesis respectively. Where simulations retain a metaphoric status (Risan 1996: 51), computer synthesis literalises the metaphor of life as information and, in accordance with the claims of 'strong' rather than 'weak' ALife, generates – through real evolution and emergence – real life, real complexity. Risan looks for and finds a resolution to the contradictions inherent in ALife in the form of irony. Claims in favour of strong ALife are then ironic, knowing and therefore permissive (104). Irony enables naturalism to re-emerge from the context of constructivism and God to re-emerge in the face of neo-Darwinism. If Dawkins's sense of irony is not immediately apparent, that of Thomas Ray is, and this may be commensurate with their claims for real evolution and real life respectively. Risan's conclusion supports the validity of strong ALife, the co-evolution of organisms and machines, not objectively, not as a witness but because the distinctions upon which witnessing is based – subject/object, nature/culture – were never sound ('we have never been modern': Latour 1993) and because the existence of artificial life forms can be proclaimed in a different ontological or epistemological register. For me, the irony which is evident in some strong ALife claims (such as Ray's) performs the same function as the denial and projection which is evident in other ALife claims (such as Dawkins's). More importantly, the validity of strong ALife is beyond rationalist assessment given the always shifting and contested criteria for life and the enmeshment of ALife in other practices and discourses of manipulation which defy even the illusory security of ontological and epistemological categories like nature or culture.

'Information Wants to be Alive!'
(Thomas Ray, *Tierra*)

In 'An Approach to the Synthesis of Life', Thomas Ray (1996) makes a point which is often echoed by other science fiction inspired ALife scientists: biology, he suggests, 'should embrace all forms of life' (Ray 1996: 111) and not just life on earth. Ideally, 'a truly comparative natural biology would require inter-planetary travel, which is light-years away' (111). Failing that, 'a practical alternative to an inter-planetary or mythical biology is to create synthetic life in a computer' (111). The aim of Ray's work is to synthesise (not simulate) life and to generate increasing levels of diversity and complexity. This aim presents him with what he calls a 'semantic' problem of redefining life 'in a way that does not restrict it to carbon-based forms' (111). There is thus something of a tautology in the construction of artificial life. In order to create artificial life, it is first necessary to redefine life as artificial.[4] For Ray, life is a facet of self-replication and 'open-ended' evolution: 'synthetic life should self-replicate, and evolve structures or processes that were not designed-in or preconceived by the creator' (112). While implying that Dawkins's Biomorph program does not (in fact) display open-ended evolution, Ray nevertheless shares with Dawkins an attempt to parallel an evolutionary phase – in his case, 'the origin of biological diversity' (113) – and a noticeable fondness for metaphor. Where Dawkins has 'selfish' genes and 'raining' DNA, Ray has CPU 'energy' and RAM 'space'. The CPU (central processing unit) of the computer is seen to be analogous to the sun as an energy resource for digital organisms. Moreover, just as organisms are selected on the basis of how well they compete for natural resources such as space, 'replicating algorithms' survive by competing successfully for 'memory space' (114). Ray argues that 'in the light of the nature of the physical environment, the implicit fitness function would presumably favour the evolution of creatures which are able to replicate with less CPU time, and this does, in fact, occur'. However, 'much of the evolution in the system consists of creatures discovering ways to exploit one another' (134). These competitive 'creatures' have no lifelike physical form – no phenotype – but consist entirely of machine instructions. They are self-replicating programs or algorithms.[5] Ray refers to the first or original creature as the 'ancestor', and the sequence of its eighty machine instructions is referred to as the 'genome'. That the process of evolution is synthesised by 'innoculating' a block of RAM (random access memory) 'soup' with 'a single individual of the 80-instruction ancestral genotype' says something at least a little biblical about Ray's desire to create life. In fact it might even be said that Ray embodies the competing strains of religious and scientific (including science fictional) discourse, with genesis in one corner and natural selection in the other. In the God versus Darwin contest in artificial life, the odds are clearly stacked, but all the secret bets are on the rank outsider. The potential pay-out is just so much higher. Ray not only creates life, but also creates the environment within which life evolves. His digital life does not live in a real computer, but in a virtual computer with virtual operating systems. This is to ensure that he retains control over the evolutionary process and guards against 'the potential threat of natural evolution of machine codes leading to virus or worm types of programs that could be difficult to eradicate due to their changing "genotypes"' (115). A somewhat sanitised digital life, then.

Ray names his virtual computer Tierra, which is Spanish for 'earth'. The act of naming is, of course, the creator's privilege and Ray then goes on to consider the nature of Tierran language and communication. This is inspired by molecular biology and the relationship between DNA, RNA and proteins. Tierran creatures are regarded as being cellular in as far as they have a 'semi-permeable membrane' of memory allocation (Ray 1996: 118). Cells, like biomorphs, appear to be female and 'mother cells' produce 'daughter cells' asexually. Because having self-replicating creatures in a fixed-sized soup would lead fairly swiftly to apocalypse, Ray declares it 'necessary to include mortality' and duly names the 'reaper' (119). The reaper is reasonably fair minded and begins to kill creatures only when the memory fills up to a specified (specified by the creator) level, and on the basis of age and/or incompetence. Ray is faced with the same issue as Dawkins when it comes to 'creating' evolution. He wants evolution to 'occur' (120) and so he makes it occur by programming ways in which genomes may change: 'In order to ensure that there is genetic change, the operating system randomly flips bits in the soup, and the instructions of the Tierran language are imperfectly executed' (120). But, Ray claims that these mutations and flaws are not necessary after a while because 'genetic parasites' which are 'sloppy replicators' and move bits of code between creatures, 'emerge' (120, 134). Ray's parasites present him with the holy grail of artificial life – the emergence of autonomous self-replicating systems ('I think of these things as alive, and I'm just trying to figure out a place where they can live . . . give life a chance': Ray in Helmreich 1998a: 3). And it was all apparently too easy and inevitable: 'It would appear then that it is rather easy to create life. Evidently, virtual life is out there, waiting for us to provide environments in which it may evolve' (Ray 1996: 135). Evidently, something is 'out there'; if not truth or virtual life, then at least the desire (and maybe the capacity) to own and (dis)embody it. Ray's project is validated by Christopher Langton in his seminal article on 'Artificial Life' (1989, updated 1996). Langton is widely referred to as the 'father' of artificial life and is credited not only with establishing the discipline but also with naming it by means of a deliberate oxymoron. Where he regards Dawkins's Biomorph as a first step in the computerisation of evolutionary processes resulting in a model of artificial selection, Langton regards Tierra as 'the final step in eliminating our hand from the selection/breeding process and setting the stage for true 'natural' selection within a computer' (Langton 1996 [1989]: 88). Like Ray, Langton sees no reason why biology should be restricted to the study of carbon-based life on earth, and, like Ray, he accepts that 'since it is quite unlikely that organisms based on different physical chemistries will present themselves to us for study in the foreseeable future, our only alternative is to try to synthesise alternative life-forms ourselves' (39). Science fiction and the promise of alien life and other worlds, perhaps in combination with the failure of NASA and other space agencies to produce them, does seem to have informed the synthetic impulse in Artificial Life. Langton himself indicates that it is more than simply an analytic method based on putting living things together rather than taking them apart (40).[6] As a synthetic approach to biology, ALife aims to do more than 'simply' recreate "the living state". It aims to synthesise 'any and all biological phenomena, from viral self-assembly to the evolution of the entire biosphere' (40). The synthesis of these phenomena need not be restricted to carbon-chain chemistry and may well lead 'beyond *life-as-we-know-it* into the realm of *life-as-it-could-be*' (40). As a generator of lifelike behaviour, Langton outlines how Artificial Life could attempt to create life *in vitro*, and how this 'would certainly teach us a lot about the possibilities for alternative life-forms *within* the carbon-chain chemistry domain that could have [*sic*] (but didn't) evolve here' (50). It is important to draw contemporary developments in biotechnology and genetic engineering (such as cloning, transgenesis and xenotransplantation) into a more broadly defined frame of artificial life. For Langton, however, the creation of life *in vitro* requires a costly and complex infrastructure and would not, ultimately, provide enough new information about possible life-forms. Computers, on the other hand, provide a relatively cheap and efficient medium for the creation of life *'in silico'*. The main proviso here is that life has to be understood or defined in purely informational terms. Once this is established, then it is simply a case of stating that 'the computer is *the* tool for the manipulation of information' and that it is capable of supporting 'informational universes within which dynamic populations of informational "molecules" engage in informational "biochemistry"' (51). The previous chapter outlined the way in which the contemporary technoscientific view of life as information is derived from molecular biology and a belief in the gene as the unit of life. Langton's definition of life combines this kind of reductionism with a high degree of functionalism:

> Life is a property of *form*, not *matter*, a result of the organisation of matter rather than something that inheres in the matter itself. Neither nucleotides nor amino acids nor any other carbon-chain molecule is alive – yet put them together in the right way, and the dynamic behaviour that emerges out of their interactions is what we call life. It is effects, not things, upon which life is based – life is a kind of behaviour, not a kind of stuff.
>
> (Langton 1996 [1989]: 53)

Whereas the definition of life as information is widely accepted within ALife and related disciplines (Boden 1996a), Langton's logically consistent claim that 'life is a property of form not matter' has had a more marked impact on the field. This reinforces the analogy between organisms and machines already facilitated by the informational concept of life, but also allows for the assertion that machine or artificial life *is* real life. Through a discussion of a project by Craig Reynolds in which the flocking behaviour of birds is simulated by 'a collection of autonomous but interacting objects (which Reynolds refers to as "Boids")', Langton (1996 [1989]: 66) asserts that whereas 'Boids are *not* birds' the flocking behaviour in Boids and birds is the same' (68). So 'the claim' is that 'the "artificial" in Artificial Life refers to the component parts, not the emergent processes' (68). If the parts are properly implemented, the processes which result are genuine. The 'big claim', then, is that artificial systems carrying out the same functional roles as natural systems 'will support a process that will be "alive" in the same way that natural organisms are alive' and that ALife is (or will be) genuine life, only 'made of different stuff than the life that has evolved here on Earth' (69). The aliens, it appears, have (almost) landed.

The philosophy and biology of ALife

ALife is an attempt to understand the essential nature of living systems through the use of computational models (Bedau 1996: 343). The working hypothesis within the field is 'that the essential nature of the fundamental processes of life can be implemented in relatively simple computer models' and that this hypothesis 'is at odds with the conclusions often drawn from the pervasive historicity, contingency, and variety of biological systems' (Bedau 1996: 343). Life, in other words, is computable – whatever it actually is. By modelling life, or its main characteristics, ALife revisits and reawakens the question of the nature of life left hanging within both philosophy and biology. The biological options, for ALife philosopher Mark Bedau, are too varied, too contradictory. He runs through the standard evolutionary, autopoietic, Gaian and astrobiological viewpoints ('we can search for [extraterrestrial] life only if we have a prior conception of what life is'), outlining the three main contending theories: 'life as a loose cluster of properties, life as a specific set of properties, and life as metabolisation' (Bedau 1996: 334). Since exponents of these contending theories have failed to rule out their challengers and since the prospect of messy ontologies and epistemologies is a reality – 'For all we know, life might be no more unified than a collection of overlapping properties from overlapping disciplines' – Bedau seeks to revitalise the quest for a unified theory by reinforcing the case for adaptation. His case for life as 'supple' adaptation (or the unending capacity to produce novel solutions to evolution on the basis of changing environmental problems) is related to, and distinguished from Ray's 'open-ended' evolution on the basis that Ray's case is anecdotal and ill-defined. In contrast, supple adaptation is an objective, empirically verifiable process (Bedau 1996: 345). Dismissing the critiques of adaptation by Gould and Lewontin, Bedau maintains that the simulated thought-experiments which characterise ALife can and do provide evidence of supple adaptation. They also bring 'explanatory order to lists of the central hallmarks of living phenomena' (354), or, more sceptically, secure conceptual and computational confinement in the face of a complex and elusive biological and philosophical concept. Pattee, responding to Langton's claim for the existence of strong ALife and for the validity of the computer synthesis as well as simulation of life, questions the technological distinction between synthesis and simulation in terms of the – as yet unresolved – biological distinction between life and not-life (Pattee 1996: 379). He, like Bedau, associates ALife more closely with computationalism than with biology and maintains that ALife is too closely tied to the problematised Platonic ideals inherent in AI. Life, in Langton's definition, is over-formalised where the compelling, and essentially biological question concerns the relation between form and matter (386). This question is fundamental to modern genetics, or more precisely, genomics where the theory, practice, philosophy and sociology of human genome research vacillates between the exploration of a simple cause and effect deterministic relation and a complex, non-linear and increasingly non-deterministic one (see Chapter 6). For Pattee, a more refined theory of both evolution and emergence is needed in order to assess the possibility of synthetic organisms existing in simulation environments. In the mean time, he argues that ALife researchers should not over-inflate the capacity of computers and 'should pay attention to the enormous knowledge base of biology', particularly in the areas of cell structure, behaviour and evolution. They should analyse 'the genotype, phenotype, environment relations, the mutability of the gene, the constructability of the phenotype under genetic constraints, and the natural selection of populations by the environment' (392). ALife, then, needs more biology, less maths.

Having examined the claims of Thomas Ray and Christopher Langton, Margaret Boden's observation that ALife 'raises many philosophical problems, including the nature of life itself' (1996b: 1) may seem to be rather understated. However, philosophical and critical analysis of ALife is severely underdeveloped;[7] Boden's work offers a useful introduction albeit from within the field.[8] Boden confirms the idea that informational concepts of life 'were widely used by theoretical biologists long before Langton gave A-Life its name' (8) and she summarises Langton's general principles of life in a list form which includes: 'self-organisation, self-replication, emergence, evolution' (8). Boden also points out that in Ray's definition of life, with its emphasis on self-replication and open-ended evolution, the existence of biochemical metabolism is 'inessential'. While such a claim is clearly controversial, 'proving that Ray's concept of life is inadequate is not easy' (12), Boden identifies functionalism as 'the philosophy of mind which defines mental states in terms of their causal relations with other mental (and environmental) states, and which assumes that these causal relations are, in principle, expressible in computational terms' (2). In other words, 'mental states' like pain, pleasure, fear and desire are not understood as 'phenomenal experiences' or even as physical events in the brain, but as 'abstract functional (causal) roles' (2). Functionalism is widely accepted within the field of AI ('most of whose practitioners define mental phenomena in informational terms') and 'many A-Life workers take a similar view of life itself' (2). This is exactly the view that Langton expresses when he says that 'life is a property of *form*, not *matter*'. Life is the computational causal relation between aspects of natural or artificial matter – its behaviour and effects but not the 'stuff' itself. Functionalism in ALife eradicates the body just as functionalism in AI dismisses experience. The body and experience are two key loci of situated and embodied knowledge which feminist work on epistemology (and science studies) seeks to promote – and which a feminist critique of ALife such as this aims to investigate. In order to do this effectively, it is important to trace – as patiently as possible – those narratives within the field which erase, and those which contest the erasure of the body as matter. Embodiment, as a goal in ALife may still be strictly informational, may refer only to the body as form and remain both functionalist and reductionist. If Langton's philosophy of life is unashamedly functionalist, it is also 'unabashedly' reductionist 'because it holds that high-level phenomena depend on simple interactions between lower-level processes' (9). Emergence, then, does not necessarily mitigate against the existence of reductionism in ALife – it depends on whether the observer is looking from the bottom-up or from the top-down. For Bonabeau and Theraulaz (1997), ALife is both synthetic and reductionist, which 'makes it quite dangerous, especially for the excited young scientists (1997: 304). Synthesis, although useful in the attempt to capture emergent properties 'implies weakened explanatory status of models, huge spaces of exploration, absence of constraints'. It is too open, too seductive, inspiring an 'irrational faith' yet ensuring that life-as-it-could-be 'is dramatically ill-defined' (305). ALife itself promotes superficial and 'short-breathed' analogies between physical and social systems and life-as-it-could-be, lacking considered constraints, necessarily reproduces and is contained by life-as-we-know-it (307). The emphasis in reductionist methodology and epistemology on internal explanations is somewhat mitigated 'by making embodiment a clear goal of all ALs' and by placing embodied creatures in some form of external environment (310). But computational reductionism – which 'stands on the idea that any phenomena that obeys the *laws* of physics can be simulated in a computer' – remains fundamental (313).

Reductionism gives rise to a central concept of ALife which is self-organisation: 'self-organisation involves the emergence (and maintenance) of order, or complexity, out of an origin that is ordered to a lesser degree' (Boden 1996b: 3). Development is regarded as being 'spontaneous' or 'autonomous', a facet of the system itself rather than the system's designer. In this respect, ALife is opposed to classical AI 'in which programmers impose order on general-purpose machines' (3). Self-organisation 'requires' a form of computer modelling referred to as 'connectionism' in which networks of simple interconnected units develop order or complexity from the bottom-up rather than from the top-down. As the units function simultaneously '(exciting or inhibiting their immediate neighbours)' they may be referred to as 'parallel-processing', and as 'they are broadly inspired by the neurones in the brain, connectionist models are sometimes called neural networks' (3). Connectionism has been taken up in some aspects of cybercultural theory (Plant 1996; Kelly 1994; De Landa 1994) where it is used as a model of autonomous development in technological and cultural 'systems'. Autonomy is another key concept in ALife and it is used to emphasise 'self-directed control rather than outside intervention' (Boden 1996b: 4). Boden stresses that it is not, however, 'an all-or-nothing property' (120) and that it is much more closely related to the idea of self-organisation than to the idea of freedom:

> To many people, the notion that computer models could help us to an adequate account of humanity – above all, of freedom, creativity, and morals – seems quite absurd. To the contrary, the suspicion is that the concepts and explanations of A-Life and/or AI must be incompatible with the notion of human freedom.
>
> (Boden 1996b: 96)

At the very least it may appear that ALife's concept of autonomy is hobbled by a kind of determinism focused on the internal environment of the system. Behaviourism, as defined by B.F. Skinner (1971), rejects the notion of freedom as 'an illusion, grounded in our ignorance of the multiple environmental pressures determining our behaviour' (Boden 1996b: 96). In ALife, those multiple environmental pressures determining behaviour are predominantly internal and pertain to the self-organisation of the system (Langton 1996 [1989]: 53) and adaptation. Determinism, especially in combination with informational or machinic metaphors of life may easily be regarded as dehumanising,[9] but ALife rehumanises informational life through the construction of artificial entities known as autonomous agents. In her book on the *Philosophy of Artificial Life*, Margaret Boden (1996a) begins to consider the implications of this research for changing cultural conceptions of life, technology and humanity. She claims that children's understanding of life has been challenged by home computer versions of Biomorph and Tierra (Boden 1996a: 28), that adults and children are developing a 'magical' view of technologies which model emergence and evolution (29) and that the way in which ordinary people see themselves is affected by what science has to say about human autonomy (95). What ALife science has to say is that 'an individual's *autonomy* is the greater, the more its behaviour is directed by self-generated (and idiosyncratic) inner mechanisms, nicely responsive to the specific problem situation, yet *reflexively* modifiable by wider concerns' (102). Prior to considering the discursive rather than ethnographic dimensions of autonomy and humanism in more detail, it is useful to consider the extent to which ALife research has succeeded in synthesising self-organised entities.

ALife's autonomous agents

One of the leading figures in what is referred to as autonomous agent, or adaptive autonomous agent research is Pattie Maes (1997), for whom an agent 'is a system that tries to fulfil a set of goals in a complex, dynamic environment' and that an agent can be said to be autonomous 'if it decides itself how to relate its sensor data to motor commands in such a way that its goals are attended to successfully' (Maes 1997: 136). Last, but not least, an agent is adaptive if it can improve its goal-oriented behaviour over time, or, in other words, learn from experience. The main goal of autonomous agents research reflects the main goal of ALife in general: to increase understanding of the principles of life-as-we-know-it and to use those principles to create life-as-it-could-be. The principles in this case are specified as 'adaptive, robust, effective behaviour' which situates autonomous agents research somewhere between ALife and AI. What moves it closer to ALife than AI is the emphasis on the 'embodiment' and emergence of adaptive behaviour. Agents are situated in an environmental context and this leads to the possibility of emergent complexity. Embodiment, in this context, refers to the 'architecture' (the tools, algorithms, techniques) for modelling autonomous agents either in hardware (as robots) or in software (as 'knobots'). Where agent architecture is uniquely flexible and there has been some success in modelling autonomous behaviour, Maes also points to some problems with 'scaling-up' to higher degrees of complexity and realising emergent potential. It is also interesting to note that she regards the predominant influence of behaviourism as a limiting factor and suggests that 'a lot could be learned by taking a more ethologically inspired approach to learning' (157). It is apparent that her own research into entertainment applications for autonomous agents is informed by studies in animal behaviour. In 'Artificial Life Meets Entertainment' she describes a number of entertainment applications including Julia, an autonomous conversing agent, and the ALIVE (Artificial Life Interactive Video Environment) project which she helped to develop. This is 'a virtual environment that allows wireless full-body interaction between a human participant and a virtual world inhabited by animated autonomous agents' (Maes 1996: 216). These agents are modelled on animal behaviour and include a puppet, a hamster, a predator and a dog. Whereas the hamster, predator and dog display reasonably conventional animal behaviours, the puppet is anthropomorphised and displays infantile human behaviour such as trying to hold the user's hand, imitating the user's actions, pouting when sent away and giggling when touched – 'it giggles when the user touches its belly' (218). What does it mean to ascribe agency to 'autonomous' lifelike forms which/who are largely, as yet, not human, not adult and not reflexively gendered? Part of this question was addressed at the Artificial Intelligence and Darwinism Symposium held at Tufts University in 1995 and recorded on CD-ROM (as *Artificial Life*). The discussion on autonomous agents took place between Pattie Maes, Daniel Dennett, David Haig, Sherry Turkle, Kevin Kelly and others – and it focused on ethics, accountability and control. Maes outlined her attempt to build agents or intelligent systems 'that perform a practical purpose and really help people deal with the complexity of the computer world by, for example, foraging, so to speak, for interesting documents for a particular user on the World Wide Web' (Maes in Dennett 1995b). These agents would watch and learn from the user and would reproduce and evolve according to their usefulness. Their existence, as Dennett points out, was anticipated by Richard Dawkins who imagined a form of computer virus which observes software use, monitors activity and reports back with data. He predicted a 'little watcher' which would move through the infosphere and proliferate' (Dennett 1995b). Some of Maes's agents do reproduce and mutations which occur in reproduction mean that offspring agents look for different kinds of documents than their parents. The documents they obtain may be more or less interesting than those of others and 'if they are less interesting then that offspring won't survive' (Maes in Dennett 1995b). Fitness is then determined by usefulness. Where Haig raises concerns about the propagation of competing agents produced by competing companies and the likelihood of 'a terrible outburst of junk mail', Turkle suggests that as 'extensions of self' agents subvert 'the notion of an identity as sort of bounded within your skin'. In answer to Haig, Maes argues that 'other companies will have to make agents that will filter that junk mail from you and maybe will make little police agents that roam around the networks to check whether other agents are doing things that they aren't allowed to do'. She dismisses Turkle's point on the basis that agents are frequently employed in the 'non-computer world' and function in similar ways by accepting delegated tasks or representing a 'client' to a third party. However, the difference for Kelly is 'that these agents are very dumb and there are lots of them'. According to him, we are faced with 'this idea of an ecology of little things out there that are somewhat representing you and have some sliver of your mind and they are out there replicating and they are mutating and they are somewhat out of control and this is a scary idea, which I find interesting'. Haig suggests that even if it is possible to control your own agents, it is not possible to control other people's and he expresses fear about the volume of information which is already being gathered about his political opinions, causes, patterns of consumption and so on. It is just not possible, Kelly concludes, to sustain accountability in an ecology of autonomous agents. What this suggests to him is that there may actually be a limit to the efficacy of evolution in AI/ALife. Reviewing the conference over all, Kelly reassess the general position on evolution:

> I think I learned that there was probably a wider agreement that evolution was a way to do things than I thought. I think that surprises me and I'm not sure that I actually, I'm still open to the possibility that perhaps it's a tool that, we may not be able to evolve everything that we want. In other words, evolution may be the way to get complex things but I wonder if we can get everything we want by evolution.
> (Kelly in Dennett 1995b)

What this means for Turkle is the need for a social practice or context with which ownership of and accountability for 'computational creatures' becomes possible: 'I mean, nobody's really had that social practise yet. And I think that it's the fact that these things are being developed without, or in seeming isolation from a social practise, that creates these ethical issues'.

It is important to emphasise the need for an ethical approach to alife as a whole and to autonomous agents in particular. Following a discussion of Rodney Brooks's research and his desire 'to build completely autonomous mobile agents' without any real regard to human implications, social and scientific applications or philosophy, Steven Levy (1992) invokes *the* cautionary tale of modern science – *Frankenstein*. Autonomous agents, as examples of self-reproducing and evolving technology, may become monstrous if a 'hands-off', sceptical or laissez-faire attitude toward ALife permits them to run 'out of our control' (Levy 1992: 334). Levy is alert to the dangerous military potential of self-reproducing war machines which may become impossible to stop. He suggests that 'dire risks' such as this are ignored for a number of reasons ranging from a sense that ALifers are mad and the project is simply impossible to the belief that it will be a long time before it is possible to create indisputably living organisms, and longer still before they threaten 'us'. The field of Artificial Life 'will therefore be policed only by itself, a freedom that could conceivably continue until the artificial life community ventures beyond the point where the knowledge can be stuffed back in its box' (339). In that case, then Chris Langton's insistence that the biannual ALife conferences address ethical issues may be 'diligent' but also, perhaps, inadequate: 'He hopes that, through frank and open discussion, the researchers would impose implicit sanctions on those who would use artificial life to arm the dogs of war. He expects the scientists to agree eventually on a framework of responsible methodologies' (Levy 1992: 339). Langton's hopes and expectations, though laudable, do not appear to have been enough to end the legacy of Frankenstein which, according to Levy (1992), has been haunting ALife conferences from the beginning (4). Despite his 'dire risk' assessment, Levy is by no means opposed to the ALife project as a whole and appears to support the central redefinition of life as information. What he is opposed to is 'culture's refusal to yield the promise of life to the realm of science' and its insistence on mystical and religious definitions (7). He is particularly critical of vitalism which, ironically, he ascribes to the story of Frankenstein. Vitalism draws on Aristotle's notion of a divine vital force or *élan vital* which exists only in living organisms, and uses it to oppose mechanistic views of life (established during the Enlightenment). The precise nature of this vital force was always uncertain but 'by the nineteenth century many were convinced that the agent was electricity, and as proof they pointed to that force's ability to twitch the limbs of the dead' (21). Frankenstein, as we know, was regenerated this way. For Levy, the persistence of 'vitalism of a sort' in us all represents 'an atavistic tendency' to refuse biological status to anything outside 'the known family of earthbound organisms'. There is, he argues, 'a particular reluctance to concede the honour of life-form to anything created synthetically', including information (22). The heretic notion that the basis of life is information may have come of age with the discovery of DNA, but predates it in Levy's account by several years.

The premise that 'the basis of life is information, steeped in a dynamical system complex enough to reproduce and to bear offspring more complex than the parent' (22), belonged first and foremost to John von Neumann. Von Neumann's self-reproducing cellular automata was made public in 1953 (Levy 1992: 45) and was to ALife what the Turing machine was to AI – a primary source of inspiration (25). Von Neumann's cellular model for a self-reproducing automaton was imagined as an enormous (actually infinite) check board on which the squares represented cells. He drew a creature on the check board and represented different cell states or activities with colour. The creature, shaped like a box with a long tail, reproduced by 'claiming and transforming territory' (43) cell by cell:

> Eventually, by following the rules of transition that von Neumann drew up, the organism managed to make a duplicate of its main body. Information was passed through a kind of umbilical cord, from mother to daughter. The last step in the process was the duplication of the tail and the detachment of the umbilical cord. Two identical creatures, both capable of self-reproduction, were now in the endless checkerboard.
>
> (Levy 1992: 45)

The idea, of course, was that the process would continue and that the cellular automata would model not only self-reproduction but also evolution, and indeed emergence. As Levy points out, questions about the control and agency of von Neumann's cellular automata – 'What happens when we set these structures free? What can emerge from them?' – became essential to the field of ALife (Levy 1992: 46).

Although von Neumann did not have the opportunity to build his model, John Conway built one very like it and named it The Game of Life. Conway simplified von Neumann's model by reducing 29 possible cell states to 2: on or off; one or zero; 'alive or dead' (Levy 1992: 51). By still using the check board idea, each cell was given eight neighbours (those which touched the sides or corners). Conway's rules were that if a cell was alive it would survive into the next generation if two or three of its neighbours were also alive. It would die of overcrowding if there were more than three live neighbours, and die of exposure if there was less than two. If a cell was dead, it would stay dead in the next generation unless precisely three of its neighbours were alive. In which case, it would be 'born' in the next generation. And that, as Levy puts it, 'was it' (52). Both stable and periodic configurations emerged on Conway's hand-built and operated grid. These were named 'much in the taxonomic style of stellar constellations, after the shapes they suggested' (Levy 1992: 52). One such shape was the 'glider', a five cell shape which shifted in each time step or generation, returning to its form after four generations having moved one cell diagonally on the grid. Conway's work on this project took place during the 1960s and by the 1970s it had been computerised and had generated a good deal of excitement in the UK and particularly US: 'It was estimated by *Time* that millions of dollars of unauthorised computer time were squandered by Life tinkerers, who even published their own newsletter listing various discoveries' (Levy 1992: 57). Despite failing to yield a self-reproducing form, Conway's claims for The Game of Life were high. He claimed that in principle it could support the emergence of all recognisable animal forms and an infinite number of new ones. He also claimed that on a large enough scale there would be 'genuinely' living configurations, 'whatever reasonable definition' of living was applied (58).

Among researchers and enthusiasts, cellular automata (CAs) are deemed 'sufficiently complex to develop an entire universe as sophisticated as the one in which we live' (58).[10]

ALife's non-vitalist vitalism

Emergence and complexity are the non-vitalists' vitalism bestowed on the ALife community by original research on cellular automata and nurtured by figures such as Langton, who 'furiously resisted any trace of vitalism in his philosophy' but 'regarded the concept [of emergence] as sort of an *élan vital* in and of itself' (Levy 1992: 107). Levy identifies a certain mysticism surrounding the phenomena of emergence which allows vitalism in through the back door of an otherwise rationalist and reductionist project. In what is ultimately a more epistemologically reflexive account, Claus Emmeche (1994) balances out both vitalist and mechanistic philosophies of life and dispels all traces of mysticism from the ALife project. By locating ALife within the framework of postmodern science, he demystifies ALife much as ALife itself, in his view, demystifies vitalism: 'It is a perspective that acknowledges that life possesses certain specific "vital" fundamentals, but that denies their mystical character'. ALife 'emphasises that it ought to be possible to imitate or remake those fundamentals artificially, in a computer, for instance' (Emmeche 1994: vii).

Spaces of dissent

Emmeche identifies the vital elements of life as it is defined within the field of ALife but takes issue with Langton's radical formalism or his assertion 'that the logical form of an organism can be separated from its material basis of construction' (Emmeche 1994: 60). He asks if it really is possible 'to completely abstract form from its material context' and turns to von Neumann for something like an answer: 'one has thrown half of the problem out the window and it may be the more important half' (62). Pattee goes further by arguing that 'one may be throwing the whole problem, that is, the problem of the relation of symbol and matter' (1996: 386). Von Neumann's equivocation leads Emmeche to reassert the interdependence of form and matter or the dual 'bio-logical' depiction of life processes. Life is, after all, 'not only digital' and 'requires a material foundation and a historical process that organises the material on higher levels and that places limitations on the logically possible ways in which life can behave' (Emmeche 1994: 63). It is clear then that ALife has an ongoing dispute which may be said to be analogous to the dispute within feminism between matter and discourse (see Chapter 7). For Emmeche, the dematerialisation of life as a concept or an epistemological category is a facet of the increasing use of computers in the construction of mathematical models of reality. The computational powers and image-making capabilities of computers supersede those of human brains and eyes (Kember 1991) and have created or revealed strange 'bewildering structures' ranging from the strange attractors and Mandelbrot sets of chaos theory through cellular automata to the nascent life forms of ALife (Emmeche 1994: 71). But what exactly is the relationship between these mathematical models and the natural world? What exactly is being modelled? It would appear to Emmeche that the mimetic function of science is being superseded by a new kind of synthetic function: 'The model no longer seeks to legitimate itself with any requirement for truth or accuracy. It creates a simulacrum, its own universe, where the criteria for computational sophistication replace truth, and only have meaning within the artificial reality itself' (1994: 160). Artificial life-forms are therefore perceived as being alienated from the physical world and because they are computational forms they are 'all too rigid, flat, without dimension' (136). Just as they may be lifelike without being fully three-dimensionally alive, they may cease to exist but never truly die. Artificial forms or states may change from being on to off (one to zero), but 'artificial life is life without death' (137). In the physical world, 'life requires death' because 'death is a process that itself involves new life' (137). In order to synthesise the decay, dissolution and re-assimilation of matter into the environment it would be necessary to build a model so complex that it ceased to be a model at all. The claims within the field that this has already been achieved are therefore overstated and alife resumes its rightful place outside of the biologics and metaphysics of life and death and 'within that domain of human construction we call language' (137). Artificial life may not deal in life and death as we know it, but even as a specialised branch of computer science which deals in 'human constructions' of life and death it has implications which may be considered profound. If Emmeche fails to acknowledge the 'material' (ideological, economic, social and political) dimensions of scientific language and discourses of life, he does point out that there are those within the ALife community who insist on the material dimensions of life itself (142, 186).[11] What this reference to an internal critique signifies for Emmeche is that ALife corresponds to a postmodern tendency towards the dissolution of copy and original, model and reality which lead to the production of simulacra. Dominant discourses in ALife conflate lifelike models with life itself and promote the existence and autonomy of virtual worlds. As a result, 'science becomes the art of the possible because the interesting questions are no longer how the world is, but how it could be' (161). For ALife to make a useful contribution to research in biology, it must to an extent denounce its autonomy and rearticulate, or contribute to the rearticulation of the artificial and natural among other traditional biological dualisms:

> The construction of artificial life can in a general way contribute to dissolving some of these oppositions, or combining them in new ways so as to shed new light on what living organisms really are; we may even use a-life to criticise the very opposition between life and death, between organic and inorganic – the essential opposition that has formed the basis for biology as a project.
>
> (Emmeche 1994: 162)

If ALife offers a means of deconstructing biological science, then the first stage is to unpack the myth of biological realism. Yet ALife offers more than an epistemological critique of biology. In so far as key biological concepts and dualisms are increasingly highlighted and contested through the convergence of biological and computer science, ALife also offers the possibility of an epistemological critique of a wider technoscientific culture. One of the issues which Stefan Helmreich's ground-breaking research on ALife outlines is that the critical potential of ALife is largely, as yet, latent within the semiotic and material dimensions of 'life-as-it-could-be'. In the discourse of ALife as it is currently manifest, 'constructions of *life-as-it-could-be* are built from culturally specific visions of *life-as-we-know-it*' (1998a: 13). In other words, ALife currently naturalises traditional biological concepts and discourses which include the reductionist concept of life as information derived from molecular biology, and the deterministic concept of human life derived from sociobiology. In *Silicon Second Nature*, Helmreich draws on the dual meaning of the Hegelian concept of 'second nature' as both mirror and successor to 'first nature'. Where first nature is the 'pristine, edenic nature of physical and biotic processes, laws and forms' (Smith in Helmreich 1998a: 11) second nature comprises both the processes, laws and forms of society which mirror them and seek to replace them. Silicon second natures then, are configured in the environments, epistemologies and subjectivities of computers, computer scientists and computer users (12) and they may or may not challenge the precepts of their ancestors. Helmreich's book seeks to force this issue by way of a 'critical political intervention into the reinvention of nature under way in ALife'. He argues 'against the digital naturalisation of conventional visions of life' and petitions for 'a greater sense of possibility in the ALife world' (15).

One way of opening up such possibilities is to recognise the existence of dominant epistemologies and subjectivities within a non-homogenous field, and Helmreich's detailed anthropological research creates a space for dissent and transformation. As well as arguing that European discussions on ALife are generally less computational than those in the US (222), he incorporates the voices of researchers who dispute the informational concept of life: 'It doesn't grab me, and it doesn't grab me for a very good reason. Real living entities not only co-ordinate information, they co-ordinate flows of matter and energy. They actually make things' (214). Helmreich draws on feminist epistemological debate in order to outline the gendered aspect of the contest for the meaning of life. Traditionally, a disidentification with the body and emotions as evidenced in the informational theory of life is associated with the masculine, rationalist epistemology of early modern science. Without seeking to reinforce any kind of essentialism, Helmreich traces the gendered Cartesian dualism of mind versus body through his interviews with both men and women, and roots it in a male dominated field of activity. The lack of women involved in ALife is attributed to socialisation, the disciplines associated with it (maths, physics, computer science) and the idea that 'in a culture in which women's professional abilities are always in question, women may be wary of attaching themselves to fringe fields' (217). The interesting question which follows from this is not the essentialist question of what ALife would be like if more women were involved, but whether the discipline would even exist if a range of feminine rather than masculine standpoints were predominant:

> Computational brands of Artificial Life may be so implicated in masculine institutions, traditions, and epistemologies that the discipline could exist without them no better than it could without computers. If there were more women in Artificial Life as presently constituted, this would probably mean more people from cell biology and ecology, in which women are more highly represented. Whether these people would find computational creatures compelling instances of life is an open question.
> (Helmreich 1998a: 218)

The call for a critical feminist and queer epistemology is concomitant with the desire that ALife should and could be reconstituted in less computational terms. Helmreich defers to women who argued that definitions of life should incorporate the value, beauty and experience of life (217) and suggests that a feminist and queer approach would, among other things question the link between sex and reproduction which currently characterises the field.[12] The heterosexual basis of ALife reduces organisms to recombining code, 'conflating reproduction and sex, and rendering uninteresting those aspects of life that do not have to do with reproduction' (218). In ALife systems, all that programs really need to do in order to be considered alive is to reproduce. One gay man interviewed by Helmreich suggested that: 'if you weeded out this philosophy that one of the intrinsic things to make the life of an organism complete is to reproduce, if you removed that, then you probably would have quite a bit of a paradigm shift over to questions about what it is about the structure of an organism that makes it alive' (219). Perhaps then, questions of metabolism would become central rather than marginalised. Another interviewee argued for a queer epistemology to critique the simulation of heterosexual selection. Modelling inheritable same sex preferences may prove controversial in the context of a society which, through forms of genetic determinism seeks to naturalise difference and discrimination, but might also indicate 'another view of life-as-it-could-be' (219).

Where Helmreich's work raises questions about the relationship between epistemology and subjectivity within a scientific discipline which normalises and naturalises biological, masculine and heterosexual discourses, there is, I think, scope for further investigation into the gaps and fissures of dominant ALife epistemology, not least through a more extensive examination of the complexity of current debate in biology, biotechnology and the life sciences. There is also much more to be said about the role of the subject in ALife since this, more than any other aspect, situates the discipline and challenges its claims to ahistorical, apolitical (evolutionary) autonomy. It is through the involvement and intervention of other subjects and subjectivities that the construction of alternative epistemologies and different configurations of power become possible. Power, as it is currently configured is monolithic. It passes from the creator to the creations of ALife worlds but is neither shared nor contested. The power of the creator is embodied in the creation through narratives which closely model those of monotheistic Judaeo-Christian religion. It is clear that ALife represents the 'god-trick' (Haraway 1991a) par excellence. This is the trick of transcendent, disembodied vision, or of seeing everything from nowhere. The notion of the god-trick which Haraway developed specifically in the context of medical imaging and other scientific visualising technologies, has greater depth than Foucault's (1987) ungendered instrumental concept of panopticism and is realised in the visual display of programs from Tierra to *Creatures* which offer users a God's eye view of artificial worlds. To my mind, ALife is also an example of another kind of god-trick which I have referred to elsewhere as that of autonomous creation (Kember 1998). The god-trick of autonomous creation is that of creating life without reference to, or dependence upon the other of the female body. Autonomous creation is masculine and it subsumes the maternal function, thereby rendering it obsolete. Most often associated with the medical and visual technologies of 'assisted' reproduction (Franklin 1993, 1997; Treichler and Cartwright 1992; Treichler *et al.* 1998) autonomous creation allows science to father itself and confers a god-like status on scientists. Helmreich locates ALife within a widespread mythical, religious, literary and scientific tradition 'of attempting to manufacture living things' which is premised on 'a sort of masculine birthing' (1998a: 5). He also demonstrates that there is more than a degree of self-consciousness in the creationist acts of at least some of the ALife Gods. Tierra is on one level, an allegory of creation. The 'seed' or 'ancestor' program is planted in the soil or earth, and constitutes the origin of life. The seed is both masculine and divine: 'In tales of procreation, males, made in the image of a masculine god, plant their active "seed" in the passive, receptive, yielding, and nutritive "soil" of females, "fertilising" them' (Delaney in Helmreich 1998a: 115). The gendering of form and matter in Langton's vision parallels seed and soil and is classical if not divine.[13] Nevertheless Langton is knowingly designated the 'father' of ALife. The 'genetic' code for Ray's ancestor is '0666god' – 'a designation that suggests that the programmer is a kind of Faustian figure, playing at a devilishly digital divinity' (117). The tightly inflated balloon of such playfully serious omnipotence is threatened less by the slow puncture of agency escaping or transubstantiating into the emergent complexity of artificial forms and environments than it is by the prick (as it were) of womb envy. As Helena says, 'Women create things, right?' and 'maybe men would like to give birth to something and here it is, this is it' (121). And no mess either. The computer is a contained, pristine birthing environment which can be turned off and walked away from at will. Power is so much more appealing without the attendant responsibility, which is why the story of Frankenstein will not cease to be of relevance within a patriarchal techno-scientific culture. Helena also echoes a familiar feminist refrain that science does not stop at mimicking the maternal or creative function; it attempts to supersede it: 'They're saying to us, "we're going to beat you guys. We're going to create entire worlds"' (121). A creationist discourse is simultaneously a colonialist one when it deals with the 'discovery', naming and controlling of new worlds by white European and American men. Thomas Ray's plan for a global Network Tierra, and a new environment to be filled with digital organisms leads Helmreich to insist on the 'terratorial metaphor' and the 'colonial imagination' with its creationist underpinnings (94). The combination of creationism and colonialism is indicative not just of masculinity but of race. Cyberspace cowboys owe much to their ancestors:

> That many Artificial Life practitioners are white men who grew up reading cowboy science fiction is not trivial. The location of SFI [Santa Fe Institute] in New Mexico, a place associated with the days of westward frontier expansion, is also fitting, and acts as a resource for imagery enabling the crafting of computers as worlds.
> (Helmreich 1998a: 95)

By initiating a postcolonial, queer and feminist critique of ALife, Helmreich offers clear directions for further work and indicates that in this particular realm of cyberspace there might be space for the decolonisation of worlds, the queering of relationships and the redefinition of life in terms that recognise that definitions matter. With reference to Judith Butler's (1993) concept of 'materialisation',[14] which attempts to bridge the divide between epistemologies and subjectivities which are regarded as being either given or constructed, natural or cultural, Helmreich suggests that 'what "life" is or becomes is "materialised"'(22). In other words it 'comes to matter (in the sense of both becoming important and becoming embodied) – in such practices as describing and fabricating machines and organisms' (22). Life comes to matter through both the description and fabrication of machines and organisms. The material dimensions of language not only are of equal significance to processes ranging from robotics to computer science to genetic engineering – but also are indivisible from them. Richard Doyle (1997) underlines this point by arguing that it is through its 'rhetorical software' that ALife produces its creationist power. According to Doyle, ALife was able to crystallise as a discipline due to a combination of three factors: the provision of cheaper and more powerful computers, the analogy (within biology) between living and non-living information processing systems and the vision made myth of Chris Langton's emergence from a near-fatal hang-glider crash (Doyle 1997: 118). His resurrection story of loose neurones and an intermittent consciousness 'sort of bootstrapping itself up' constitutes a powerfully productive religious allegory and origin story for ALife (119). Langton's emergence into consciousness and life is experienced out-of-body and reported in 'transcendental terms' (Doyle 1997: 118) which evoke the epistemological and ontological status of the discipline they inform. The material and metaphorical existence of gliders link together Langton's biography and his biography of ALife in which a 'glider' pattern is identified as a propagating information structure in CA. For Doyle, gliders and other spatial metaphors like emergence 'situate A-Life as a transcendental place from which to view traditional, carbon-based life' and resonate with Haraway's (1991a) concept of the god-trick or 'gaze from nowhere and everywhere that characterises the disembodied objectivity of technoscience' (Doyle 1997: 121).

N. Katherine Hayles's (1999a) analysis of the disembodied and dematerialising narratives of ALife is contextualised by her discussion of the embodied status of narrative itself and the materialisation of scientific concepts in literature. She concentrates on Tierra in which the organism/machine analogy (or, anthropomorphism of computer code) is not – as in the case of Dawkins – 'incidental or belated' but 'central to the program's artifactual design' (Hayles 1999a: 228). In other words, Ray does not 'discover' creatures in his computer, rather, he designs an artificial computer in which creatures can 'appear' – so naturalising the organism/machine analogy. Hayles shows how the visualisation of Tierra (in a video produced by the Santa Fe Institute in order to publicise its work) serves to naturalise the analogy, disguising its status as a narrative construct:

> In the program, the 'creatures' have bodies only in the metaphorical sense . . . These bodies of information are not, as the expression might be taken to imply, phenotypic expressions of informational codes. Rather, the 'creatures' *are* their codes. For them, genotype and phenotype amount to the same thing; the organism is the code, the and code is the organism. By representing them as phenotypes, visually by giving them three-dimensional bodies and verbally by calling them 'ancestors', 'parasites', and such, Ray elides the difference between behaviour, properly restricted to an organism, and execution of a code, applicable to the informational domain.
> (Hayles 1999a: 229)

Through a detailed examination of Ray's verbal and visual narratives, Hayles seeks to demonstrate the complex interplay between metaphor and materiality and between embodied subjectivities and disembodied epistemologies. Ray's program instantiates his investment in the metaphor of life as information and in arguments which privilege form over matter. Hayles's critique centres on the strategic exposure of elisions and erasures within these arguments (1999a: 12) and on the 'rememory' of embodiment.[15] Her methodology centres on literature and on narrative as embodied sites which facilitate the traffic between science and culture. Hayles is concerned with 'narratives about culture, narratives within culture, narratives about science, narratives within science' (22). While maintaining that literary texts are not merely passive vehicles for scientific ideas but actively shape what science and technology signify in cultural contexts, she also suggests that they 'embody assumptions similar to those that permeated the scientific theories at critical points'. These assumptions include 'the idea that stability is a desirable social goal, that human beings and human social organisations are self-organising structures, and that form is more essential than matter' (21).

The future of ALife – consolidating (digital) naturalism

A critical cultural analysis concerned with examining the science (and) fiction of ALife might usefully employ all of the forms of narrative listed by Hayles (1999a). Significant among the narratives within (ALife) science are those which constitute conference proceedings. Both Helmreich and Risan have offered insightful anthropologies of the field, but one question which remains concerns the extent to which ALife discourse has itself 'evolved', not least through the agenda-setting emphasis of conferences. Early proceedings already considered include Langton's *Artificial Life. The Proceedings of an Interdisciplinary Workshop on the Synthesis and Simulation of Living Systems* (1989) and Varela and Bourgine's *Toward a Practice of Autonomous Systems. Proceedings of the First European Conference on Artificial Life* (1992). But clearly, ALife is no more a static than a homogenous field. It is, perhaps, easy for interdisciplinary projects – such as ALife and, indeed, cultural studies – to be accused of lacking or losing direction, and there may currently be only limited allegiance to Langton and Ray's original claims for strong ALife interlaced, as they are, with a strong narrative current of science fiction. Life-as-it-could-be still stands in a relation of possibility to life-as-we-know-it but, at least at one end of the art and science continuum, that possibility may be being constrained by the increasing emphasis on naturalism and on biologism. Risan (1996) reports that the proceedings of the third European conference were introduced by associating increased biologism with the very survival of ALife: 'It is our opinion, and that of many others, that the future *survival* of ALife is highly related to the presence of *people from biology* in the field' (Moran *et al.* in Risan 1996: 44). In the preface to *Artificial Life V*, Langton and Shimohara (1997) balance the turn to biology with a continued emphasis on engineering goals. A decade after Langton's first workshop, he argues that 'Artificial Life should not become a one-way street, limited to borrowing biological principles to enhance our engineering efforts in the construction of "life-as-it-could-be"'. He maintains that 'we should aim to influence biology as well, by developing tools and methods that will be of real value in the effort to understand "life-as-it-is"' (Langton 1997: viii). This suggests something of a balanced, reciprocal relationship between biology and engineering, weak and strong ALife which salvages his original goals and enlists engineering in the service of this, if not other planets: 'It is now clear that we are responsible for the state of "Mother Nature" here on Earth. It is time to take that responsibility seriously, by vastly extending the time-horizon in which we think about our scientific, engineering, and social goals, from years and decades to centuries and millennia' (viii). Here, of course, space and the quest for alien life is elided by computerised evolutionary time and the quest for sustainable human life.

In his introduction to the highlights of *Artificial Life VII* (a special issue of the journal *Artificial Life*), Mark Bedau (2001) points out that all of the papers address fundamental questions about living systems while 'most build bridges to concrete biological data and generate experimentally testable predictions' (2001: 261). He includes papers on the evolution of multicellularity and genetic codes; on neuro-robotics; cancer and the evolutionary growth of complexity. The conference theme, 'Looking backward, looking forward', coincided with the new millennium and sought to assess the achievements of the field to date, as well as set out – and to an extent prescribe – the remaining open questions. The agenda-setting aims of the conference are realised in a concluding paper which explores three broad questions: 'How does life arise from the nonliving? What are the potentials and limits of living systems? How is life related to mind, machines and culture?' (Bedau *et al.* 2001: 263). These questions pertain primarily to biology, evolution and evolutionary psychology and though not seeking to be exclusive, they counteract 'the centrifugal force tending to pull apart interdisciplinary research activities' (263). While acknowledging the dimensions of a field broad enough to encompass the discovery and/or creation of new and unfamiliar forms of life, and practical enough to guide the use of new technologies for extending and manufacturing life (including drugs, prosthetics, the Internet, robotics), Bedau *et al.* (2001) clearly state that 'Artificial Life is foremost a scientific rather than an engineering endeavour' (364). ALife is to be, at least at the outset of the new millennium and a new generation of researchers, subsumed within biology: 'Given how ignorant we still are about the emergence and evolution of living systems, artificial life should emphasise understanding first and applications second' (364). Their list of open problems reflects this priority, and the extent of the (re)turn to biology is perhaps best illustrated here by the author's willingness to include the speculative and partially non-scientific concerns related to evolutionary psychology. Casualties of the list include the nature of life itself and those areas in which ALife plays a significant role, namely robotics, games and art (365). Philosophy, entertainment and art alongside engineering are therefore subjugated in the name of biology. The biological tasks highlighted include:

> Generate a molecular proto-organism in vitro; achieve the transition to life in an artificial chemistry in silico; determine whether fundamentally novel living organisations can exist; determine what is inevitable in the open-ended evolution of life; determine the predictability of evolutionary consequences of manipulating organisms and ecosytems; develop a theory of information processing, information flow, and information generation for living systems.
> (Bedau *et al.* 2001: 365)

A number of these tasks are dependent on the lessons learned from genomics (366, 367) and all take the indirect, synthetic route to a clear definition of life. The tasks relating to evolutionary psychology include: 'Demonstrate the emergence of intelligence and mind in an artificial living system; evaluate the influence of machines on the next major evolutionary transition of life; provide a quantitative model of the interplay between cultural and biological evolution; establish ethical principles for artificial life' (365).

One of the substantive issues here is the extent to which life and mind are intrinsically (that is, biologically) connected. One of the methodological issues is whether it is more beneficial to analyse mind and intelligence when they are embodied in living systems. Both issues inform the approach within ALife to autonomous agents and any progress in the task of demonstrating the emergence of intelligence and mind in artificial living systems would inform the controversy surrounding the computability of life (Bedau *et al.* 2001: 372). The evolution of machines follows, teleologically, the evolution of culture, which is at most semi-autonomous from biological evolution. The rate of technological change suggests 'that machines might play an unprecedented role in the next major evolutionary transition' (373) and the task described is to predict and explain this role. If autonomous agents become established in the natural world then this role will be central (373). At the very least there is a supporting role for technology in the provision of a suitable infrastructure for evolutionary change (373). Perhaps the most ambitious task set out by Bedau *et al.* (2001) in this category is that of quantifying the interplay between cultural and biological evolution – and in effect resolving the debate, or rather the science wars, surrounding evolutionary psychology. Culture, in this account does not change, it evolves, and evidence of evolution can be found in the emergence of economic markets, the growth of technological infrastructures and the development of scientific beliefs (374). Changes which might otherwise be described as socio-economic or philosophical therefore become biological. ALife here adopts the principles of cultural evolutionism set down in sociobiology, evolutionary psychology and memetics. Where sociobiology and evolutionary psychology analyse how cultural traits evolve due to their effect on biological fitness, memetics looks at how cultural traits evolve in their own right (374). The distinctions and similarities between the genetic and psychological transmission of information are fundamental to the establishment of a quantitative model of the interaction between biological and cultural evolution. Finally, the authors return to the ethical challenges held in suspense since Langton's initial injunction and highlighted by ALife's adoption or adaptation of a concrete biosocial analogue.

Four main ethical issues are indicated: the sanctity of the biosphere; the sanctity of human life; responsibility towards new forms of artificial life and the risk entailed in exploring the possibilities of artificial life. It is interesting that all of these issues are similarly indicated in genomic discourse, suggesting that ALife and genomics are informed by a single, fundamentally humanist, bioethics which functions, if it functions, somewhat retrospectively as an evolutionary brake. The practice of engineering life explicitly includes 'natural or artificial' systems and the biosphere or ecosystem explicitly incorporates 'indispensable' elements such as the Internet: 'Existing computer viruses wreak havoc as it is, but imagine how much worse they would be if they spontaneously evolved like artificial life systems' (Bedau *et al.* 2001: 374). Concerns about the sanctity of human life inform much of current ethics and ALife, 'like the theory of evolution, will have major social consequences' including 'our future increasing dependence on artificial life systems' (375). Increasingly sophisticated artificial life-forms will require public protocols analogous to those covering the treatment of human and animal research subjects and the uses and applications of such forms demands an assessment of costs and benefits. The existing development of military and commercial applications (Chapters 4 and 5) of ALife research, particularly autonomous agents, has not yet been explicitly assessed. Bedau *et al.* conclude that where the ethical issues for ALife do resemble those concerning genetic engineering, animal experimentation and AI, 'creating novel forms of life and interacting with them in novel ways will place us in increasingly uncharted ethical terrain' (375). It is clear from this that the engineering goals of life-as-it-could-be are not wholly displaced by the biological imperative which primarily secures the prospect of a(lien)life within the realms of science rather than fiction, naturalism rather than constructivism.

To summarise, this examination of the field of ALife has highlighted the contradictory and gendered dynamic of creationism and evolutionism in which self-proclaimed gods create and name artificial worlds and feminised entities which reproduce asexually and give rise, through emergence, to pristine prelapsarian life. In the conflict between creation and evolution, God and Darwin, the role of religious mythology and particularly Christianity is emphasised. ALife allegorises creation and the fall and is attuned to the narratives of *Frankenstein* and *Faust* through which it connects with molecular biology, especially genomics. It is my contention that Western Christian mythology overwhelms those gestures towards Eastern mythology which inhere in ALife's Californian countercultural past. The process of emergence is at the heart and soul of ALife's reappropriation of creationist power by means of evolutionary machines. Eschewing spiritualism, emergence is presented as a means of securing the purity and promise of digital naturalism in the face of the evident and more lowly constructivism of 'artificial' life. In the retreat from base engineering, the increasing turn to biological hegemony in ALife is commensurate with the quest for digital naturalism. And as alien life is re-presented in more familiar, less threatening almost archetypal forms and frameworks, ALife follows essentially the same pattern as related discourses in evolutionary psychology and genomics – the containment of unnatural, unlawful and potentially unwholesome kinds within the sanctified if eternally mythical boundaries of Nature.

What has also been highlighted here is the ethical debate surrounding the generation of autonomous agents and their origins in von Neumann's cellular automata and Conway's The Game of Life. These mathematical experiments in how cellular forms can give rise to complexity inspired the largely disavowed resurgence of vitalism in ALife. Vitalism and creationism in ALife are repackaged and re-presented as emergence and evolutionism respectively. They are the absent presence of a technoscientific discipline and discourse which always seems threatened by the power of its own mythology.

ALife's internal contradictions are manifested partly in dissenting debates on the relationship between form and matter; debates which might usefully be considered alongside those on the relationship between matter and discourse within contemporary feminisms (Chapter 7). These debates concern the philosophy and politics of the (female) body of nature and have a direct bearing on the ontological and epistemological status of posthuman ALife identities. To what extent are they (to be understood as) both facts of nature and artefacts of culture?

It is my main contention that a cyberfeminist critique of ALife, affiliated to postcolonial and queer critiques through the demand for decolonised worlds and denaturalised relationships and identities, must recognise and work with such internal dissent by risking the renunciation of oppositional rhetorics. These serve only to reinstate the dichotomy of nature and culture problematised across a broad spectrum of technoscience but rendered clearly obsolete in artificial life. Cyberfeminism cannot *resist* the increasing biological hegemony in artificial life cultures, but it can enter into a productive dialogue with the 'enemies within' (Segal 1999).

Chapter 4
CyberLife's *Creatures*

This chapter combines a detailed analysis of the ALife computer games *Creatures* and *Creatures 2* with an investigation of the games' producers (CyberLife Technology Limited [CTL] and CyberLife Research Limited [CRL]) and consumers. Textual analysis is therefore combined with an extended interview discussion (over two years) with ALife engineer Steve Grand (originally of CTL, then of CRL and the designer of *Creatures*), and with a survey of *Creatures* (series) Internet-based user groups. The *Creatures* products are analysed in the context of other ALife computer games and programs and in the context of the 'Sim' games, including *SimLife*, which incorporates ALife principles. The aim of this contextualised approach is to highlight the specificity of CyberLife's *Creatures* in relation to other computer games and to use it as a case study which illustrates a dynamic relationship between the games' producers and consumers, between ALife science and engineering and between science and culture. Through *Creatures*, CyberLife claims to have (co)evolved life in the computer, and to have enabled autonomous agents to proliferate in the digital ecosystem.

The media effects debate is concerned with the positive or negative effects of the mass media,[1] particularly television and video but also new media including computer games. It tends to concentrate on the effects of screen violence on the audience (Keepers 1990). Where the media are regarded as being all-powerful, violence in the media is said to have a direct effect on the audience and to cause violent behaviour. The audience in question is constructed as being passive and susceptible, and technology is given a deterministic role. At the other extreme is the active audience which filters or totally subverts media messages and is premised on a humanist model of the rational, autonomous subject (Henriques *et al.* 1998; N. Rose 1999; Blackman and Walkerdine 2001). Historically, the media effects debate has alternated from one extreme to the other and has by no means been confined to academic circles alone. The mass media are rather preoccupied with their own effects, and particularly with the effect of screen violence on young, impressionable or 'vulnerable' minds. Blackman and Walkerdine (2001) have argued that the main determinant of vulnerability is class as well as age.[2] The argument that the media have negative effects is echoed in psychology research on early video and computer games. In his research on video games in arcades, Mark Griffiths offered a definition and a model of addiction which, he argued, was equally applicable to computer games (Kember 1997). The addicted, isolated game player corresponds with the notion of a passive audience and contrasts with the (gendered) individual who actively projects his desires for mastery and control (Turkle 1984; Kinder 1991) onto the game, and with the formation of subversive computer game subcultures (Panelas 1983; Buckingham 1997, 2000; Sefton-Greene 1998).

CyberLife's *Creatures* demonstrates a consumer involved (rather than producer dominated) commercial enterprise which suggests that the ALife in ALife products is neither statically defined or the sole activity of remote and specialised scientists. ALife engineers designed the products according to their perception of the scientific principles and game users playing the role of genetic engineers (creating a new species of artificial 'creatures') contributed to the development of the design. The game users who are involved in the user groups and in the correspondence with *Creatures* producers are a self-selecting group of North Americans and Europeans who do not constitute a subculture – or counterculture (Ross 1991) – because there are no consistent signs of subversion or rebellion against the 'parent' culture of science and/as commerce. Rather, the game users took the genetic engineering lead from the game narrative and objectives, and extended it, making it their own via a much greater facility with computer technology and Internet culture than had been anticipated by the producers. The results were then fed back to the producer/parents who in turn restructured the game to allow for greater user involvement. This does not amount to a dialogue (about ALife) between scientists or engineers and predominantly young people of school or college age, but it does imply a dialogic structure of communication which disrupts binaristic and hierarchical models of power in computer game and new media theory, and possibly even in effects-based theories about the relationship between science and culture. This chapter aims to give one illustration of the idea that it is not useful to debate the 'effects' of science on culture or vice versa. It is arguably more productive to highlight the specific interaction between science and culture, producers and consumers which challenges these distinctions and changes the terms of the debate.

The chapter is divided into four sections and the first, 'SimWorlds – ALife and Computer Games', offers a brief analysis of *Tierra*, *SimEarth*, *SimLife* and *SimCity*. These simulations are seen to offer mirror rather than alternative worlds, reproducing hegemonic discourses of origin and evolution without (with the possible exception of *Tierra*) being instances of origin and evolution – or examples of artificial life. The Sim games are primarily ecosystem simulations designed for educational as well as commercial purposes. The 'Creatures' section incorporates detail about the games, their design and consumption and includes interviews with Steve Grand plus samples from the Internet user groups. 'CyberLife – Selling ALife' examines the science and philosophy behind CTL – producers of *Creatures*. It looks at the marketing and promotion not only of the games but of ALife itself, and considers other applications in the military, banking and retail. Finally,

'CyberLife Research Ltd – or "real" ALife' examines the relationship between AI and ALife in the context of blue-sky research into the synthesis of consciousness in the computer. This section includes Steve Grand's project to develop a situated robot (Lucy) and the question of whether or not it is really possible to capture life and consciousness in your computer.

SimWorlds – ALife and computer games

Stirring the primordial soup

In the computer program *Tierra* (Spanish for 'Earth') Thomas Ray claims to have synthesised natural selection. The documentation for *Tierra* (Ray and Virtual Life 1998) provides exhaustive detail about the program and how to operate it, and Ray's ambition for the program is established at the outset. Ray is an ecologist with particular interest in the Cambrian explosion of diversity (600 million years ago) when the dominance of single-celled organisms gave way to complex multicellular life-forms. If, as he puts it, 'something comparable can be made to occur in digital organisms' it would be instructive not only for natural science but also for computer science or 'for evolving software to make the transition from serial to parallel forms'. *Tierra* creates a virtual computer and operating system within the user's machine, and the architecture of the operating system is designed so that the executable machine codes are evolvable. Ray is not evolving computers but claims to evolve computer codes as if they were organic. It is clear from his contribution to the ALife debates that he does not uphold a clear distinction between natural and artificial organisms, and the *Tierra* documentation consistently employs biological terms and metaphors. The RAM (Random Access Memory) of the virtual computer is referred to as the 'soup' in which the machine codes or 'self-replicating' algorithms 'live' as 'creatures'. Natural selection and evolution are facilitated by genetic processes of 'mutation' (random flipping of bits) and 'recombination' (swapping segments of code between algorithms). The virtual computer provides the program user with the means to control and monitor the evolution of computer code creatures. Evolutionary control is offered through the provision of three different mutation rates, disturbances, the allocation of CPU time to each creature, the size of the soup and the spatial distribution of creatures. The observational system is described as 'very elaborate' and it includes a record of births and deaths, code sequences of every creature and a 'genebank' of successful 'genomes'. It is also possible to automate the ecological analysis and record the kinds of interactions taking place between creatures. Observation and control is facilitated by the provision of two user interfaces which display information in predominantly numerical rather than pictorial form. The interface is detailed and information is basic only in so far as it is not visual.

Tierra consists of a complete virtual world or ecosystem governed by Darwinian evolutionary laws. While being enclosed and self-contained, this virtual world is also completely accessible to the program user, who is offered panoptic vision and other god-like powers. The user is able not only to 'see everything from nowhere' (Haraway 1991a: 189),³ but also to stir the primordial soup and create life through evolution. The god of Judaeo-Christian religion and the 'god' of evolution are brought together and combined in the role of the user. God and Darwin play together from the outset, making *Tierra* as much an ALife fantasy as an ecological playground. Running the program involves initiating the virtual computer 'world' and 'innoculating' the soup with the 'ancestor' creature – a single cell with a specified genotype. The program then runs for 500 generations in a soup of 50,000 instructions. Runs can be restarted and genomes can be created, replicated or extracted. The price of being god of *Tierra* is a pre-conversion to ALife's particular brand of creationist evolutionism combined with the technical ability and will to navigate the interface beyond the quite innocuous surface (giving cell numbers and sizes) to the ever deepening and (perhaps to some minds) impenetrable depths of computer code.

The Creation part 1: making worlds

SimEarth: The Living Planet (Maxis Inc. 1995, 1997) is a planet simulator based on the Gaia theory of James Lovelock. Gaia is a holistic Darwinian evolutionary theory of life on earth, and it is associated with environmentalism. For Lovelock, Gaia is 'a theory that sees the evolution of the species or organisms by natural selection and the evolution of the rocks, air and ocean as a single tightly coupled process' (Maxis Inc. 1997). While the (CD-ROM) instruction manual acknowledges the specificity of this world view, the game itself – or 'system simulation toy' – clearly sets out to reinforce it. *SimEarth* has a strong educational dimension and emphasises the notion of responsibility towards the planet and its occupants. *SimEarth* is populated by 'SimEarthlings' ranging from single-celled plants and animals to intelligent species. Intelligence may be a feature of any animal anywhere on the evolutionary scale, but there can be only one intelligent species on the planet at any one time. Evolution is placed within a hierarchical framework even if humans are not necessarily placed at the top of the hierarchy. The player or user may well view all SimEarthlings from a 'satellite's point of view' complete with two levels of magnification, but also embodies a Gaian god who has the welfare of the SimEarthlings and the SimEarth 'in your hands'. As a Gaian god, the player is invited to play with the tools of the game, but not with the rules. As, strictly speaking, a 'software toy' rather than a game, *SimEarth* offers more flexible and open-ended interaction. In this case, the toy incorporates a number of different planets with which the player can experiment. *SimEarth* is described more specifically as a system simulation toy and 'in a system simulation, we provide you with a set of RULES and TOOLS that describe, create and control a system' (Maxis Inc. 1997). Each planet is a system, and it quickly becomes clear that where the tools are for playing with, the rules are to be learnt. The challenge is to understand how the rules work and then apply them by using the tools. The rules in question apply to the management of global 'factors' including chemical factors (atmosphere, energy); geological factors (climate, continental drift, earthquakes and other disasters); biological factors (formation of life, evolution, food supply, biological types and distribution) and human factors (war, civilisation, technology, waste control, pollution, food supply and energy supply). The tools in question provide players with the ability to create, modify and manage a planet: 'Create a planet in any of four Time Scales. Physically modify the landscape of the planet . . . Nurture a species to help it evolve intelligence' (Maxis Inc. 1997).

The rules and tools of *SimEarth* combine to form a simulation with acknowledged technical and epistemological limitations. The simulation is described as 'a rough caricature – an extreme simplification' in so far as even the most powerful super-computers cannot yet accurately simulate climate and weather patterns let alone other (for example, biological) aspects of the game. The evolutionary model is limited in that it resembles that of Earth, and other possibilities are not explored. Intelligence may not be confined to humans but its evolution follows the same path as it is perceived to have done in human civilisations. The assumption is also made that intelligence is an evolutionary advantage, which 'might be flattering ourselves' (Maxis Inc. 1997). Gaian theory is also acknowledged to be a controversial, not a universal theory. It is clear then that *SimEarth* constructs a mirror world rather than an alternative one, and that it reflects (albeit in a semi-transparent way) existing hegemonies and epistemologies. Gaia may not be a universally acknowledged science, but it appears as a universal within the game in that, despite the illusion of choice, the player is offered no other view of how life-supporting planets might work. Playing by these rules, the player then chooses between seven planets, each with its own scenario and goal. One planet involves a demonstration 'of life on Earth as a self-regulating whole' using populations of daisies, and another tells the whole evolutionary story from origins to cosmic terraformation – from a scientific viewpoint which is evidently inspired by science fiction. There is a macho evolutionary option – 'choose and help a particular species gain mastery of the planet' – and a more softly focused fantasy of designing, managing and maintaining 'the planet of your dreams' which may be high tech or low tech ('where the biosphere is never endangered'). Players identify their own skill level, name their chosen planet and have to pick from a world in the geologic, evolution, civilised or technology 'stage'. Two main windows are displayed on the computer screen (as with all other Sim games): one showing an atlas of the world, and the other a more detailed close-up of an area which can be directly manipulated. Information about specific events – such as the evolution of a particular type of organism – appears as a message at the top of each window. The menu on the atlas window gives access to a number of new windows controlling the variables for geosphere, atmosphere, biosphere and civilisation.

Information is graphically displayed so that the biosphere window, for example, displays a number of deer-like creatures whose simulated reproduction and mutation levels can be altered. The main window displaying a detailed area of the planet enables the player to view, add to, and move life-forms and aspects of the environment but reinforces the linear, teleological process of Gaian evolution.

Planetary evolution should progress from the formation of continents to the appearance of life, civilisation and technology in that order. If civilisation is introduced too early the player receives a message saying 'that is not allowed in this time scale'. Complex life-forms will die out if they are introduced too soon, and so intelligence can only evolve at a certain stage – regardless of the particular species it graces. The teleology is based on the evolution of life on earth up to and including advanced technological (or Westernised) civilisation governed by the notion of progress. The necessary use of fossil and nuclear fuels underlies the conflict between technological progress and the environment. The conflict is resolved by the narrative fact that a successful civilisation is one which leaves in a spaceship to explore planets elsewhere. So much for the prospects of life on earth. In this simulation, Gaian theory privileges 'the living planet' over specific inhabitants, representing it as a whole organism which reacts to the player's decisions. In the 'Gaian Window', earth is anthropomorphised as a face which graphically displays what it thinks about what the player does – so aligning Gaian gods with ordinary mortals and underlining the environmental lesson.

The Creation part 2: making life

SimLife. The Genetic Playground (Bremer 1992), like *SimEarth*, is an ecologically complete simulated world, but with more emphasis on the genetics and evolution of plant and animal species. The Gaian element of the game is important but overshadowed by the focus on Artificial Life. Ken Karakotsios, who designed the game, gives 'inspirational thanks' to Richard Dawkins and Chris Langton, and in the introduction to the user's manual, Michael Bremer states that 'SimLife is an Artificial Life laboratory/playground designed to simulate environments, biology, evolution, ecosystems, and life' (Bremer 1992: 2). The main purpose and main feature of *SimLife* is the 'exploration of the emerging computer field of Artificial Life' (2). ALife, like *SimLife* 'creates a laboratory in a computer, where the scientist can completely control all environmental factors' (6), but 'life as we know it' emerges from the bottom up. Life is defined from a Dawkinsean (ultra-Darwinist) perspective as 'a gene's way of making copies of itself' (16) and as 'anything that exhibits lifelike behaviour, including: adaptive behaviour, self-replication and the ability to extract order from the environment' (17). One of the specified roles for the user is an ALife experimenter, and this combines with the other named roles of Charles Darwin, 'normal human being trying to play and win a pretty complex computer game' and 'a being with amazing powers who creates worlds' (12). The user builds, names and populates a world, having the power to confer life and death. Using the 'life button' (complete with the image of a double-helix) the user populates her/his world with plants and animals but also accesses the 'smite' icon (a black lightening bolt) which takes out any animal and leaves behind a pile of bones. Life, death and sex have specific sound-effects (fanfare, 'oooo' and 'oo-la-la' respectively) and evolution occurs through natural selection and survival of the fittest. Each creature, or 'orgot', in *SimLife* is a composite (head, body and tail) of three known species and the user can change each of the three segments or create a creature from scratch. Genes can also be accessed via the life button and manipulated at will. Each gene codes for a specific function (flying, walking, swimming and so on), feature, size and gender, and genetic engineering is one of the two highest levels of complexity in the game. The other involves manipulating the laws of physics. Time can also be controlled (evolution can be made to occur at different speeds or the whole game can be paused) and the map and edit windows offer further levels of monitoring and observation (including self-observation or evaluation). As in *SimEarth*, however, the user manipulates the tools but not the rules of the game (despite the illusion of omnipotence) and no matter how many variables are altered, evolution must progress and the ecosystem must be balanced and diverse:

> What can be considered the ultimate goal of SimLife is to look beyond the game, to understand that the real world with its millions of species with their combined billions of genes are all interrelated and carefully balanced in the food chain and the web of life, and that this balance can be upset.
> (Bremer 1992: 3)

SimLife is ultimately an environmentalist and Darwinist educational tool. It provides a lab book designed for school children and an experiment ('Splatt') to observe evolution through natural selection. What is interesting about *SimLife* (apart from the strangely mythical composite creatures) is the extent of its claim to create life in the computer. SimLife-forms 'easily meet' the definition of life offered in the manual. They 'metabolize energy from your wall socket. They require the proper environment – the SimLife program – to survive. They react to stimuli in the environment. They evolve' (Bremer 1992: 83). But they are not complex life-forms and are also said to be alive in a similar way to viruses, requiring 'a host – the computer – to live in and with' (83). In this sense they may be on a similar level of complexity to Tierran creatures – albeit in embodied form.

Cain's creation

In *SimCity* (Bremer and Ellis 1993) the player is invited to 'enter' and 'take control' by being 'the undisputed ruler of a sophisticated real-time City Simulator' (Bremer and Ellis 1993: 4). It is possible to be the 'master' of existing cities such as San Francisco, Tokyo and Rio de Janeiro or to 'create your own dream city (or dream slum) from the ground up' (4). The player is given the prescribed roles of Mayor and City Planner or a simulated city populated by 'Sims' or Simulated Citizens. Sims, 'like their human counterparts' (4), live life (exclusively) within the city, building houses, condos, churches, stores and factories. They also complain about things like taxes and mayors and the player's success depends on how Sims react to her/his government.

SimCity claims to have been the first of the new type of entertainment/education software referred to as system simulations. Rules and tools are provided to create and control the city system, the rules being based on a particular kind of (naturalised) city planning involving human, economic, survival (strategies for dealing with natural disasters, crime and pollution) and political factors. The tools of the simulation enable the player to plan, lay out, zone, build and manage a city and the goals relate to effective economic growth and management. The overall aim of *SimCity* is to balance the eco(nomic)system in order to facilitate progress – or the evolution of an urban capitalist economy. The tools of the game – including discrete residential, industrial and commercial zones as well as transport and civic facilities – reduce and reproduce the values and norms (and the architecture) of late-twentieth-century North American cities. The manual recommends that the number of industrial (external market) and commercial (internal market) zones combined should roughly equal the number of residential zones since this will ensure optimum employment and thus minimise migration. Development of the residential areas is dependent on land value and population density, and quality of life is a measure of the relative attractiveness assigned to different zone localities: 'it is affected by negative factors such as pollution and crime, and positive factors such as parks and accessibility' (43). Progress is prescribed by the inevitable growth of internal markets and 'acceleration' of the commercial sector which 'can turn a sleepy little town of 50,000 into a thriving capital of 200,000 in a few short years' (43). Crime rates are related to population density and land value and a recommended long term approach to lowering crime 'is to demolish and rezone (urban renewal)' (45). The introduction of a police station may be effective in the short term provided that funding is adequate. The player controls the budget for police, fire and transport departments and raises money through tax. Sims are very likely to revolt if tax rates are too high, and the player may be faced with 'an angry mob led by your mother'. The player is offered a bird's eye view through the two windows on to the city (map and edit) and can control the speed with which time passes. But there is no direct control over the Sims who inhabit and make use of the structures provided. Sims are not visible to the player and are represented only statistically.[4] They function as solitary invisible units of production and consumption who commute to and from industrial and commercial zones, but do not travel to other residential zones or communicate with each other. In the manual, Cliff Ellis provides a 'History of Cities and City Planning' which includes sections on the 'Evolution of Urban Form', 'Transition to the Industrial City' and a guide to 'Good City Form' (Bremer and Ellis 1993). *SimCity* normalises the good city form of late-twentieth-century North America by making it impossible to build anything else. Different historical and geographical scenarios are neutralised in a regulated socio-economic system which simultaneously erases and is conflated with a self-regulating natural system or ecosystem. Nature is external to the city system (there is no day and night, seasonal change or weather) and takes the form only of disasters such as tornadoes – which can be switched off by the player. As in all the other Sim games, the software system or environment is a self-contained if massively simplified world.

'Sim' worlds are by no means alternative. They are mirror worlds which reproduce – through simulation – hegemonic discourses of origin and of natural/socio-economic evolution. Are they then *instances* of origin and evolution? Only *Tierra* and *SimLife* claim to be. The other games are more prescribed and lack the necessary 'emergent' technology (bottom-up or parallel processing). Thomas Ray claims that *Tierra* will exemplify the laws of evolution because his algorithms reproduce and mutate.[5] Since Ray's creatures are not embodied and are manifested only as algorithms,[6] his claim depends entirely on the legitimacy of strong ALife claims (notably that life is a property of form not matter). *SimLife* also has genetic artificial organisms of very limited complexity (Cliff and Grand 1999) and is arguably most effective as an ecosystem simulation much like *SimEarth*. The Sim games are distinct from *Tierra* which has specialist ALife appeal (rather like Craig Reynold's Boids and John Conway's The Game of Life CA) and does not aim to be educational as much as illustrative. The Sim games are educational and deal primarily with the balance and development of natural/socio-economic systems. They are ecological in emphasis (including *SimCity*) and have a very different appeal than the faster, more competitive and more violent computer games (such as the 'beat 'em-ups'). In Sim games, violence has a different form (such as the 'smite' feature of *SimLife*) and has evolutionary consequences. It is not as cathartic as, for example, *Tekken 3* or *Mortal Kombat* and the games are not as popular.[7] They are clearly at the more cerebral end of the market and, arguably, because Sims and orgots are not particularly complex, they have not attracted the kind of user investment that the *Creatures* games have. In these, players are engaging with virtual – often anthropomorphised – pets of considerable complexity. What Sim games do most effectively is naturalise genetic and evolutionary determinism in an environmentalist educational scenario and – in the case of *SimLife* – introduce ALife in to one area of popular culture.

Creatures

> Put some life into your PC!
>
> (*Creatures* publicity slogan)
>
> Norns have no mental lives and hence cannot be conscious . . . but they are alive.
>
> (Steve Grand, interview 1999)

In 'Creatures: An Exercise in Creation' Steve Grand (1997a) describes his design for producing lifelike autonomous agents whose biology and biochemistry is sufficiently complex as to be believable. Grand's agents are designed with a large simulated neural network and a basic biochemical model which creates diffuse feedback in the network and represents reproductive, digestive and immune systems. Moreover, the structure and dynamics of the neural network, the structure of the biochemical model and numerous morphological features of the agents are defined by a simulated genome. The genome is of variable length and appropriate for open-ended evolution. The article is presented from a practical viewpoint and goes on to state that the design has been implemented in the form of a commercial computer game product. Underlying Grand's practical description are clear assertions about the value of biological metaphors and the importance of emergent behaviour in creating intelligent agents.

The design for *Creatures* is based on Grand's point of view as a computer engineer rather than a scientist, and for him, this entails a rejection of reductionism in favour of a holistic approach to generating artificial intelligence and artificial life. 'Why', he asks, 'do we create neural networks that have no chemistry' when organisms are heterogenous rather than homogenous systems? Most attempts to generate intelligent or lifelike agents are based on single mechanisms, and while there are good reasons for this (from a scientific or research perspective it is often necessary to simplify the object of study in order to learn in detail how it works),[8] there is also the risk that this methodology 'will fail to deliver the emergent richness that comes from the interactions of heterogenous complexes' (Grand 1997a: 19). Put more simply, 'the fact that organisms are combinations of many different processes and structures suggests that most of those systems must be necessary, and we should heed this in our attempts to mimic living behaviour' (19). What follows is a complicated description which captures the complexity of simulating a biologically whole organism complete with its own genome. The design specifies a class of 'gene' for each kind of structure in the organism, and determines whether it applies to males, females or both. It also controls when, in the creature's life cycle, a specific gene (for example, governing the reproductive system) switches on. Creatures may then experience puberty, although it is noted that 'genes do not code for behaviour, but for deep structure – the behaviour is an emergent consequence of this structure' (23). Genes are assembled into a single 'chromosome' and when creatures mate, chromosomes are crossed over to produce offspring that inherit their complete definition from both parents. 'Mutations' and 'cutting errors' (involving dropped or duplicated copies of genes) produce variations in the gene pool, allowing 'our creatures' to 'truly evolve' (23).

Grand's design for an autonomous intelligent creature capable of learning, and situated in an artificial 'world' is completed by the addition of a simple speech mechanism.[9] It is realised in the *Creatures* computer program 'that allows people to keep small communities of little, furry, virtual animals as "pets" on their home computer' (24). Creatures can be taught to speak, rewarded with a 'tickle' or punished with a 'smack'. They eat, play, travel and 'learn how to look after themselves' (24). Users care for their creatures and diagnose and treat illness when it occurs. Eventually, creatures get old and die, but if they live to puberty, users can encourage them to breed and reproduce: 'He or she can then swap those offspring with other *Creatures* enthusiasts over the World Wide Web' (24). From this, relatively early description of the program, it is already possible to identify quite a range of roles for the prospective user: pet owner, parent (of an anthropomorphised creature), medical scientist, breeder/genetic engineer, trader and computer (games) enthusiast. An important aim of this chapter is to highlight the dynamics of user involvement as a means of challenging the effects model of new media communications, and more significantly of science (producers) and culture (consumers).

The first *Creatures* computer game was released on CD-ROM in Europe in November 1996. In 'The *Creatures* Global Digital Ecosystem', Dave Cliff and Steve Grand state that more than 100,000 games were sold within a month and that they attracted a great deal of media attention (1999: 77). Later releases in the US and Japan in 1997 produced global sales of over 500,000 games by 1998. *Creatures 2* was released globally in 1998 with an initial shipment of 200,000. The technology in *Creatures* is directly drawn from ALife research,[10] and the official guide to *Creatures 2* contains a brief justification of bottom-up (ALife) as opposed to top-down (AI) programming (Simpson 1998). The working definition of ALife offered here is as follows: 'Artificial Life concerns itself with capturing lifelike behaviour by creating small systems (called autonomous agents) and allowing these small systems to interact with eachother to create more complex emergent behaviour that none of the individual systems are aware of. Because it is lifelike behaviour we are after, we call it artificial life' (Simpson 1998: 155). The emphasis is on a quest for emergent (intelligent) behaviour, and where other examples are offered (John Conway's *Game of Life* and Craig Reynold's *Boids*) the *Creatures* producers (CyberLife) are quick to claim that their product is leading the quest:

> All life is fundamentally biochemical, and CyberLife believes that to capture human-level intelligence inside a machine, you should create a complete functioning human. This could be achieved by modelling the individual cells and arranging them in the same way that they are arranged in a real human. The result should be a human that is a human, brain and all. There is still a long way to go, but *Creatures* and *Creatures 2* are substantial steps in the right direction.
>
> (Simpson 1998: 159)

From this it is clear that CyberLife is interested in more than commercial success in the computer games market. This is perhaps simultaneously an end in itself and a means to an end of realising one of the key aims of ALife research. There is then a reciprocal relationship between ALife science and engineering in this context that belies the separation of these categories.[11]

Creatures involves a virtual 'world' (Albia) populated by 'species' of autonomous agents which in the second version of the game, include norns, grendels and ettins. Norns are the focal species with which the player is encouraged to interact, and they possess artificial neural networks, biochemistry, genes and organs. A norn's behaviour 'is generated by its "brain", an artificial neural network that coordinates the norn's perceptions and actions, according to a set of behavioural "drives and needs"' (Cliff and Grand 1999: 79). There are a total of seventeen drives (ranging from hungry and thirsty to amorous and lonely) all of which can be monitored during the game (Simpson 1998). There is a short-cut in the simulated neural network in as far as it 'is not *directly* involved in either perceiving the environment or generating actions' (Cliff and Grand 1999: 79). The task of accurately modelling the physics of sound, vision, smell, taste and touch is currently considered to be too technologically complicated and advanced. Similarly, it is not possible to compute a neural network with individual neurons controlling individual 'muscles' in the norn's 'body'. In place of this, 'there are a fixed number of predefined action scripts (e.g., "move left", "push object"), written in a higher-level language' (79). These action scripts (and the drives) correlate with the simple language which norns can learn with the aid of 'learning machines' (artificial computers) situated in their environment. Using two separate machines, norns learn the words for drives and concepts expressed, respectively, through adverbs and adjectives ('intensely hungry') followed by verbs and nouns ('get food') (Simpson 1998: 69, 70). The game player, or user, must encourage the norns to learn language and 'talk' to them by positioning a virtual hand near an object and typing the object category on the keyboard. Apparently, 'Norns can distinguish only between categories; they can't distinguish between the specific objects within those categories' (Simpson 1998: 76). Users must then familiarise themselves with the norn system of classification. Because norns inhabit a virtual environment, they are referred to as 'situated' autonomous agents. Their autonomy is figured in their capability of co-ordinating actions and perceptions over extended periods of time without external (human) intervention: 'Interactions with the human user may alter the norns' behaviour, but the human can only *influence* a norn, not control it' (Cliff and Grand 1999: 79). The virtual hand of the user (operated by a mouse) can be used to reward (tickle) or punish (slap) a creature and to pick up and drop certain objects. It cannot be used to pick up a norn unless it is in the process of drowning, but it can be used to push it away from (or indeed towards) danger. The virtual hand can be made visible or invisible to the creatures (Simpson 1998: 64). Each individual norn's neural network is affected by its biochemistry. Specific actions such as eating 'can release reactive "chemicals" into the norn's "bloodstream", where chain reactions may occur' feeding back and altering the performance of the network. Norns may eat toxic foods in their environment, and will therefore need medical intervention and care from the user. The interaction between neural network and biochemistry also 'allows for modelling changes in motivational state such as those that occur when a human releases hormones or ingests artificial stimulants such as caffeine or amphetamines' (Cliff and Grand 1999: 80). Norn behaviour is, in other words, biologically and biochemically produced, albeit emergent rather than directly programmed. Albia contains 'bacteria' which can harm the norn's 'metabolism' and be counteracted by eating some of the available 'plant life'. The way in which bacteria affect norns and are affected by certain plants is genetically encoded for each strain of bacterium, 'so there is an opportunity for coevolutionary interactions between the bacteria and their hosts' or between the environment and the individual (80). Artificial genes govern the norns' brains, chemistry, morphology (physical appearance) and stages of development from birth, childhood, adolescence, adulthood to death. Genes are passed on through sexual reproduction (sex is represented by a 'kisspop' in the game and is strictly 'behind closed doors' and heterosexual) and are indirectly related to behaviour: 'the net effect is that new behaviour patterns can evolve over a number of generations' (80).

The creature's brain, organ system, genetics, immune system, respiration and cardiovascular system, digestive system and reproductive system are all modelled and described in detail. They can be monitored and (in most cases) manipulated by the user through the provision of 'applets' or kits such as the Health Kit (basic health care including a selection of medicines), the Owner's Kit (used to name norns, take photographs of them and study their family tree), the Breeder's Kit (covering reproductive systems and providing aphrodisiacs if needed), and the advanced Science Kit (detailed monitoring of organs, DNA, biochemistry and complete with syringe and chemical mixtures for treating illness and injury) and Neuroscience Kit (for experimenting with brains). The Science Kit and the Neuroscience Kit are 'pick-ups' which can be accessed only by persuading one of the norns to activate them. There is also an Observation Kit (to monitor population figures and details such as age, gender, health status), an Ecology Kit (to monitor the environment), a Graveyard and an Agent Injector which allows new objects in to the environment. These objects include norns imported from Internet websites. The main aim of the game, or 'toy' as *Creatures* is also referred to,[12] is 'to give birth to some Norns, explore Albia, and breed your Norns through as many successive generations as you can' (Simpson 1998: 19). However, it is clear that there are advanced features and users, different levels of involvement, and a variety of user roles.

Playing the game

The virtual world in *Creatures* is called Albia and this has both a mythology and a colonial history. Albia is a rich natural and technological environment originally inhabited by the ancient race of Shee who left behind a temple and laboratories for engineering plant and animal life. Grendels are the Shee's monstrous mistake and norns are their crowning achievement. Having created life on Albia, they set off in a rocket to colonise space. The tropical volcanic island then experienced a natural disaster necessitating the resurrection of the Shee's favoured species among the remnants of bamboo bridges, forts and intact underground laboratories.[13] Albia has a balanced ecology with plentiful natural resources and danger in the form of poisonous plants, violent disease-carrying grendels and oceans deep enough to drown unsuspecting creatures in. In keeping with traditional Western techno-scientific perspectives, nature is constructed as a resource for exploration, observation, experimentation and exploitation. The presence of both danger and biological potential is seen to justify and reward technological intervention. Albia has its own geography of east and west (as a disk it has no meaningful north and south) but its landscape of exotic plants, animal life and buildings encodes it as being Eastern, oriental, other. The technoscientific and colonial enterprises are therefore conventionally associated.[14] The flourishing underground terrariums and laboratories of the Shee represent a technological Eden, or tamed, enhanced nature where norns can be safe and enjoy life (at least until the user discovers the genetic splicing machine). Above ground, the barren volcanic area remains an untamed and hellish realm of molten lava fit only for the uncivilised grendels to inhabit. In Albia, the next generation of biotechnological scientists are the direct descendants of the ancient Gods of Shee.

Game playing begins in the hatchery area, 'a warm, cosy, friendly place' (Simpson 1998: 10) with an incubator. The player selects one of six male and female eggs, 'each of which contains its own unique digital DNA' (22) and places it in the incubator. Some 5 to 10 seconds later ('depending on the complexity of the DNA') a norn is 'born'. The player is informed that 'baby norns are a little like two-year old toddlers' and advised to name and track them prior to teaching them to speak and eat (25). The naming of norns is a fairly emotive experience (especially when they reciprocate) and from here onwards, drowning them through lack of adequate supervision can be quite upsetting. It is also very easy to do since the player has no direct control over these cute but wilful creatures. It is possible to spend too much time placing photographs on headstones in the graveyard or making futile attempts at resuscitation at this stage of the game. The parent-educator role of the player quickly gives way to that of general medical practitioner and trainee medical specialist. This is due to the norns exploring their environment and encountering natural hazards, and to an increase in numbers which makes them even more difficult to control ('the most interesting objects in Albia to Norns are other Norns') (Simpson 1998: 81). This is despite a whole barrage of monitoring and surveillance technology and the player's overall panoptic perspective on the world. One interesting feature is the Creature's View option which 'lets you see what the selected creature is looking at' (53). As well as being a necessary mechanism for naming objects, this ability to adopt the creature's point of view is perhaps a further expression of kinship and a displacement of the surveiller–surveilled dichotomy. Surveillance offers no guarantee of control in Albia and this is most evident in the context of reproduction. The Breeder's Kit allows the user to monitor the reproductive system of individual creatures and displays their sex, age, image, life stage and estimated fertility (using a graph of hormone levels). Since 'female Norns are fertile for several minutes, whereas males are capable of getting a female Norn pregnant most of the time' (43) it follows that close monitoring of female reproduction may be necessary. In a somewhat familiar scenario, if a female becomes pregnant 'the picture of the Creature's body is zoomed in, and it shows a little egg that grows gradually over time' (47).[15] Norns are blessed with a 20 minute pregnancy after which time 'the cycle resets and a little baby egg is hatched!' (47). Just like that. The player-programmer's eyes are shielded from this rather effortless process which is signalled by the appearance of an egg in the bottom right hand corner of the screen. Once the player has reached the stage at which norn eggs begin to appear the 'natural' way rather than via the hatchery, the role of parent gives way to that of breeder and ultimately overseer of an evolutionary process which can be influenced but not controlled. Norns will breed without intervention, and naturally occurring eggs do not need to be placed in the incubator. Despite a default setting of sixteen creatures, the population of Albia may start to escalate, making further interventions (such as birth control and the exportation of norns) necessary. The breeding of successive generations of norns (complete with their digital DNA) leads of course to evolution which is – given the underlying ALife philosophy – the true agent of the game and more powerful than the god-like Shee and their player-programmer descendants.

Creatures is by no means simply about interacting with virtual pets. Rather, the player is ultimately positioned as the overseer of a process of evolution involving artificial life forms with which he or she has a degree of kinship. Norns are like children: a new generation. The introductory tutorial to the original game states quite clearly that 'our new-born Norn is alive and like any child she has her own personality'. More than that, as representatives of the ALife project, norns are the next stage in evolution – a new species. It is clear, within the narrative of the game, that 'we', the players, have responsibility for this new species, but as overseers of the evolutionary process we are evidently not in control of it. We may observe, interact, participate and intervene in the process, but ALife carries forward a firm belief in the sovereign power of evolution. Human agency is at best secondary to the primary force of nature. The role of the player in *Creatures* is god-like in so far as it involves bringing a new, genetically engineered species in to existence. But the power of the creator is compromised by the fact that neither the individual creatures or the process of their evolution are controllable. In contrast with conventional video and computer games, the player does not play a character on the screen and does not control their actions. Creatures appear to make their own decisions about what to do and where to go. They learn how to behave and how to survive, and the player can only attempt to teach them through punishment, reward and communication. It is necessary to activate a surveillance camera in order to track the whereabouts of the creatures and the only real control players have is the pause button or exit option which places the whole game in suspended animation. The overseer of *Creatures* is a kind of parent-god whose omnipotence is exchanged for kinship. Kinship is represented on the level of narrative and interactivity and is underlined by a deeper level or principle of connection. The game is based on the principle that norns, as artificial life-forms are alive or possess the same essential life criteria as humans, including autonomy, self-organisation and evolution. As an example of ALife, *Creatures* represents the connection between organic and artificial life forms, and through the role of the overseer it loosens the relationship between vision and control which characterises technological forms of visualisation in science and at the intersections between science, art and entertainment.

Kinship with creatures, or the connection between organic and artificial life-forms, is something which is simultaneously underlined and disavowed in the game which encourages players to play with the idea of it through increasingly enhanced genetic engineering features. The most basic level of genetic engineering is breeding. Then comes the genetic splicing machine, a 'great gadget' that 'allows you to breed creatures together that don't normally match – such as a Grendel and a Norn' (Simpson 1998: 203). The device for creating such 'half breeds' is 'essential for any budding Dr. Frankenstein' (142). Creatures must be lured into the room, locked in to the machine and both donors will be 'lost' in the transformation. The appeal is that 'you never quite know what will come out' (143). Interest in these transgenic organisms is enhanced by a feature which allows them to be exported and imported via the Internet. Users can swap or trade their monsters and – through demand – can download a Genetics Kit which 'lets you view each and every gene in a Creature's genome and edit its properties . . . You can even create a whole new genome from scratch!' (203).[16]

Creatures on the Internet

Using a phrase coined by Thomas Ray, Cliff and Grand (1999) claim that what distinguishes *Creatures* from other ALife games (apart from the combination of simulated neural network and biochemistry) or products based on autonomous agents, is the occurrence of 'digital naturalism' in communities of users and the possible occurrence of culture in communities of artificial agents (1999: 81–83). *SimLife* is mentioned as 'one of the first pieces of entertainment software explicitly promoted as drawing on alife research' (81). But *SimLife* deals with digital organisms which are nowhere near as advanced as those in *Creatures* (82). Similarly, '*Dogz, Catz, Fin-Fin* and *Galapagos* are all presented as involving ALife technologies, but none of them (yet) employ genetically encoded neural network architectures or artificial biochemistries as used in *Creatures*' (83). Neither (as a consequence) do they allow for the possibility of culture emerging in communities of artificial creatures. Cliff and Grand argue that it is likely that 'rapid and productive evolution' will occur in CyberLife systems and that, although it is 'highly unlikely' in the current (second) version of *Creatures* it is 'tempting to speculate' about the emergence of social structures:

> Given that the norns can communicate with one another, and that supplies of some environmental resources (such as food or 'medicine') can sometimes be limited or scarce, it is not inconceivable that simple economic interactions such as bartering, bargaining, and trade occur between norns, allowing for comparison with recent work in simulated societies such as that by Epstein and Axtell.
>
> (Cliff and Grand 1999: 85)

Epstein and Axtell's (1996) work will be discussed in the following chapter, as will the development of artificial societies in environments governed solely by the principles of genetic determinism and evolution. Given the relatively advanced organisation of game 'users' rather than 'agents' it seems more appropriate at this stage to look more closely at the idea of 'digital naturalism' in *Creatures* on the Internet.

Cliff and Grand compare *Creatures* with Ray's *Tierra* but, again, claim that the agents are significantly more complex: 'In colloquial terms, if the agents in *Creatures* are similar to animals in their complexity of design and behaviour, then the agents in Tierra are similar to bacteria or viruses' (Cliff and Grand 1999: 85). Yaeger's *PolyWorld* is considered to be the closest comparable program in so far as it, like *Creatures*, 'attempts to bring together all the principle components of real living systems into a single artificial (manmade) living system' (85). Cliff and Grand point out relatively small differences in technology but larger differences in the aims of the projects since *PolyWorld* is primarily a tool for scientific inquiry into the issues addressed in *Tierra* and a test-bed for theories in evolutionary biology, behavioural ecology, ethology or neurobiology (86). *Creatures* is clearly an entertainment application with technology designed to be adapted in industrial engineering. It bears greater comparison with Ray's *NetTierra* which has been in development since 1994. A development of *Tierra*, *NetTierra* is intended to run on the entire Internet rather than on a single computer. It would run on spare processor time and on as many machines as possible, migrating organisms across the network in search of idle computers: 'typically on the dark side of the planet, where the majority of users are asleep' (Cliff and Grand 1999: 86). Ray argues that the program will create a 'digital ecosystem' supporting diversity and self-organising evolutionary processes. Industrial applications of *NetTierra* are possible and it would be necessary, according to Ray, for 'digital naturalists' to observe and experiment with the evolving life-forms 'possibly removing promising-looking "wild" organisms for isolation to allow "domestication" and subsequent "farming"' (Cliff and Grand 1999: 86). Cliff and Grand maintain that, 'without any prompting', the sizeable community of *Creatures* users with their independent newsgroups and some 400 websites, 'appear to be engaging in exactly the kind of digital naturalism that Ray foresaw the need for in *NetTierra*' (87).

The original producers of *Creatures* anticipated a slightly different response to their product than the one they received and, according to Steve Grand, its appeal stems from the producers' receptiveness not just to user opinion, but to user involvement in the design and development of the product. It had been thought that the major issue for game players and observers would be 'the philosophical question of whether the norns are truly alive' (Cliff and Grand 1999: 87), but the major issue appears to have been the practice rather than the philosophy (or ethics) of genetic engineering. Interest in breeding and exchanging norns exceeded initial expectations, but the 'rapid appearance of users *reporting the results* [my emphasis] of "hacking" genomes, producing new "genetically engineered" strains of creatures' took the producers of the game by surprise (87). When asked about the extent to which *Creatures* was designed to create a community of users on the Internet, Steve Grand reveals how the design was adapted to accommodate a community of technically astute users which had formed semi-autonomously:

> The practical key to making it work was to make the program open, so that people could alter it and add to it. Partly by design, and partly because of the building-block nature of biological systems, there turned out to be a number of ways that people could enhance the product and hence their enjoyment of it. Within a few days of launching the game, there was a new species of creature and also a couple of freeware add-ons available over the Web. People had developed these by hacking into the code and working out how parts of the genetics and script languages worked. Infact they were so good at this that I decided to save them the trouble of working it all out for themselves, and simply published the necessary documentation so that they could get on with it.
> (Interview, September 1999)

Computer hacking plays an interesting and perhaps novel role in the ownership of software in this case in so far as it is not, strictly speaking, an example of either theft or subversion. It is noted that the results of hacking genomes was 'reported' – not just to other users but also back to the producers who, in turn, responded by making an 'open' program still more accessible. The stereotype of the hacker is of an isolated asocial or antisocial computer 'criminal' (Ross 1991) who is probably between the ages of 13 and 30 and almost certainly male. The users represented in *Creatures* user groups are predominantly of school or college age and include both men and women.[17] The question of gender might stimulate an analysis of potentially different types of use, but the focus here – and arguably the more pertinent focus – is on the dynamic interplay between CyberLife producers and consumers as they embody one index of the continuum between science and culture.[18] Hacking is generally associated with subcultural or countercultural activity (Ross 1991) which does not appear to be relevant here. Hackers do not generally share their spoils with targeted groups or individuals. While the economic and copyright ownership of *Creatures* is not in question, the sole ownership of the technology and of successive 'generations' of the product is. While crediting CyberLife's web design team and developers' forum initiative (set up to support individuals producing 'add-ons' or new elements of the game), Steve Grand also acknowledges that *Creatures* products have been moulded 'not only through suggestions but also through direct help from some very experienced and smart Creatures users' (1999).

How then might the relationship between producers and consumers, science and culture, capital and labour be characterised in this context? How does the interplay between designers and users of *Creatures* contribute to an understanding of the dynamics – the flows of ideas and resources – of digital culture and the digital economy? Clearly it does not fit a model of either exploitation or subversion, nor does it reduce to the familiar concept of cultural appropriation or assimilation which depicts 'the bad boys of capital moving in on underground subcultures/subordinate cultures and "incorporating" the fruits of their production (styles, languages, music) into the media food chain' (Terranova 2000: 38). For Terranova, the complexity of labour in late capitalism is characterised by the concept of 'free labour', which, in the context of the Internet incorporates the construction of websites, modification of software, participation in mailing lists and inhabitation of virtual spaces (33). She lays stress on the argument that the virtual reality of the Internet is not a form of unreality or any kind of ideological *tabula rasa* since it is structured by the cultural and economic flows which characterise network society as a whole (34). It might be added that this early sense of possibility linked to unreality is also countered by a return to naturalism in digital culture. Critical of Richard Barbrook's concept of the 'gift economy' in which 'gifts of time and ideas' serve to overturn capitalism from within (Barbrook in Terranova 2000: 36), Terranova maintains that although new forms of labour may not be produced by capitalism in any direct sense, they have, however, 'developed in relation to the expansion of the cultural industries and are part of a process of economic experimentation with the creation of monetary value out of knowledge/culture/affect' (38). It is, she suggests, too easy to employ a model of capitalism against expressions of Internet utopianism – or rather, evolutionism – centred on self-organisation and the depiction of the Internet as a hive mind (collective intelligence) or free market (44). Free labour does signify a degree of co-evolution or collectivism which is however not natural, not ideal. Terranova's main complaint against what she terms Internet utopianism, and I term evolutionism, is its tendency to neutralise the operations of capital (44). It is nevertheless effective – through the stress on self-organisation and collective intelligence – in capturing the existence of 'networked immaterial labour' (44). This, at least in part, has been encouraged by the existence of the open source movement – fundamental to the technical development of the Internet – in which software companies make their program codes freely available to the public for modification and redistribution (49). The company then gains its returns through, for example, technical support, installation, upgrades and hardware (50). The open sourcing of Netscape in 1998 rejuvenated debates on the digital economy and led Barbrook to underline his claims that 'the technical and social structure of the Net has been developed to encourage open cooperation among its participants' (50). Conversely it was regarded as an alternative and successful strategy of accumulation based on encouraging users to spend more time on the Internet (50). With no direct investment in access, this raises the question of what, on its own scale and in the gaming rather than telecommunications context, CyberLife stood to gain by open sourcing *Creatures*. The answer, as suggested in Steve Grand's comments, might well lie in pragmatics. That is, there was not a great deal of choice. The *Creatures* code was hacked, altered and returned because, in a sense, it was already open. It was already open by virtue of its design and distribution, and Grand therefore simply realised what might be termed a digital form of co-evolutionary, cultural and economic interdependence.

For Cliff and Grand, the main indicator of digital naturalism is the emergence of unexpected user activity. The first example they give explains the presence of the genetic splicing machine in *Creatures 2*. The creation of hybrid 'Grenorns' was a user initiative – 'we didn't think this was possible, since grendels had been deliberately made sterile . . . to prevent them from overrunning the world' (Cliff and Grand 1999: 88). Users initially tampered with the norn eggs by manually inserting a genome from a grendel in place of one 'parent'. The result was a random cross between the two species and the topic of 'much newsgroup discussion' (88). Of the 'naturally' occurring mutations, one is the 'Highlander Gene' which results in an immortal agent, and another is the 'Saturn Gene' which causes norns to shiver continuously and results in a 'rather morbidly popular phenotype' (88). What Cliff and Grand refer to as 'digital genetic engineers' have modified individual genes spreading popular mutant strains via the web. One of these, the 'G-defense gene' turns the creature's fear response into an anger response, making it more aggressive (89). A pattern begins to emerge. Aspects of anthropomorphism and sentimentality, stressed by the game producers, gives way to a certain amount of sadism connected with a diminishing sense of kinship. On the one hand, a 'save the grendels' campaign attracted significant support, as did a European drive to ease the language difficulties for migrating norns (unsurprisingly, by increasing standards of spoken English). Apparently, an Australian family emailed Grand a norn that was deaf, blind, insensitive to touch and generally not getting much out of life at all. Grand diagnosed a mutated brain lobe gene, and after corrective modification, rest and relaxation, the norn was sent home (90). On the other hand, there have been disturbing reports of organised norn torture and abuse by an online character named Antinorn:

> There's a guy whose pseudonym is Antinorn, and he runs a website devoted to ways of being cruel to norns. Because of his wicked sense of humour, there are plenty of 'battered norns' around, and so people have set up adoption agency sites to look after them and find them new homes! The newsgroup has several times exploded into a frenzy over this topic and I think it's very healthy.
>
> (Interview, September 1999)

The newsgroup (alt.games.creatures) was monitored at around the time of this interview (August/September 1999). Amidst some chat about school classes and extracurricular activities apart from *Creatures*, Mae, Julius, Cati, Indigo, Patrick, Dave, Kate, Freya and others discussed Bastian's problem with a dead genetically engineered norn – 'flying around my world in circles' and resisting the dead creature remover – alongside Antinorn's antics. The debate about norn torture was, in fact, linked with the question of whether or not they can be considered to be alive. Aliveness was measured against human and animal criteria and connected to an ethical debate about rights. According to Bastian, all of the 'no-torturing stuff' is based on an association between norns and human life, and the fact that humans, unlike animals, have 'mercy for the weak ones'. But Bastian thinks that 'we should not use this attitude on other creatures – especially not on digital ones' because they will weaken the evolutionary chances of our own young: 'If we are to [*sic*] merciful with our C2 norns, they take the place you could use for stronger newborns'. Norn torture is then sanctioned on the basis that it is not as bad as the kind of medical interventions often practised on people: 'if someone wants to torture norns, so let him . . . what some other players do to their norns (let them live ages with injections and then let them starve very slowly) would be even much more cruel'. The debate about what can be considered alive picks up on philosophical, biological and ALife references (including references to the life status of viruses) and there is a sense, for example from Julius and Indigo that norns are only really 'technically' alive. Dave agrees and brings the subject back to norn torture, AN's (Antinorn's) website and the debate about whether or not this has a right to exist. Kate makes a plea for free speech and xOtix supports this by adding:

> I don't believe that simulated torture of a simulated computer-generated <u>model</u> of a limited low form of life (didn't we once agree on spider intelligence as the max? Remember that thread?) is 'bad' or 'evil'. I only have one concern. I wonder what engaging in this kind of activity does to the person who is doing it?

Bastian summarises the debate by agreeing that norns are not 'really' alive but then adds that there will, in future 'be much more complex digital "life forms" that will have the same rights as we . . . and therefore cannot be tortured'. This appears to constitute a closure, and the discussion wanders off course until Tom appears:

> Hi
> Does anyone know how I can 'kill' my Norn??
> I don't know what's her problem, but she suffers, is lonely . . . all the time. I've tried almost anything: I gave her food, injections, other Norns, . . . but she lies there and cries.
> She doesn't listen to me anymore. Perhaps it is a genetic defect.
> So I want to kill her SOFTLY. I don't want to export and delete
> My Norn.
> Please help me.
> Tom

When asked for his view on norn torture and abuse, Steve Grand replied: 'I think it's wonderful! Well, I guess I feel sorry for the norns (although not terribly sorry – it's no more cruel than stepping on an ant)'. For him the concept of cruelty demonstrates that his project has been successful and that people are engaging with the ideas on which it is based. What annoys him is the way in which US publishers censored the 'slap' and 'tickle' feature of the game for fear of a moral panic among parents and teachers. The issue of norn cruelty forced the game producers to change the original title and 'tone down the yelp' that norns emit when slapped. This feature almost had to be removed altogether. Such censorship does not appear to have deterred the dedicated, and a thriving website (run by Antinorn) entitled Tortured Norns co-exists with literally hundreds of other independent sites, and with the official Creature Labs set up by CyberLife.

The Creature Labs website includes information on new products (including *Creatures 3* with enhanced brain power and social behaviour for norns), news and general self-promotional material designed to attract investment: 'We create games and online virtual worlds that are pushing back the frontiers of simulation and modelling to create powerful solutions in Entertainment and on the Internet' (http://www.creaturelabs.com). It is clear that the company wishes to 'take our products even further onto the Internet' by, for example, developing online games designed for both the 'loyal Creatures user community' and new audiences. A chronology of the company's development notes that in 1998 the Creatures brand website was launched with a weekly hit rate peaking at 1.6 million. In 1999, *Creatures* and *C2* combined sales exceeded 1 million units and new games *Creatures Adventures* (for children) and *C3* were launched. Also in 1999 CyberLife launched the Creatures Development Network (CDN) to support 'third-party' developers. The site states that over 1000 people signed up in the first week. Details on the CDN are given in the Creatures Community category (which also contains news that CyberLife has completed and published details of the recently mapped Norn Genome). CDN is a free developer program giving users access to the tools and technology in *C2* 'plus a chance to make money by selling your add-ons and objects through the CyberLife website e-commerce system'. CDN has a forum for discussing technical issues with other players and producers. According to CyberLife 'you are free (subject to our terms and conditions) to use knowledge gained here to produce free COBs [creature objects] for yourself and others or you may want to offer them to us to distribute or even sell on your behalf (for a portion of the sales). Information and materials on CDN remain the copyright of CyberLife but COBs created by 'third-parties' are copyrighted to them. The Creatures Community section of the main Creature Labs website also has a Community Center with links to independent websites, webrings and virtual worlds. It is designated as a place for exchanging news about the creatures community and for adopting norns from the main database or arranging 'a date for a lonely heart (norn, of course)'. The site explains that webrings are co-operatives of website owners who link to and from each other and provides information on how to access the Creatures webrings (some 228). Titles revealed by a search included All Creatures Great and Small (which wasn't to do with the games), Middle Agers for Norns (which was), Norn Sanctuary and Kickass Creatures. The latter contained the Tortured Norns site complete with pitiful images, a tortured norns forum and downloaded dialogue from the newsgroup (http://www.TorturedNorns.homecreatures.com). On 23 July 2000 AntiNorn received a stern warning from the SPCN (Society for Prevention of Cruelty to

Norns), and on 1 May 2000 s/he (according to Grand 'he's definitely a "he"') appeared to receive information from a disgruntled CyberLife employee. To a certain extent, AntiNorn fulfils the traditional hacker role, subverting the 'cutesy' and educational aspects of the game and the producers' claims to have created life in a computer. Certainly, AntiNorn's cannibal norn challenges sentimentality and it is interesting to note that the promotional literature for *Creatures 3* claims that norns behave 'as though' they were alive because they 'almost' are.

CyberLife – selling ALife

Creatures is produced by CyberLife Technology Limited. The company was established in 1996 (although it had been operating under other names since 1989) and prior to a reorganisation in 1999, it comprised three departments: Creature Labs, CyberLife ALife Institute and CyberLife Applied Research. The Institute was responsible for advanced long-term high-risk research into 'artificial lifeform technology' (Grand interview, September 1999). Connections to the wider ALife and AI community was made via the Institute (through publication and conferences) which was financially supported by the other two departments. The Applied Research department developed widespread applications for the blue-sky research, and Creature Labs focused on specific products, notably computer games. In 1999 the company split to form CTL (now officially Creature Labs Ltd) with a focus on games software and online entertainment and CRL chaired by Steve Grand and comprised of other members of the earlier CyberLife ALife Institute.

The original CTL established the company's commitment to the biological method and metaphor of computer modelling and its awareness of the marketability of key concepts and processes such as emergence, evolution and adaptation. The website states that 'CyberLife is designing and building a new generation of Living Technology' (http://www.cyberlife.co.uk) and it advertises biologically inspired modelling. What biologically inspired modelling involves is bottom-up rather than top-down programming and an attempt to simulate complexity through emergence: 'rather than developing massive rule bases, we choose to model life forms from the bottom up in order to capture the intelligence and subtlety of human and animal behaviour, using properties such as emergence'. Biological modelling produces life-forms which 'grow' inside the computer and are whole or complete. They are more than bits (or bytes) of artificial intelligence programming designed for specific albeit advanced or complicated tasks. They are synthetic organisms which can reproduce, evolve and adapt as a 'species' and have (at least initially) lower level but wider possible applications. CyberLife promotes norns as just such a species which may or may not extend its function in the future: 'Some of their offspring, or their cousins, may learn to do useful jobs for people, or simply to keep people entertained until the day comes when we know how to create truly intelligent, conscious artificial beings' (http://www.cyberlife.co.uk). CyberLife is ultimately 'concerned with the re-vivification of technology' by creating lifelike little helpers 'who actually enjoy the tasks they are set and reward themselves for being successful'. The reward is artificial 'natural' selection and survival of the fittest in a Darwinian evolutionary environment which supports and mirrors the economy within which it operates. In a commercial context, biological ALife modelling and psychological AI modelling have different methods and metaphors but similar goals; to compete successfully and survive in the information economy. ALife products have, it could be said, evolved as a result of the failure of the AI project to deliver HAL to an expectant market:[19] 'Artificial intelligence is not achieved by trying to simulate intelligent behaviour but by simulating populations of dumb objects, whose aggregate behaviour emerges as intelligent'.

CyberLife's biological simulation engine is called Origin (originally Gaia). Origin models artificial life forms and environments by mirroring the function of cells within a living structure. Living cells function autonomously and in parallel with other cells to create a unique organism: 'There is no "guiding hand" controlling cells; likewise, each of the objects in Origin can detect the information they need and determine their own state without needing to be controlled or co-ordinated from above' (http://www.cyberlife.co.uk). The major advantage of this cellular modelling structure is that just like real biological systems, it can handle complexity, scale up from small to larger populations of cells and is highly adaptable (Grand interview, Sept 1999). As a software architecture, Origin can be customised to suit specific needs and provided to customers as a toolkit. Robert Saunders assesses the value of Gaia/Origin and CyberLife Technology in general as a basis for innovative or creative design computing (Saunders 2000).[20] Creative design is based on the introduction of new knowledge into the design process through, for example, emergence. Saunders links emergence and creativity with reference to interesting and unexpected properties of a system which requires an observer 'with a set of expectations' (3). After cellular automata and 'evolutionary systems' (such as GAs), emergence and creativity are demonstrated in Gaia's system architecture of cells, chemistry, neural networks and genetics. For Saunders, creativity, like life itself, is a difficult concept to define and 'attempting to create a creative entity is much like trying to create an autonomous agent which can be said to be alive' (5). The real test of a system is whether or not its behaviour matches the observer's expectations of something that is creative or alive: 'Creativity then is an emergent property of a complex system measured in relation to the model of creative agents which people derive from experience' (5). Where Gaia's simulated organisms, such as Creatures, behave in a sufficiently lifelike manner within the confines of their own artificial environment, the question for Saunders is whether they could be equally believable in another 'limited domain' (5).

Apart from the entertainment and games industry, CyberLife technology has been applied (but not as yet developed) in the contexts of the military, medicine, banking and retail. As well as modelling the part animal, part human-like norns in *Creatures*, CyberLife claim that 'we can produce human-like virtual entities in a wide range of simulated environments' (http://www.cyberlife.co.uk). These entities will demonstrate, for example, how the layout and design of retail space affects human behaviour, and they could prevent the pharmaceutical industry from having to carry out research *in vivo*. The 'privately owned' company has supplied products and technology to organisations such as Motorola, NCR and the UK Ministry of Defence. The US company NCR wanted to research customer behaviour in a high-street bank 'so it commissioned CyberLife to breed surrogate people that could wander around inside a virtual bank and test the layout of machines and services – without the time and expense of real-life tests' (Davidson 1998: 40). The software agents were programmed with certain 'drives' (queue, leave, deposit or withdraw cash, seek financial adviser) and behaved in similar ways to real customers in real banks. Their competing drives serve as a test of the bank layout since a successful design enables the agent to complete its transactions 'before it is overwhelmed by the urge to take its business elsewhere' (40). CyberLife advertises similar forms of modelling for other businesses such as shopping malls and theme parks on its website. The contexts are interchangeable because of the adaptive modelling system, and the principles remain the same. Modelling begins with a drawing of the physical space into which human-like agents are placed, 'each with their own set of needs and drives'. After that, data are added from customer records, such as age ranges and patterns of consumption. A picture of consumer behaviour is built up as the agents move around the retail space. A change in physical space, such as the relocation of particular shops, will demonstrate changes in consumer behaviour.

In March 1998, CyberLife Technology Ltd announced to the press that it had signed a contract with the Ministry of Defence research organisation DERA (Defence Evaluation Research Agency) to construct an artificial pilot capable of flying a simulated military aircraft. Both European and US defence agencies are interested in the prospect of Unmanned Air Vehicles (UAVs), and CyberLife's project was one of the first to attempts to fly an aircraft with an autonomous agent rather than through ground-based pilots using remote control. Artificial pilots have the advantage of being in the aircraft and also of being artificial and therefore relatively invulnerable. Potentially, artificial pilots would supersede the capacities of human pilots and provide a more cost-effective service involving smaller, faster and less detectable aircraft.[21] As with other applications, CyberLife used data from the relevant source – in this case actual flight data – to generate the simulator. The simulated aircraft was akin to the Eurofighter except that it required no human control and could sustain flight, pursue enemy aircraft, evade attack and make 'reasoned decisions in order to complete its mission requirements' (http://www.creatures.co.uk). The artificial pilot is based on, and uses essentially the same technology as CyberLife's Creatures; namely a complex simulated neural network and biochemistry controlled by binary genes. The idea is that the pilot can learn from experience and reason in the face of novel situations (Davidson 1998: 39). DERA recognised the potential in the science behind the computer game and commissioned the research in order to provide something more realistic than the computer-generated opponents ('following rigid, rule-based systems') being used to test pilots in flight simulations. In keeping with ALife principles, CyberLife attempted to create this ideal opponent using biological methods including evolution and emergence. Populations of 40 pilots were subjected to 400 generations of evolution. The success of different flight strategies was tested and success was determined by how long each pilot remained in the air and how well it tracked or evaded the enemy. Only the best pilots from each generation were selected and allowed to reproduce (passing on their top-gun genes). Unlike norns, the pilots were not programmed with specific drives, and so behaviour was (even) less predictable. What emerged was some human-like and some distinctly non-human behaviour and techniques. Like human pilots, the artificial ones banked the aircraft in a turn and rolled over before attempting a steep dive: 'Humans roll over before diving to stop the blood rushing to their heads. The synthetic pilots don't suffer such physical constraints. They have developed this tactic because it helps to keep targets in their sights for longer' (Davidson 1998: 43). More unusually, according to Steve Grand, one rather successful breed flew the aircraft in a constant tight roll which, 'as far as we could tell' significantly increased the stability of the planes. This 'strange behaviour' evolved 'because the synthetic pilots didn't care in the slightest which way up they were, and had stomachs like cast iron' (September 1999). The lack of physical constraints suffered by artificial pilots (especially g-force) could lead to a radical redesign of real aircraft which would not have to be constrained by human dimensions and could be made to turn and accelerate much faster. Perhaps one of the reasons why this test-bed project has not, so far, been developed further is that such far-reaching implications did emerge and that in some ways it exceeded (and therefore failed to meet) its brief. Artificial pilots are not 'realistic' opponents. There is also the possibility that ALife could produce something much simpler and more 'organic' than a synthetic human in combination with another kind of machine:

> We think there is a lot of potential for far more advanced pilots, although it is hard to tell whether we should really be talking of pilots in aircraft or synthetic eagles, where the neural network is the brain and the aircraft is the body. So far we haven't had the opportunity to do the much more advanced and difficult research needed to find out.
> (Interview, September 1999)

For Grand, the most important applications of CyberLife technology are still to be developed. He lists space exploration alongside entertainment, military, medical and other commercial uses. However, it is clear that the technology is still undergoing a crucial stage of development, namely the incorporation of consciousness or imagination (which are considered to be linked, though not synonymous, attributes). Consciousness is a subject of debate and controversy in AI and related fields and it is acknowledged to be hard to define (and therefore simulate) (Velmans 2000). But for Grand it is associated with the ability to imagine or plan possible actions – without having to carry them out – and the ability to make novel connections between thoughts and ideas. These imaginative abilities are likely to have real benefits, 'allowing organisms (or machines) to be creative, to plan and to speculate' (Grand 1999a). As a consequence, CyberLife (Research Ltd) is concerned with new kinds of neural networks 'which have the facility to decouple themselves from simple sensory-motor interaction with their environment and develop an internal mental model of the world, which they can manipulate "at will"' (1999a). Whether or not this constitutes consciousness may be hard to determine: 'we're rather hoping the machines will tell us!'. What it does seem to constitute is the production of agents with increased agency. Saunders (2000) evaluates the imaginative abilities of norns and finds them limited (Grand declares that norns are alive but not conscious). However, their ability to learn from experience and generate expectations or hypotheses 'which can be tested and re-evaluated in new situations' is the key to being able to develop agents 'capable of accomplishing a limited range of creative tasks' (Saunders 2000: 6). In the context of design computing this development could produce sophisticated agents capable of tackling design tasks autonomously. Success could be rewarded in the normal way (genetic reproduction) leading to the prospect of able artificial assistants (already 'gaining a great deal of acceptance in the wider computing field') which are ultimately collaborating with each other in self-organising social systems. Specialised agents in a simulated environment collaborate to solve problems which none could do alone: 'This would mirror the real world approach to design and may make for interesting comparisons in the way work is distributed between individuals in self-organising social systems' (8). By this logic, imaginative or conscious autonomous agents pass through an object or instrumental stage to become microcosms of human-like cultures and societies in which human agents invest anthropological, psychological or sociological concerns. In other words, they become mirror worlds (Helmreich 1998a) offering novel opportunities for narcissism.

CyberLife Research Limited – or 'real' ALife

CyberLife's research on consciousness is currently being undertaken by CyberLife Research Ltd (CRL), a completely independent company chaired by Steve Grand. CRL is developing the blue-sky projects initiated by the original CTL and its relation with commerce and industry is more open-ended in as far as the focus is not on the development of specific products or product lines.[22] A biological engineering company with close links to the ALife research community, CRL demonstrates what might be referred to as 'real' ALife.[23] Its practical and philosophical commitment to ALife is premised on the belief that whereas AI programming failed to work (or pass the Turing Test), ALife engineering is working towards the creation of artificial intelligence in artificial life-forms. The fusion of biology and technology is central to the success of ALife engineering in as much as it produces machines which are more 'robust', 'adaptive', 'intelligent', 'flexible' and 'friendly' than, for example, HAL: 'Many of us grew up with Dan

Dare comics, Star Wars movies and Kubrick's 2001. We believed in a world full of androids, cyborgs and intelligent robots. We also believed in a world of constant warfare and cold-blooded mechanical logic. Only part of that story will be true' (http://www.cyberlife-research.com). Biologically inspired or 'sentient' technology is less abstract than the 'smart' systems developed using conventional AI techniques. It is more robust because, unlike smart systems, it is not necessary for every eventuality to be programmed in to the design and so it can more easily cope with the unexpected, learn from mistakes – and adapt to its environment. Sentient technology is more intelligent in the way that a mouse is more intelligent than a chess computer: 'A mouse will always lose at chess to a chess computer, but try throwing them both in a pond and see how they fare'. Less clever in a narrow sense or specific context, sentient technology produces a broader based more flexible intelligence than non-biological forms. It aims to be friendly by having a brain which is similar in one fundamental way to the human brain – conscious. In CRL's website publicity, consciousness is clearly associated with imaginative and creative powers, the 'special features' of human brains. So the chief characteristic of its artificial neural network 'is that it will have an imagination, and it will be able to use this imagination to visualise possible futures, make plans and act them out within a complex and messy external world'. The basis of the biological approach to artificial intelligence is a belief that complexity can be synthesised but not analysed, built but not broken down. The analytic method characterises traditional AI science more generally. It is based on the (reductionist) idea that in order to understand (and recreate) something, it is necessary to break it down into its constituent parts. But intelligence is, arguably, more than the sum of its parts and has eluded the AI project. If intelligence cannot be modelled directly (or programmed and controlled), then the building blocks (cells) from which intelligence emerges can be. The transition from top-down to bottom-up computing is characterised as a shift from 'command and control' to 'nudge and cajole', and as a rejection of a 'macho attitude to intelligence and software design' (Grand 1999a: 74). In the context of AI and ALife, emergence is characterised as a feminine process based on a nonlinear, non-deterministic model of connection or communication between multiple rather than individual units. Captured in a computer, this (irrational) process mirrors the creativity of the brain ('since it is the simultaneous interaction of many parts that creates behaviour at new levels of description'). The key to artificial intelligence then, is a holistic, co-operative, cell-based biological approach which (until recently) has eluded 'blinkered, domineering, chess-playing computer nerds' and saved them from the dubious honour of creating machines which think like them: 'We've carefully avoided creating HAL, and have therefore saved humanity from hearing those awful words . . .: "I'm sorry Dave, I'm afraid I can't do that". Robots will never take over the world now, and it's all thanks to us' (74).

Because CRL's approach to the development of artificial intelligence is organic, it depends on the creation of life-forms which are 'embodied' and 'situated'. Embodiment, in this context, refers to the creation of whole (optimally) three-dimensional entities which demonstrate the advantages of rendering emergent behaviour visible and learning by putting things together rather than taking them apart (Grand 1998c). Situatedness, in this context, is about being embedded in, and responding to a real environment (Grand 1999a), and it is based on the idea that intelligence cannot exist in a vacuum. Artificial organisms, or systems, 'need to be grounded in some kind of rich, noisy environment' (Grand 1998b: 3) and preferably one inhabited by other systems. This definition would appear to fall only slightly short of suggesting that intelligence (which is not actually defined by CRL) can develop only in a *social* context. CRL's stress on embodiment and situatedness is contrasted with a tendency, within the ALife community, to place too much emphasis on abstract and mathematical processes of simulated evolution such as CAs and GAs. Simulated evolution is regarded as an 'ecological specialisation' which is stifling the development of ALife and may not be the panacea which it appears to be. Modern computers can accelerate the process of simulated evolution, but perhaps not fast enough to attain even low levels of animal intelligence and, while scientifically informative 'is of limited use to an engineer' (Grand 1998a). While it is good, evolution is 'not *that* good' and it contributes to impoverished ideas about the meaning of life itself. Referring to the 'currently fashionable' definitions of life which revolve around self-replication, evolutionary potential and other 'fairly mechanistic criteria', Grand acknowledges that 'at a reductionist level' evolving systems can be considered to be alive or lifelike. He also points out that just because evolutionary systems fulfil some of the conditions for life, the 'tendency to assume that this is all of life' is a syllogistic fallacy:

> Artificial life, one of the most holistic, synthetic fields in science, is falling foul of *reductio ad absurdum*. In practical terms, this means that research into morphogenetic, chemical and neural mechanisms, communication, perception and intelligence is being eclipsed by the overwhelming shadow of evolution.
>
> (Grand 1998a: 20)

In an effort to dispel the overwhelming shadow of evolution, Grand reemphasises the importance of the term 'organism' in the study of life and seeks to shift the focus away from 'pure science' to technology. What this results in, ideally, is a meeting between biology and cybernetics. Cybernetics is associated with a less Newtonian, less materialist view of the universe 'in which everything is regarded as software, and the world is studied and classified, not in terms of "things"', but in terms of the relations between things. In this scenario, 'atoms, mind and society are seen as essentially the same kind of non-stuff' and 'everything is irreducibly connected to everything else' (Grand 1998c: 72). This is the kind of holism expressed in connectionist theory (Plant 1996; Kelly 1994; De Landa 1994), and it is made possible by the emphasis on form over matter (Langton 1996 [1989]). This may not sit as uncomfortably with the notion of embodiment as would at first appear, since embodiment refers here to cybernetic or cyber-biological systems rather than anything quite as material as physically and socially situated subjects.

ALife's stance on the concept of matter is informed by the idea that (natural or artificial) intelligent systems are too 'messy' and complex to fit with the 'reductionist, materialist and mechanist' approaches of classical physics which has informed much of contemporary science. ALife originally constituted something of an epistemological challenge to physics which, almost by definition, is concerned primarily with matter, and established an 'obsession with stuff' (Grand 1999b). Grand points out that this obsession with stuff is an understandable result of the way in which human senses respond to the environment; organising a mass of coloured dots picked up by the visual senses and converting them into discrete physical objects which are then classified and mapped back onto the external world. As well as organising the world, the sense are blind to non-material phenomena – such as minds. So there is a tendency to think of tangible things as real and intangible things as unreal, and to assign a value to the difference. These values are constructed in language so that 'material facts' are good and 'immaterial' means irrelevant. 'Tangible assets' are superior to intangible ones, and 'substantial' is a positive attribute where 'insubstantial' is not: 'Even the word "matter" carries emotive baggage when we discriminate between things that matter and things that don't!' (Grand 1999b). CRL's philosophy is based on rejecting the distinction between form and matter and the appearance that the world is divided into discrete objects. It carries forward Langton's (rather Saussurian) credo that things are not as significant as the relationship between them. In fact, Grand denies that matter exists at all. What exists, he argues, is a hierarchy of 'ever more sophisticated persistent phenomena' ranging from photons of light and subatomic particles through atoms, molecules and cells to whirlpools, bridges and bodies. All phenomena are metaphorically software rather than hardware since 'everything is made from the same (non)stuff'. Moreover, it is subject to a small number of basic mechanisms such as modulator mechanisms which allow one flow of cause and effect to modify another: 'In electronics modulators are called "transistors" while in biology they are referred to as "synapses" or "catalytic reactions"'. The identification of 'cybernetic elements' such as modulation provides engineers with 'a universal LEGO set' from which 'we can create all kinds of organisations, including intelligent living ones' (Grand 1999b). The idea that people are not material objects but persistent phenomena is clearly a bold and controversial one from a sociological and scientific perspective, and it constitutes one of the more problematic aspects of ALife for feminism. Grand brings the idea down to earth a little with an illustration about memory. In 'Three Observations that Changed my Life' Grand (1997b) invites the reader to recall an episode from childhood, explore it briefly and then question how it is possible for 'you' to have had this experience when 'you' were not there at the time: 'Probably not a single atom that's in your body now was there then. You still consider yourself to be the same person, yet you've been replaced many times over'. This certainly puts cell division and replacement in a new light. The upshot for Grand is that 'whatever you are, therefore, you are clearly not the stuff of which you are made' (1997b: 14).[24]

The deconstruction of form and matter in ALife (which seems to vacillate between a challenge to hierarchical epistemological dualism in a hegemonic scientific philosophy and an attempt to invert the binary in favour of form) has immediate implications for the reconceptualisation of contested categories such as 'mind' and 'consciousness' as these have traditionally fallen under the overarching category of non-matter (Velmans 2000). The exponents of ALife science necessarily stand in opposition to current trends which seek to redefine consciousness as simply a property of matter.[25] For Grand, this is a 'mechanistic' argument ('the mechanists are always trying to reduce mind to physics') which is counterposed by a 'dualistic' one. Dualists maintain the distinction between mind and matter 'but then go and spoil it all by trying to reify spirit and "promote it" into some magical substance' (Grand interview, August 2000). For dualists, mind and matter are different but mind is essentially 'some special kind of stuff'. For materialists, there is only matter and mind 'is some kind of illusion (simply the product of nerve impulses'). So in order to shock people out of what he refers to as 'those dogmas', Grand offers his own assertion that 'there is only mind' – or rather, 'there is only form' (Interview, August 2000).

In *Understanding Consciousness*, Max Velmans (2000) characterises reductive materialism as a belief that consciousness is only a state or function of the brain. In his opinion, this conflates the generally accepted idea that consciousness has neural causes and correlations with the idea that it is 'ontologically identical' to a brain state. Velmans (2000: 32) characterises emergentism as a belief that consciousness is a higher-order property of brains which cannot be reduced to neural activity. Where there is no clear definition of consciousness it is, he argues, distinguishable from mind, self-consciousness, wakefulness or thought. If it is not possible to say what it is, it is possible to say what it is like or rather 'pick out the phenomena to which the term refers' (6). An ostensible definition of consciousness refers to the awareness of experience or 'what it is like to be something' (Nagel in Velmans 2000: 5). By regarding the brain as an open, not closed system in continued interaction with the environment, Steven Rose (1999b: 15) argues – after Marx and Nietzsche – that 'consciousness is fundamentally a social phenomenon, not the property of an individual brain or mind'. It is especially not the property of an individual brain metaphorically linked to a computer. This brain as computer metaphor is, he suggests, flawed because brains deal with meaning and not with information: 'I have argued that brains and minds deal with meanings imposed by their "hard-wired" ontogenesis, and by the historical personal development of the individual and the society and culture in which that individual is embedded' (6). The brain as computer metaphor, rather like the life as information metaphor has generated a good deal of controversy and polarised arguments represented, for example, by Roger Penrose (1999) and Igor Aleksander (1999). For Penrose (1999: 155), the human quality and conscious phenomenon of 'understanding' is 'not something of a computational nature at all', and is something of a test case for AI, or the 'fundamentally lacking' computational model of mind (156). Conversely, Aleksander (1999: 181) is committed to the task of engineering a conscious machine and, citing Penrose, is keen to join the 'acrimonious' battle among 'the great and the good' for capturing the high ground or revealing the brain's deepest secret. Aleksander claims to be more willing than his colleagues to situate current theories of consciousness in established philosophies from Aristotle to Wittgenstein, and replays the ascendancy of ALife over AI by arguing that in order to approach consciousness, 'one needs to study the properties of a neural net which enable it to be a dynamic artificial organism whose learned states are a meaningful representation of the world and its own existence in this world' (Aleksander 1999: 185).

Grand's allegiance to AI and to strong ALife means that he believes that machines are capable of being conscious just as computer software is capable of being alive (1997b). Whereas norns were 'alive, but not conscious', Grand's next project – Lucy ('a robot baby orang-utan') – might ultimately be both. In order to understand consciousness (or 'that sense of "being" – the "I" that I find inside my head'), Grand aims to synthesise it: 'but if I can ever do such a thing it will be by deliberate accident' (Interview, August 2000). In other words, he hopes that it will emerge through simulated 'subconsciousness' and its underlying neural substrata. For Grand, one of the main phenomena to which the term consciousness refers is imagination, so 'I'm trying to create a robot that can make plans and rehearse them in her head – i.e. she will have an imagination' (Interview, August 2000). The cybernetic mechanism which provides the basis for his model of imagination is the servomotor (electronic motors used to move the wing flaps in radio-controlled aircraft). Servos have inputs which provide information about where the aircraft is and where it should be – one represents an actual state and the other represents a desired state. The circuitry inside the aircraft tries to move it in order to minimise the difference between desired and actual state: 'If you think of a billion such servos side by side, and think of the desired state as being a "mental" state, then you have an imagination machine' (Grand interview, August 2000). The basis of this imagination machine is the representation of the imagination 'as if' it were raw sensory data. Aleksander depicts a similar machine which 'given a state of the input stimulus (the perceptually available state of the world), can mentally imagine . . . the effects of available actions' (1999: 193). Such a machine may be able to represent emotions such as love as abstractions devoid of personal knowledge and experience (197) and is based on a view of consciousness as something rather simple 'not something that escapes computation, but something that is the ultimate masterpiece of iconically adapted firing patterns of parts of the brain, something which the advances of neural computation allow us to approach, study and imitate, something which is just too important to be smothered by an assumed complexity engendered by taboos rather than science' (199).

Lucy will have a simulated brain, but what is the status of this brain and could it really give rise to anything which might truly be called imagination or even consciousness? Steve Grand summarises his claim as follows:

> A raw computer simulation of a phenomenon is not an instance of that phenomenon, no matter how much it looks like it (an algorithm directly simulating the motion of a particle does not itself have mass and inertia). However, a metaphenomenon built from such simulated building blocks is fundamentally indistinguishable from the same metaphenomenon built from so-called 'real' building blocks – they occupy different universes, but are equivalent.
>
> (Grand 1997b: 17)

Allowing for the moment that the brain is a persistent stable phenomenon then it follows that a simulated brain is not a brain (or an instance of a brain), but the metaphenomenon of a simulated brain – or the way in which that brain behaves – may be the same (if the simulation is sufficiently complex) as the metaphenomenon of a 'real' brain – or the way in which the 'real' brain behaves. So if consciousness emerges from real brains it can emerge from computer simulated ones (so denying the distinction between natural and artifical [meta]phenomena). Does this mean that computers are, or can be, conscious or alive? Grand's answer is 'no', but the things built inside them can be: 'I conclude, therefore, that a computer cannot be alive or conscious, nor indeed can a computer program. On the other hand, things built inside computer programs can' (1997b: 17). Mary Midgley (1999) concludes *From Brains to Consciousness?* (S. Rose 1999a) by situating this sensitive concept – 'a term used to indicate the centre of the subjective aspect of life' – within the science wars and her own argument for ontological unity and epistemological diversity (Midgley 1999: 249). While there is a place for causal, physical explanations of consciousness, science also searches for the connection between cause and effect and 'if this can be found at all in the case of consciousness – which is still not clear – the search for it must certainly involve reference to a much wider context which takes both aspects seriously as a whole' (250). This more rounded approach runs counter to scientific specialisation and incorporates biosocial aspects of life itself.

Chapter 5

Network identities

The concepts and artefacts of artificial life are not confined to or contained within a specific technoscientific discipline or the artistic, commercial and institutional projects it stimulates. They traverse and exceed the boundaries of 'science' and 'culture' becoming elements within the process of globalisation which pertain to the ontology and epistemology of posthuman network identity. Significantly, within this context, posthuman identity is less instrumental than that of the previous race of cold-war cyborgs. Posthuman identity, informed by the discourses of artificial life, centres symbolically on the humanisation of HAL, raising bioethical questions about life-as-we-know-it and life-as-it-could-be.

Artificial agents

HAL

> In the 1980s, Minsky and Good had shown how neural networks could be generated automatically – self-replicated – in accordance with any arbitrary learning programme. Artificial brains could be *grown* [my emphasis] by a process strikingly analogous to the development of a human brain. In any given case, the precise details would never be known, and even if they were, they would be millions of times too complex for human understanding.
> (A.C. Clarke, *2001. A Space Odyssey*, 2000 [1968]: 98)

In this description of the HAL 9000 series, the 'highly advanced' computer or 'brain and nervous system' (Clarke 2000 [1968]: 97) of the spaceship *Discovery* would seem to bear a closer resemblance to what would now be termed artificial life rather than artificial intelligence. HAL ('Heuristically programmed Algorithmic computer') was modelled on the emergent potential of parallel rather than serial processing which is analogous to brain development. He was more 'grown' than built. As such, he may have been slightly misplaced at the fictional centre not only of popular fears about the development of intelligent machines which turn out to be disastrously disobedient towards their creators, but of professional pronouncements of the failure of Artificial Intelligence as a project. HAL may, after all, have been a premonition of the new era of essentially unpredictable, non-deterministic bottom-up rather than top-down machines which embody post-Newtonian concepts of chaos and complexity and are always already out of control. However, the HAL 9000 series was clearly at a pioneering, experimental stage and HAL had been given clear instructions to monitor and support the principal life-forms aboard the spaceship. He could talk in 'perfect, idiomatic English' (99) but as Andrew Leonard asks of more recent 'chatterbots', did he really have much to say? (Leonard 1997: 32). Chief astronaut Dave Bowman evidently pondered this question and recognised that the computer's intelligence, or at least his consciousness, was largely the result of projection and anthropomorphism: 'He knew, of course, that HAL was not really *there*, whatever that meant . . . But there was a kind of psychological compulsion always to look towards the main console lens when one addressed Hal' (Clarke 2000: 146). Anything else would, naturally, be rude. HAL was also infamously defensive in the face of a challenge to his information processing accuracy: 'I don't want to insist on it, Dave, but I am incapable of making an error' (147). Not unheard of in lifelike society, but clearly not a good sign. HAL, at this point, begins to sound ominously inflexible. Shortly after this statement he jettisons Dave's colleague into deep space: 'Too bad about Frank, isn't it?' (156). In 'The Year 2001 Bug: Whatever Happened to HAL?', Steve Grand (1999a) argues that, as a fictional example of artificial intelligence, HAL's act of murder and his own demise was not the result of unfriendliness but the fact that, although superficially or technically intelligent, he wasn't really too smart at all. HAL, according to Arthur C. Clarke (2000 [1968]: 99), 'could pass the Turing test with ease', but during a space odyssey in which his role was primarily to serve the astronauts, he was asked (by Mission Control) to keep the true purpose of the trip to Saturn from them. His major malfunction, then, was not a result of maliciousness but of a failure to deal with conflict – with a task which was not clearly right or wrong, black or white, master or slave, binary. The 'conflict between truth, and concealment of truth' undermined his integrity (albeit that the threat of disconnection effectively finished it off) (162). HAL did not understand the purpose of little white lies and could not (affectively) cope with his own. His failure, for Steve Grand, was the failure of the Turing Test as a measure of intelligence. The basis of the Turing Test is a concealed computer being able to pass as a human during a dialogue.[1] Turing's confident prediction that by the year 2000, computers would routinely pass his test has failed to be realised and, broadly speaking, ALifers such as Grand argue that this is the fault of top-down as opposed to bottom-up processing. Computers programmed to perform high level functions in a limited context – space flight or chess – have only the appearance of intelligence. Put another way, the intelligence is in the programming, not in the computer. Moreover, intelligence in this context is misconceived as a linear command and control (information) process. Turing's 'organised machine' (Grand 2000: 16) – the forerunner of the digital computer – models a mind which 'gives us the impression that it is top-down (employs a chain of command), serial (only one mind per brain, operating one step at a time) and procedural (works in terms of logical procedures to be followed, as in a recipe)' (17). The bottom-up alternative or 'self-organising machine' is based on a more biological conception of the brain as a complex parallel processor lacking a central control mechanism. Intelligence is then an emergent phenomena of the lower level functioning of brain neurones, the precise details of which 'would never be known' (Clarke 2000: 98). Grand's book, *Creation. Life and How to Make It*, details the seemingly impossible, contradictory task of using organised machines to create self-organising ones – to generate chaos out of order – and so achieve the goal that Turing sought.

If HAL was ultimately too robotic to achieve self-organising status – Dave eventually pulls out his plugs, causing a mechanical breakdown not unfamiliar in the station announcements at East Croydon during the UK's late 2000 floods: 'The train now at platform three . . . has been terminated at Victoria . . . a buffet service of drinks and light refreshments will be serv . . . please stand back . . . please stand back . . .'[2] – the ETs in *2001* would seem to qualify. These represent the robust, adaptive, flexible and friendly symbiosis of mind and machine. They are disembodied hyper-intelligent entities who turn out to have had a benevolent hand in human technological evolution and who complete this process by assimilating and immortalising an astronaut in/as cosmic space. The transcendent star-child Dave happily inhabits 'formless chaos', the still 'unused stuff of creation, the raw material of evolutions yet to be' (250). And if a further analogy between the evolution of cosmic and computer space were needed, 'Here, Time had not begun; not until the suns that now burned were long since dead, would light and life reshape this void' (250). Worryingly, when Dave returns to Earth, he brings apocalypse.

Situated and autonomous robots

The goal of nouvelle AI – which is sometimes otherwise known as ALife – in 2001 is to turn the principles of classical AI on their heads and to generate friendlier and more humane machines than HAL, either as software or as hardware. The key to engineering this artificial humanity is seen to lie more in biological than in the more traditional psychological methodologies. Kevin Kelly (1994) traces the history of the desire for artificial humanity and its evolution in the early 1990s: 'What we want is Robbie the Robot, the archetypal being of science fiction stories: a real free-ranging, self-navigating, auto-powered robot who can surprise' (Kelly 1994: 34). Ideally, Robbie the Robot should be a 'mobot' (mobile robot) as opposed to a 'staybot' and so dispense with the need – and indeed the risk – of electrical plugs: 'Any robot is better if it follows these two rules: move on your own; survive on your own' (34). The ongoing quest, it would seem, is for the lifelike, walking, talking, independent (in)visible friend with attitude, but not aggression promised in comic books, television serials (*Lost in Space* springs to mind) and *Star Trek/Wars* sagas over the years. The longing for son – or daughter – of R2D2 would appear to be widespread if not universal, and this is indicated in the media and industrial interest invested in MIT's research in robotics, and in particular, in the work of Rodney A. Brooks. Brooks, like Langton, is credited with a seminal paper of the late 1980s in which he paved the way for a generation of downsized, dumb mobile robots capable of convincing NASA of their collective efficacy as limited autonomous agents who could get jobs done. In 'Fast, Cheap and Out of Control: A Robot Invasion of the Solar System' (Brooks and Flynn 1989), Brooks' cosmic pragmatism evolved, according to Kelly, into NASA's self-guiding, solar-powered microrover and a generation of off-the-shelf miniatures whose rationale 'is upside-down to the slow, thorough, in-control approach most industrial designers bring to complex machinery' (Kelly 1994: 37). Brooks realised 'bottom-up smartness' (Kelly 1994: 38) through Genghis, a literally brainless model cockroach whose ability to walk emerged out of the collective behaviour of the twelve motors and twenty-one sensors distributed across its network (38). Subsequent mobots built on the Genghis model by means of 'a universal biological principle that Brooks helped illuminate – a law of god: *When something works, don't mess with it; build on top of it*' (39). In the evolution of natural systems, improvements are overlaid on behaviours which have already been successfully test driven. These remain unaffected by and unaware of their superiors. Brooks' mobots learn to move through an environment by building a hierarchy of behaviours ranging from avoiding obstacles, to creating an internal map to making, and then modifying travel plans (40). Each layer in the hierarchy takes care of its own task and higher level behaviours subsume lower levels in order to gain control. Brooks' 'subsumption architecture' is, for Kelly, almost universally viable as a description and prescription for the organisation of biologies and societies. It becomes part of his naturalised political economy of information systems. For Kelly, the federal structure of the US government is a subsumption architecture of (relatively autonomous) towns, counties and states which amounts to significantly more than the sum of its parts. Centralised command and control may be 'the most obvious way to do something complex, such as govern 100 million people or walk on two skinny legs' (42), but as the economy of the former Soviet Union demonstrated, it is inherently unstable and impractical:

> Central-command bodies don't work any better than central-command economies. Yet a centralized command blueprint has been the main approach to making robots, artificial creatures, and artificial intelligences. It is no surprise to Brooks that braincentric folks haven't even been able to raise a creature complex enough to collapse.
>
> (Kelly 1994: 42)

Apart from subsumption architecture, the other key to Kelly's political economy of information systems is (evolutionary) competition. Whether in minds, bodies or economies, 'the dumb agents in a complex organization are always both competing and cooperating for resources and recognition' (44). Successful agents persist, unsuccessful agents do not. Information systems are therefore 'ruthless cutthroat' ecologies (44) analogous to market economies. Brooks' vision of an ecology of artificial agents (in which 'many hands make light work, small work done ceaselessly is big work' and 'individual units are dispensable' (49)) is both situated and embodied. In other words, it promises the co-existence and co-evolution of natural and artificial beings. His mobots qualify as beings in as far as they are not programmed tools but autonomous 'things' that exist in the world and interact with it, pursuing multiple goals.

Before Brooks' insect-like mobots appeared in the late 1980s, successful navigation came in the form of an internal map complete with the location of all obstacles, and software which could continually relocate and recalculate position and direction. Brooks' beasts have no such map or program, but a collection of miniprograms representing simple behaviours such as 'move front right leg' or 'turn left'. As Freedman (1994) points out, 'all of these behaviours competed for control of the robot at any point in time' and the winner was determined not by a central control system but by interaction with the environment:

> In effect, the insect robots were not thinking but reflexively reacting to their surroundings. This happens to be how real-life animals do it. You don't have to continually ask yourself where you are on a map in a room in order to calculate your next step. You just look and move.
>
> (Freedman 1994: 2)

These robots were quicker, faster and more efficient (requiring much less power than 'map-reading, step-calculating computers') than their predecessors which is why, after initial scepticism from the AI community, they caught on (2). In 'Bringing Up RoboBaby', Freedman (1994) elucidates Brooks's motivation and what might appear as self-contradictory interest in scaling up directly from insect to humanoid robots. Partly in order to stay at the forefront of AI and complete a big project before he became sidelined by younger researchers, and partly in acknowledgement that building humanoid robots is the AI researcher's ultimate quest, Brooks went for gold with Cog. Cog (derived from 'cognitive') is a widely publicised and documented humanoid torso situated at MIT's Artificial Intelligence Laboratory[3]. It represents a purist investment in growing intelligence from the bottom up, and is developmentally anthropomorphised as a baby. Cog has arms, hands, a moveable head and three basic senses – vision (provided by cameras), sound and touch. It employs a subsumption architecture or hierarchy of behaviours which are developed without the aid of a ready-made program: 'Instead of playing the piano, "conversing" with visitors, or welding car parts, as other robots have been programmed to do, Cog will groan and coo, stare at colourful objects, and flail its arms in an effort to grab things' (Freedman 1994: 3). But since the project began in 1993, it has, arguably, become less purist in so far as it entails directed evolution or development. The aim is not simply to evolve greater complexity but human-like complexity and this has necessitated the input of developmental psychology and elements of behavioural programming. What seems to emerge from this is the insight that intelligence exists in – even stems from – the eye of the beholder. As Hayles states in relation to Karl Sims's simulated creatures: 'Invariably viewers attribute to these simulated creatures motives, intentions, goals, and strategies' (Hayles 1999b: 1). However, the meaning of alife software and hardware stems from the interaction between the various human and non-human actors involved, and is more than the sum of projection or anthropomorphism. In order for Cog to develop anything close to human intelligence it must be socially as well as physically embodied in an environment which centres on human interaction. In order to stimulate human interaction, Cog's emergent babyness has been booted up (Graham-Rowe 1999) through the development of Kismet, 'a disembodied Cog-type head' complete with large expressive eyes, ears and jaw plus 'built-in drives for social activity, stimulation and fatigue'. Kismet can express feelings and 'like a baby, it can manipulate a softhearted human into providing it with a companionable level of interaction' (Beardsley 1999: 2). What is learnt through Kismet will be transferred to Cog (and this may not be that different from watching children play with animated child-like toys, or parents interacting with young children). Development – within the remit of developmental psychology – is a factor of behavioural cues and responses which, to a certain extent, are in-built or pre-programmed.[4] Just as the machines in ALife are humanised, the humans are, to an extent, roboticised. One of the unexpected findings of the Cog-Kismet project is, according to Graham-Rowe, the fact that giving them the appearance of infant actions and reactions is more than just a superficial trick: 'The robots, like children, will not develop unless their carers read more into their behaviour than is actually there' (Graham-Rowe 1999). Carers project meaning onto what are in fact only a few innate expressions, interpreting them as indicators of an infant's emotional state. Signs of contentment lead to a maintenance of existing levels of interaction, whereas signs of discontent lead to alterations or intensifications of interaction. An infant will learn from the consistency of the carer's responses how to manipulate them and so optimise the rate at which information is absorbed. As Cynthia Breazeal, Kismet's principal researcher, puts it: 'Kismet takes advantage of the way we are programmed to interact with small children' (in Graham-Rowe 1999). Within this behaviourist model of robot development, the emergence of intelligence depends on a quite mechanistic process of communication between believable (rather than autonomous) agents and users defined as carers.

While Cog qualifies as (expensive) blue-sky research,[5] there is only a short step between Kismet and the bot-end of the toy industry into which Brooks's company, the *iRobot Corporation*, has entered with an animatronic doll named My Real Baby. Initially set up to exploit the commercial potential of Genghis-like mobots, the company builds machines for the research community 'providing budding roboticists with hardbodies to reprogram' (Davis 2000: 3). When Brooks found that his mobots appealed to the young, as well as the young at heart, he found, as Davis points out, his entrance to consumer robotics. My Real Baby, produced in conjunction with toy manufacturers *Hasbro*, inherits a tradition of automata stemming from the eighteenth century and Jacques de Vaucanson's clockwork duck, via the doll industry of the nineteenth century to the more recent exponential growth of high tech toys stimulated by Furby and a pack of robot dogs. In this 'backstabbing, nail-biting, copycat arena' economic if not evolutionary competition is fierce and Brooks's doll is up against Caleb Chung (father of Furby) and his Miracle Moves Baby, produced by *Matel*. Further competition is provided by *MGA*'s My Dream Baby which, though less animatronic is able to grow and eventually walk. At a cost of approximately $100, these dolls at least signal a technological transformation in the toy industry as it absorbs biological and evolutionary paradigms to generate products which appear to exist between the non-categories of life and lifelike. Brooks's baby is based on a subsumption architecture and pragmatically adapted to the target market. A combination of sensors, processing and internal behavioural models, it is 'designed to trigger the nurturing play of little girls' (4) by exploiting what Chung refers to as 'genetically hardwired bonding responses' (Davis 2000: 6). An interesting question which arises from this industry and from related research is just where a product or artefact needs to be on the continuous line between life and lifelike in order to generate at least some of the same responses as a real human or other animal. Do robot toys just 'hardwire the animated qualities that kids already lend their toys through their imaginations', and who is actually being 'hardwired' – the toy or the user? (6) Is it enough to simply give the cues to autonomy because the gaps will be filled in by human genetics or imagination? In 'The Uncanny' (1919), Freud's discussion of the primitive animist instincts stimulated by automata would seem to confirm this, so why proceed and attempt to obscure the boundary between simulated and actual autonomy? Perhaps because some toy researchers and industrialists are invested in the more metaphysical aspects of strong ALife. For Chung, the industry's products go 'way beyond toys' and 'are the next iteration of our attempt to re-create life' (Davis 2000: 6). It is also interesting that the quest for humanised machines is figured, in these examples, through the interaction of programmed humans functioning not altogether unlike transitional objects along the path to posthuman selfhood.[6] This suggests a more dynamic interaction between humans and machines than is indicated in deterministic statements about 'hardwired' responses. It suggests that posthuman identity is co-evolved in the feedback loops of a distributed cognitive system within which humans and machines are not entirely discrete entities. What binds us together, according to Hayles, is not (only) biotechnological complexity but the ability to generate a narrative of it: 'Spliced into a distributed cognitive system, we create these narratives not by ourselves alone but as part of a dynamic evolutionary process in which we are coadapting to other actors in the system' (Hayles 1999b: 8).

Erik Davis (2000: 7) points to some dissent within the robot community over the desirability of humanised as opposed to utility machines, mobot friends as opposed to mechanical servants, but while this question is of technological and ethical interest, it is clear that both are in the process of being developed and commercialised. Where *Wired* outlined the characteristics of a new generation of 'Robosapiens', declaring that 'at least for now we and the robots are in this together' (Boutin 2000: 1), *New Scientist* headlined with 'Jobs for the Bots', declaring that 'they're mobile, they're autonomous . . . After years of hype and crushing failure, robots are ready to start serving us in our homes' (Graham-Rowe *et al.* 2001: 27). These robots, such as pool cleaners and lawn mowers, are single-purpose machines which operate in relatively 'structured' – or predictable – environments. They have become commercially viable due to the increased capacity and decreased cost of microprocessors and sensors, and they are not particularly intelligent: 'None use cutting-edge AI' (34). However, smarter, general purpose domestic robots are in development in the form of *Gecko Systems*' Carebot and *Probotics*' Cye. Both are 'PC extensions', meaning that instead of having artificial brains, they utilise the processing power of computers via a wireless link. Both have vacuum attachments but like their more limited outdoor colleagues are quite likely to mow down children, pets and other large obstacles which test their limited navigational abilities (31). The imperfections of domestic robots are, however, more acceptable than the imperfections of military and industrial robots and, according to Brooks, the fact that they may still be categorised as toys is not necessarily a problem since the expectations placed on toys is lower and their main task is to entertain (Graham-Rowe *et al.* 2001: 35).

Life is not so easy for the autonomous roboticists at the UK's Centre for Computational Neuroscience and Robotics (CCNR), based at the University of Sussex. With only a fraction of the funding available in the US, Phil Husbands describes (in an informal interview) the research as more theoretical than engineering based, and centred on the finer points of designing artificial brains for useful (cheap, fast, out of control) mobots. CCNR combines the work of the Evolutionary and Adaptive Systems Group with that of the Sussex Centre for Neuroscience in order to foster the synergy between real and artificial neural systems. Computational models are able to provide insights into the process of neurotransmission in the brain, and in particular into the spatial and temporal dynamics of the gaseous transmitter nitrous oxide. These models have inspired a new area of research in robotics – GasNets – in which elements of the transmission of information by diffusion are incorporated into artificial neural networks. Members of CCNR also work on evolutionary robotics and evolutionary electronics, and there is a widespread use of Genetic Algorithms (GAs) as a methodology for designing complex systems of various kinds (Harvey *et al.* 1996). Algorithms are basic computational operations (strings of code), and genetic algorithms employ Darwinian principles of evolution in order to increase the fitness of successive generations of algorithms, where fitness is a measure of success in solving specific computational problems. Variation is provided in a population of trial solutions through generating genotypes of artificial DNA. The process of artificial selection favours the fittest algorithms which survive and reproduce at the expense of the less successful (Emmeche 1994: 114). In robotics, evolutionary computing techniques such as GAs are used to develop artificial nervous systems and in electronics, evolutionary techniques are used to optimise electronics design. Apart from project specific research grants, funding is provided by links to companies such as BT (British Telecom) and MASA (Mathématiques Appliquées S. A.) – a French group with a campus site which offers an employment route for members of CCNR and which specialises in generating and hybridising evolutionary algorithms for commercial optimisation. Optimisation represents a solid commercial application for research undertaken at CCNR and has been applied to routing and scheduling systems. MASA also has links to the games, defence and space industries and has developed a wargame using autonomous agents – a version of which will be made available for military training in France.[7] Nick Jakobi at MASA UK was responsible for the Evolved Octopod exhibited as part of the European Conference of Artificial Life in 1997 and featured posing on Brighton beach on the front page of the *Guardian* (28 July 1997). Billed in the paper as 'Frankenstein's successor: a purple spider with a mind of its own that likes to go walkabout', the Evolved Octopod is autonomous in as far as its ability to walk is self-taught and self-sustained; evolved not programmed. Yet Jakobi is wary of the semantic baggage carried by the concept of autonomy, especially when it is applied to software 'agents', which, as algorithms, are mathematically able but semantically challenged, informational but not particularly meaningful. In other words, such agents are unable to understand the meaning of anything much, least of all what they are.

Bots

Bots are the precursors of autonomous agents and 'the first indigenous species of cyberspace' (Leonard 1997: 8). They are to software what robots are to hardware; algorithms rather than animated machines governed by rules of behaviour. Written by teenagers and computer scientists alike, these strings of code 'are variously designed to carry on conversations, act as human surrogates, or achieve specific tasks – in particular to seek out and retrieve information' (7). Bots are the friendly robots of cyberspace, designed to assist computer users by limiting the information overload of the Information Age. Mailbots, for example, filter email for junk and 'spam' advertising, but the first 'natural' habitat of bots – which flourished in the 1990s – is the World Wide Web. Leonard celebrates the range and diversity of this new species (in evolutionary terms the apes to human-like agents) which, as part of the unnatural order, highlights the process of classification as one of (arbitrary) categorisation rather than nomenclature, and also highlights the contested role of anthropomorphism in the successful mediation between digital and biological entities (22). Bots, for Leonard, may not be alive but they are 'cool', content to exchange agent for alien status and thereby inhabit the grey area between fiction and reality in computer space and in the minds of computer users: 'They stoke our imaginations with the promise of a universe populated by things other than ourselves, beings that can surprise us, beings that are both our servants and, possibly, our enemies' (10). Bots are impure (semi-programmed) life-forms which achieve autonomy only when released into their unnatural environment where they operate 'out of direct control' (21). And autonomy is 'the crucial variable, the dividing line between dull computer clay and rich digital life' (21). If true autonomy remains the goal of agent research, the chatterbot experience offers some lessons in mistaking autonomy for the appearance of autonomy. In 'Artificial Life Meets Entertainment', Pattie Maes (1996) examines Michael Mauldin's chatterbot 'Julia' which resides in a text-based multi-user simulation environment (MUSE). Julia's behavioural repertoire incorporates discourse with players, and she has attitude. She is able to converse in a lifelike manner where lifelike is defined as 'non-mechanistic, nonpredictable, and spontaneous' (Maes 1996: 213) and she tells a sexually predatory character to 'take a long walk through an unlinked exit, Space-Ace' (214). Julia's conversation module is based on a range of potential responses and as a chatterbot she is, according to Maes, more advanced than Eliza, the original chatterbot of 1966, in as much as she has more tricks and a more sophisticated memory (213). Leonard is not as certain of chatterbot progress. He describes the initially inflated responses to Eliza – 'the first computer program that could carry on a conversation with a human being' (Leonard 1997: 33) – which created the illusion (now quite unconvincing) of intelligence by discoursing within a limited context, or predefined script. Initially, a number of different scripts were available, but the one which achieved notoriety was the doctor script in which Eliza operated as a Rogerian therapist. Carl Rogers was a prominent psychologist in the early to mid-1900s, and in Rogerian therapy the therapist is confined to questions and 'content-neutral' statements which encourage the patients to clarify their own thinking: 'The Rogerian model solved a number of problems. Eliza did not have to generate her own content, answer questions, provide information, or do anything other than rephrase incoming statements. By controlling the context, Eliza could pretend to be in a position of unchallengable authority' (34). Julia inherited the context trick which enabled Eliza to become part of computer science folklore, but a clever trick is still a trick and Leonard states that 'No bot has passed the Turing Test' (43). Symptomatic of the failure of top-down expert systems, chatterbots (or natural language processing systems) illustrate a major fault line in classical AI – limiting the context is an intelligence trick and being able to talk does not necessarily make you clever. Chatterbots are the HALs of real AI; in the process of being replaced by more biologically inspired models and the 'profound understanding that we can't dissect intelligence piece by piece' or 'make absolute sense of the world' (46). While emergence replaces the context trick with something at once more modest and more magical in autonomous agent research, bots are arguably restricted to more instrumental roles. Or perhaps they simply share their intelligent agent status with their newfound environment – the Internet.

The distinction between bots and agents is not absolute, especially as information and communication industries begin to realise the consumer marketability of useful, user-friendly agents. Where academic researchers might need to be clearer about the classical or nouvelle AI qualifications of their products, corporate researchers (as the MASA example illustrates) have no qualms about creating hybrids if hybrids can get the job done. Leonard points out that the growth of the Internet has led to a boom in claims (from academic and corporate researchers) about intelligent agents, and argues that there has been a commensurate growth in the agent technology industry (1997: 54). At BT, agent and AI research in general is co-ordinated through *BT exaCT*, established in July 2000 as the company's advanced communication technologies organisation. Like MASA, BT adapts intelligent systems research to scheduling work, including the award winning 'dynamic workforce scheduler' which was developed for, and is used by BT's 30,000 strong 'front-line' workforce. The scheduler automates the deployment of BT's engineers. The same core technology is also used in the company's billing systems and booking systems (for broadcast services). The Agent Research Programme is investigating and developing different aspects of agent technology including: Collaborative Agents, Personal Agents, Information Agents and Mobile Agents (http://www.labs.bt.com/projects/agents.htm January 2001). ZEUS is BT's agent toolkit or architecture, designed to collaborate with and relate to other agents, and modelled as a social entity or actor. BT's 'Intelligent Personal Assistant' or 'electronic butler' is described as an adaptive system which 'manages information, communication and time for its user by learning their preferences and interests and using these to carry out proactive information seeking, filtering and activity planning'. The fundamental component of these personalised interface agents is a 'user model' or personal profile which enables them to learn the user's 'true requirements and needs' and to track changes in these requirements and needs over time. Instead of involving the user in the 'tedious' process of informing the agent, BT argues that the agent must be free to acquire the information it needs by accessing the various systems that the user interacts with. This places a great deal of emphasis on trust: 'These agents have access to much personal information about the user and in some cases they are performing actions on behalf of the user. This issue is perhaps the most crucial obstacle facing personal agents if they are to become widely accepted in the real world'. Personal agents are the most clearly anthropomorphised, not just as social actors but as individuals capable and desirous of earning trust. Again, there is a sense of the transfer of agency from the delimited if not dehumanised 'user' to the 'agent' which (or who) must be trusted to function correctly (for example, by storing information in the right place), to refrain from 'improper suggestions' (such as scheduling meetings that the user has no desire to attend), to present information in an appropriate and timely fashion and to refrain from disclosing personal or sensitive information to a third party. Agents must then incorporate 'privacy levels' (a level or so short of understanding the concept of privacy) and expertise ratings specified by the user to prevent unnecessary information overload. By declaring him or herself an expert on a given subject or interest, the user is provided with less information than that required by a non-expert. According to BT, a better way to do this, however, would be to combine the notion of expertise with the knowledge gained through access to the user's habitual patterns. The transfer of control is combined with the transfer of agency.

BT's Information Agents project involves extending the capacities of ZEUS to incorporate retrieving and processing information and its Mobile Agents are biologically inspired – like Genghis – by insect-like animals rather than human-like ones. Mobile agents are here described as 'a flexible software tool suited to the demands of a network environment', and the goal of this project is 'to achieve self-organising adaptive agent systems, which maximise network and machine resources, with minimum intervention'. These are perhaps the closest to artificial life-forms understood not as individuals but as component parts of a larger distributed agency: 'Clear analogies can be drawn with living cells within organisms or the members of social insect colonies' whose modus operandi is co-operative and whose complexity is emergent. BT's brush with the blue sky is balanced with – and contingent on – the delivery of business solutions which give the company a greater fitness in the marketplace. As Nader Azarmi, BT's AI technology manager succinctly puts it: 'I think we have a goldmine here and that is my next target – to turn this programme into a major revenue generator and cost saver for the company's operations'.[8]

Although for Leonard (1997: 12), bots are the real product of the Net and its parallel distribution, it is clear that both bots and agents are being designed to adapt to the Network ecology. As well as increasing the believability of bots, the Internet enhances agent status as it 'lowers the level of suspension of disbelief necessary to engage in a conversation with what might be a machine' (1997: 52). But Turing Test apart, the Net is regarded, in the context of ALife *culture*, as a suitable environment in which agents can grow, evolve and perhaps ulimately migrate and settle in diasporic societies. Artificial, intelligent or autonomous agents are an increasing part of the academic and corporate research in AI/ALife and migrate into the wider culture through media publicity and industry products. Similar agents are produced by artists with various degrees of allegiance to the nebulous, non-homogenous interacting spheres of Artificial Intelligence and Artificial Life.

ALife aesthetics

Nick Jakobi's Evolved Octopod was exhibited as part of Brighton Media Centre's *LikeLIFE* exhibition which was associated with the 1997 European Conference on Artificial Life (ECAL). *LikeLIFE* was 'a collection of installations, robots, creatures and artworks inspired by living things' (ECAL '97: 1). More specifically, *LikeLIFE* was inspired by a concept of Artificial Life which centres on elucidating 'the processes underlying living things – evolution, development, self-organisation – and reproducing them artificially' (1). Among others, the exhibition featured the work of Karl Sims, Norm White, William Latham and Nell Tenhaaf. It also showed the first *Creatures* computer game designed by Steve Grand, and Stelarc's cyborgian video documentary excerpts (*Psycho/Cyber*, 1997) of his electronic semi-autonomous third arm. Karl Sims's computer animation *Co-Evolved Virtual Creatures* (1992) illustrated the evolution of form and behaviour in a Darwinian scenario featuring simple three-dimensional shapes, while Norm White's *Helpless Robot* (1987–96) used a synthesised voice in order to persuade users to move it to a more desired position. The goal of this object with limited visual but strong behavioural appeal, was to learn about and hence predict human behaviour (in a limited context). Linking the work of Sims and Latham is the concept of artificial evolution:

> Instead of deciding what the finished form will look like, the artist defines rules for how a form will grow. The unplanned and unexpected result emerges from the interaction between many different independent elements. The artist then adjusts the developmental rules, allowing some forms to survive and mutate, and driving others to extinction: survival of the most beautiful rather than survival of the fittest. The final image is thus due to a dialogue between artist and computer.
>
> (ECAL '97: 2)

This dialogue between artist and computer has a history as long as accessible, reasonably user-friendly machines, and for Paul Brown (artist in residence at Sussex University's CCNR and School of Cognitive and Computer Science during 2000) dates back memorably to the 'Cybernetic Serendipidy' exhibition at the ICA in 1968 (P. Brown 2000). *LikeLIFE* staged William Latham's *Mutation X*, *Evolution of Form* (1989) and *Biogenesis* (1993) and featured images generated by his Mutator computer programme – a development of Dawkins's *Biomorph*. *Biogenesis* is a video which depicts the evolution or emergence of complex forms from a simple form designed by the artist. These more complex forms are unexpected and emerge spontaneously from the initial blueprint. They are visually quite spectacular and are like the images generated through the use of fractals and chaos theory. Latham states that fractals are used, for example, to add 'slight irregularities that give forms a more natural appearance' (Todd and Latham 1992: 208), but adds that they are used in combination with the rules of his computer program and not as an art form in their own right. Sherry Turkle (1997) states that artificial life was 'encouraged by chaos theory, which appeared to demonstrate that mathematical structure existed underneath apparent randomness and that apparent randomness could generate mathematical structure' (1997: 152). Latham states quite clearly that his artworks are a form of artificial life, and claims that they comment creatively and critically on the role of both science and art. They are a parody of genetic engineering (Todd and Latham 1992: 208) and have revised the role of the artist. Latham claims to have developed 'evolutionism' as a new style which divides the artistic process into two stages or parts: 'in the first part, the artist creates an artificial world by defining systems and structures for form and animation generation; in the second part, he works as a gardener within this world, using aesthetic judgements to breed artworks' (207). The role of creator and gardener may be performed by different people, and artistic autonomy is inevitably shared with the computer.

Latham's work is incorporated in Peter Bentley's (1999) comprehensive survey of evolutionary design, a development of computer aided design (CAD) and analysis in which evolutionary programmes – notably GAs – facilitate the emergence of novel designs across a range of contexts. Evolutionary design 'allows a designer to explore numerous creative solutions to problems (overcoming 'design fixation' or limitations of conventional wisdom)' (2) and has been applied by Bentley in areas ranging from floor-planning to music ('Nature', BBC Radio 4, 14 May 2001). Referring to Dawkins's *Biomorph* evolutionary design and selection process, Nelson (1999) describes the inspiration for applying genetic algorithms to the growth of musical organisms: 'the process immediately reminded me of the means that composers use to search the hyperspace of musical constructs to find just the right choice for a particular moment in a piece'. Bentley (2001: 1) displays a non-ironic level of enchantment with computers which, in his mind, act as 'greenhouses for a new kind of nature' or 'digital biology'. Significantly, this digital biology encompasses cultural processes and forms such as design and music as well as entities 'just the same as you' (2). Digital biology may be more, but is rarely less than digital naturalism (Macgregor Wise 1998).

While Bentley approaches artificial life as a computer scientist, Nell Tenhaaf approaches issues of autonomy, emergence and self-organisation from a biological perspective. In her two artworks of 1995, *Apparatus for Self-Organization* and *Orphaned Life-Form*, she reflects on a central debate within biology: 'should we understand living form – including our bodies – as being dictated by a molecular genetic blueprint, or as the self-organised result of the dynamics of life itself?' (ECAL '97: 5). Here Tenhaaf addresses the internal conflict between autopoietic and genetically deterministic views of the biological organism. In her ECAL '97 conference paper she declares herself to be a 'critic of the genetic imperative and convinced of some conspiracy acting against the interests of the rest of the living organism' (Tenhaaf 1997: 1). Nevertheless, she continues to represent 'the tension between genetic code and organic matter' which is, for her, a gendered tension between 'the key to the mystery of life on the one hand, the (masculine) determinant of human fate, and on the other hand, the (feminine) burden of flesh to be mastered and shaped' (1). In her exhibited artworks she seeks to undermine what she calls 'bio-power politics' by using electronic devices which suggest both coding and the properties of matter itself, and by encouraging the viewer 'to *intuitively* understand how the self is shaped in relation to one's biological substrate' (3). Influenced by complexity and cultural theorists alike, Tenhaaf attempts to find a convergence between a 'biological aesthetic' and an art language which has its own morphogenesis. Regarding ALife modelling as a potential point of convergence, she draws on Humberto Maturana and Francisco Varela's concept of autopoiesis as the basis for the development of a bio-aesthetic. Cautious of the epistemological consequences of naturalising scientific paradigms, Tenhaaf reframes the autopoietic organism within a post-structuralist epistemology and ontology where it stands as a metaphor for embodiment which is precisely bio-logical; real and symbolic; matter and code.

Tenhaaf's faith in the convergent potential of ALife modelling is further explored in the Life 2.0 International Competition where she chaired the jury of pooled expertise in cultural theory, philosophy, electronic media art practice and ALife research. The jury assessed thirty projects from eleven countries and awarded joint first prize to *La Cour des Miracles* by Bill Vorn and Louis-Phillipe Demers of Montreal, Canada and *Tickle* by Erwin Driessens and Maria Verstappen from Holland. *Tickle* is a small autonomous robot with rubber-studded wheels which navigates the surface of the body, tickling as it goes. Its smartness lies in its ability to avoid steep slopes and in its 'tongue-in-cheek acknowledgement of the kinds of accoutrements that are the promise of an encroaching but ever-elusive cybernetic future' (http://www.telefonica.es/fat/ajury.html 16 Janury 2001). *La Cour des Miracles* is a dystopia of dysfunctional robots analogous to a medieval 'cripples and beggars' court', and third prize winning *Bomb* (Scott Draves, San Francisco) 'is an instance par excellence of the capacity of alife algorithms to computationally generate imagery in such a direct way that the user can experientially grasp some alife principles without even knowing it'. What ALife aesthetics emphasises then, and what MIT's Cog-Kismet project is realising in a different way, is the *affect* of lifelike behaviour and the experiential existence of intelligent or autonomous agency in the eye/I of the observer. This aesthetic may be as old as the history of automata but it is currently expressed as a factor of biological machines which by definition share agent status with human-animals, but which in the discourses and practices of autonomous agent research, derive agent status from human-animals. As well as constantly regenerating pixel patterns based on sound or other input, *Bomb* is also a 'visual parasite', 'growing versions and offshoots, downloadable onto virtually any platform and with its source code available to other programmers'. These and other Life 2.0 projects incorporate ALife research in hardware and software, but differ from each other in how explicitly ALife methods are used. Of the ten finalists, approximately half employed ALife techniques, and the remainder adopted more metaphorical approaches. One-third of the competition entrants were women, and the work submitted 'begins to describe the possibilities for an expanding cross-fertilisation between cultural and scientific artefacts'. During the public presentation of Life 2.0 in February 2000, the organisers screened a number of groundbreaking works in ALife and awarded a special mention to *TechnoSphere* by Jane Prophet, Gordon Selley and Mark Hurry. This is described as 'one of the first examples of an online ALife ecosystem'.

TechnoSphere consists of a virtual environment of mountain ranges and plains covering a 16 square kilometre area, and two classes of animals – herbivores and carnivores. These artificial creatures have a life-span of up to five TechnoSphere months and a cycle ranging from foetus, child, adult to veteran. Herbivores graze on virtual grass and carnivores graze on herbivores and each other. They exhibit a hierarchy of behaviours with foraging, sleeping, mating, procreating, playing, evading, relocating and rotting at the higher levels. Unlike the creatures in *Creatures*, those in *TechnoSphere* consist of only one sex and there is no gender: 'Animals in TS have no gender as such, but its easy to perceive them as taking on one role or another at different points in their life' (www.technosphere.org.uk). Strictly speaking, the creatures are hermaphrodites in that they can become pregnant themselves or make other creatures pregnant. Contra Darwin, 'the creature which initiates the mating is the one which becomes pregnant'. Mate selection appears to be not merely species specific but based on physical appearances; an animal looks for a mate with the same body as itself. Moreover, 'consummation doesn't take that long and conception is rare'. Parenting is effectively dispensed with and 'the "mother" has no obligations to her children', albeit that the children inherit their 'names' or passwords from the 'mother' not the 'father'. Undoubtedly, this is a slightly reflexive, reconstructed, somewhat tongue-in-cheek ecosystem whose goal is more about the interactive process than about the biological and evolutionary criteria of the product itself. As a non-participant observer this is only really apparent in the, at times, unfortunate appearance of the creatures themselves, consisting as they do of a strange combination of eyes, heads, bodies – and wheels. A participant in *TechnoSphere* is able to construct their own creatures from the available body parts which include compound eyes, eyes on stalks, 'a bunch of choppy crushy things' (in the words of Sigourney Weaver in *Galaxy Quest*) for mouths, and spoke wheels. Having constructed and named some creatures and provided an email address, participants receive an email asking them to confirm their creations and 'bring them to life' by clicking on the relevant link. Each creature is then given an identity number and enters *TechnoSphere*. The ID number can be used to access the creature's family tree. A few days after receiving confirmation of Errol and Sydney's entrance to *TechnoSphere*, it was reported that Sydney had managed to get Riley pregnant with baby Riley and to conceive baby Sydney with father TurdMonster. Sadly, Errol did not do as well, and was killed by a predator within ten seconds of arriving. Sydney lasted rather longer and was succeeded by some meaner creatures named Hal and Hannibal.

A real time three-dimensional version of *TechnoSphere* – in which it is possible to see the creatures rather than just hear from them – is installed at the National Museum of Photography, Film and Television in Bradford, UK (and has also appeared in California and Sydney), and as of February 2001, the total number of creatures 'alive' in the ecosystem was in the region of 200,000 with over 300,000 sadly deceased. It is possible to access and compete for a creature's hall of honour ('fiercest', 'greediest', 'most potent'), and the extent of user involvement has in the past been sufficient to compromise the entire project. As co-author, Jane Prophet pointed out: 'The interaction between users and between us and users is as central to *TechnoSphere* as the underlying ALife engine which defines the interaction between creatures' (Interview, 14 February 2001). The idea for the project came together in 1995 when 'the web was a grey place with pages of black text and blue text links' dotted with the occasional image. Excited by its potential for connectivity, Jane Prophet also wanted a project combining interaction with landscape which had so far failed to attract funding: 'The web was a challenging environment for us to work in . . . and it was free'. While relinquishing any strong ALife claims to emergent behaviours, Prophet does claim that the project 'counters biological determinism and rigid Darwinian evolution'.

Nerve Garden is a virtual ecosystem with a straighter face than *TechnoSphere*, since it claims to provide an online 'collaborative ALife laboratory' for the purpose of education and research. For Bruce Damer, the project's initiator and CEO of Digital Space Corporation, 'the whole idea is to create a common space on the Net where all kinds of weird forms can emerge' (in Kuchinskas 1999). Inspired by Karl Sims's *Co-Evolved Virtual Creatures* and Thomas Ray's *Tierra*, *Nerve Garden* is the first product of a special interest group, Biota.org, which was 'charted to develop virtual worlds using techniques from the Artificial Life (ALife) field' (Damer *et al.* 1999). Biota.org is part of an organisation, also founded by Bruce Damer, called Contact Consortium, which 'serves as a focal point for the development of on-line virtual worlds and hosts conferences and research and development projects' (2). Launched at the 1997 SIGGRAPH conference, *Nerve Garden 1* allowed users to germinate, mutate and seed three-dimensional plant models and to move through the ecosystem utilising the viewpoint of a flying insect. The immersive three-dimensional interface, combined with sound emanating from different objects 'encouraged participation and experimentation from a wide group of users' (5), but the project was found to lack the sophistication of 'true ALife systems like *Tierra*' and to require upgrading. The virtual plants failed to interact with each other or the environment and 'there was no concept of autonomous, self-replicating objects' (6). *Nerve Garden II* addresses these faults by utilising a high performance general purpose cellular automata engine called Nerves: 'Nerves is modelled on the biological processes seen in animal nervous systems, and plant and animal circulatory systems' (6) and it facilitates the introduction of autonomous entities termed 'polyvores'. Nerves implements biological processes in artificial systems or 'embodies' the behaviour of artificial forms, and yet embodiment would not appear to be a priority in Damer's cosmic key-note speech at Digital Biota II, the Second Annual Conference on Cyberbiology. In 'Why is Life trying to Enter Digital Space?', Damer (1998) employs both genes and memes in a teleology of life-as-information which extends from terra firma to cyberspace and even to outer space. The reason why life (itself) attempts to enter digital space is to cheat not only physical but planetary death: 'Through us, our genes can now know about their ultimate end. The inevitable end will happen in a distant eon as our red giant sun, starved of hydrogen, consumes all genes and memes still dwelling down in this gravity well' (Damer 1998). Transcendence, within the framework of Western Enlightenment epistemology, is achieved through disembodiment – 'massless organisms can escape gravity's well easily, and traverse the solar system in mere hours' – which has the added advantage of lending itself to much faster (if less rounded) evolution:

What advantages are there to be gained by evolving to live in such an ephemeral, narrow and arid ecosystem? One advantage is super charged selection; freed from the slow speed of chemical replication and limited supply lines of atoms, the essential genetic machinery can be copied relentlessly rapidly.
(Damer 1998)

Artificial cultures

Damer's science fictional vision of the future of ALife from a gene's eye view serves to contextualise computational models of human cultures and societies. Where these offer a method and epistemology for the study of life-as-we-know-it, they do so within a narrative framework where life-as-we-know-it is in the process of being superseded by life-as-it-could-be. This is not as much an apocalyptic as an evolutionary scenario in which the next stage in the evolution of life is digital life – and the aliens are (be)coming. Within this evolutionary scenario the concept of culture seems to regress from a social to a bio(techno)logical context from which it is expected to re-emerge.

In 'Artificial Culture', Nicholas Gessler argues that 'we now have at our disposal a new, untapped and rich paradigm for building theories of cultural evolution', and this paradigm, inspired by artificial life, is computational anthropology (1994: 430). Computational anthropology provides a language for, and a means of experimenting with human cultural processes and 'it can help us investigate the deep relationships between culture and biology, the individual and population, the social and physical environments, and the natural and artificial worlds of artifacts and technology' (http://www.sscnet.ucla.edu). Within the paradigm of computational anthropology, culture is viewed as a computational system – 'not in the weak metaphorical sense' but literally – and as a manifestation of the ubiquitous evolutionary process of information exchange (Gessler 1999: 1). This then, is a memetic view of culture which Gessler realises in a software model. *Artificial Culture* (AC) is a computer program which functions as a test bed for the theory of cultural evolution, and it builds on foundations which already exist in ALife. Rudiments of artificial cultures, according to Gessler, already exist in ALife environments and ecosystems populated by autonomous agents, but the crossover between anthropology and artificial life has been minimal: 'so foreign' do the practitioners of ALife appear 'that ironically AL may more frequently be the *object* of ethnographic field studies than it is the *paradigm* for their understanding' (Gessler 1994: 431).[9] Artificial culture enacts a theory of culture which is evolutionary and emergent. Gessler argues that within anthropology, culture is at least 'operationally' defined 'as a corpus of shared traits held in common by a social group'. Without wishing to deny that traits are sometimes shared, it seems to Gessler 'to be the differences in these traits among participants that motivates societal evolution and change' (432). Evolution operates through cultural variation and the emergence of behavioural patterns from individual local rules:

> As cellular automata illustrate, individual actors, operating under individual local rules, can automatically produce collective global patterns of behavior that emerge solely through their mutual interactions. Importantly, these global patterns of behavior are not programmed into the simulation. They have no existence within the individual actors themselves. Rather they come into being only as the entire system operates. I suspect that many of the same processes are operating in culture.
>
> (Gessler 1994: 432)

Within cultural studies, this view of culture approximates to a structuralist as opposed to culturalist argument (Hall 1981) and is anti-humanist, denying agency to the individual and conferring it instead on the system. Based on a desire to develop the incipient cultures of ALife programs like *Tierra*, *SimLife* or perhaps *Creatures*, the aim of *Artificial Culture* is to create a population of evolving mobile autonomous agents which are both embodied and situated and which may contribute to the so far elusive theory of cultural evolution. Economic as well as scientific advantages are acknowledged outcomes of any successful attempt to elucidate the 'patterns governing fad and fashion' (433). In the mean time, Gessler details, and continues to develop the architecture for the program which includes a terrain, virtual objects (including resources), 'personoids', an ethnographer and God.[10] Personoids are intelligent agents which interact in the social environment of other agents and in the physical environment of artefacts and objects. In the simulated physical environment, time is represented through an analogy with film where 'the experimenter or director always specifies the starting world' but can then stand back and watch the story emerge infront of the camera: 'In a simulation, the programmer is the director of a self-organising improvisational production' (Gessler 1999: 3). Space is represented by a grid of cells analogous to a chess board, and extending 100 cells per side. Resources (food and water) may be rendered renewable or non-renewable and the agents are 'actors' with a basic physiology, metabolism and state of health related to energy levels. Agents reproduce, age and die and, as in TechnoSphere, they are sexually undifferentiated: 'In this world of sexual equality, constructed to avoid the complex issues of gender in an already complex world, reproduction is a unisex affair: agents are sexually undifferentiated. Each agent can have sex with any other' (1999: 4). While rooting his view of culture in biological processes, Gessler tampers with biological processes in order to accommodate certain cultural truths; in this case that homosexuality and reproduction are not necessarily mutually exclusive. In this utopia, as in others (notably, Haraway's Cyborg Manifesto [1991a]), gender is neutralised in order to resolve the problem of equality. Somehow this always implies that the problem of equality is unsolvable in gendered cultures and societies which AC does not mirror but re-create. As originally envisaged, the ethnographer is an embodied, necessarily partial observer 'operating within the physics of the virtual world and perceivable by its objects' (Gessler 1994: 434). It enacts a reflexivity denied in the figure of an omniscient god which is 'disembodied and objective' but 'exists for us only in a less-than-ideal state' (434). In an updated account, these two figures appear to combine in the role of the 'investigator' which represents the user in a virtual ethnography (Gessler 1999: 6). Gessler anticipates both ethical questions and epistemological criticisms. He points out that the ethical questions are already raised in science fiction; the rights of aliens and the rules of behaviour in a realm where the distinction between nature and artifice has collapsed (1994: 435). Computation, as a form of model-building is, for Gessler, a universal scientific method based on building increasingly reliable representations of the world. It is also a universal facet of mind and matter; the key to all cybernetic systems. In line with the current trend towards synthesis rather than analysis, Gessler maintains that in order to understand cybernetic systems it is necessary to grow them. Within the ALife community, reductionism is associated with the analytic rather than synthetic approach; with physics rather than biology. Moreover, for Gessler, 'the charge of reductionism . . . is usually made from the illusory high ground of privileging discourse over computation' and 'those who make that claim need to be reminded that natural language is reductive too, that its structures are sentential, highly sequential and that it takes great writers to represent a world of multiple actions happening all in parallel' (1999: 6). Where AI failed to match the representations of great writers, ALife is about 'staying simple' – and believing in emergence.

Viewing Nicholas Gessler's *Artificial Culture* (1999) from the viewpoint of Raymond Williams's *Keywords* (1983),[11] it is clear that Gessler's concept of culture is already biologically regressed (atavised to the petri dish of a computer program) as well as non-theistically emergent. This cultural evolutionism (also manifested in evolutionary psychology and the theory of memetics) is characteristic of a new epoch which has been variously named as the Information Age (Castells), the Biological Age (Grand) and the Neo-Biological Age (Kelly). Whatever the name, Grand argues that the third great age of technology 'in which machine and *synthetic* organism merge' has begun (Grand 2000: 8). In this biotechnological epoch it is not simply the computer, but increasingly the Net which defines the parameters of culture and identity. The Net does this partly in as far as it is regarded as an ecosystem for emergent artificial life forms *and* as an entity or intelligent life form in itself.

Kevin Kelly's vision of 'Network Culture' incorporates 'all circuits, all intelligence, all interdependence, all things economic and social and ecological, all communications, all democracy, all groups, all large systems' (Kelly 1994: 25) in a co-evolved single organism which is analogous to an emergent hive mind. It is a decentralised distributed intelligent entity which assimilates and elides identity: 'As we wire ourselves up into a hivish network, many things will emerge that we, as mere neurons in the network, don't expect, don't understand, can't control, or don't even perceive' (28). Individuals are signified as bees in the hive, neurons in the network, cogs in the wheel which is more than the sum of its parts. Derided as a key example of the Californian Ideology (Thomas and Wyatt 1999: 695) developed on the pages of *Wired* magazine, Kelly's vision of the early 1990s is nevertheless strangely echoed in a millennial issue of *New Scientist* dedicated to the role of the Internet as 'Global Brain'. Here, Michael Brooks (of Sussex University) examines the claim of Francis Heylighen (of the Free University of Brussels) that the global brain will grow out of attempts to manage the store and flow of information on the Internet. These attempts are more sophisticated than 'simple-minded' search engines and websites put together by people who are oblivious to the needs of individual users. As part of the Distributed Knowledge Systems (DKS) project at the Los Alamos National Laboratory in New Mexico, Johan Bollen (former student of Heylighen) has built a web server called the Principia Cybernetica Web which continually rebuilds the links between pages in order to adapt them to the user's needs. New links are added when the server 'thinks' they may be useful, and old links are closed down when they are no longer used. Brooks points out that this dynamic process of strengthening and weakening links is analogous to the synaptic connections which grow and fade in the human brain.

The Principia Cybernetica Web may have dramatic implications, but its mechanism is relatively simple. The server uses 'smart cookies' (small strings of data) to recognise individual users, and it tracks their routes through the site. It can then make specific recommendations and tailor-make its structure or pattern of hyperlinks. As well as strengthening and weakening links (which in conventional websites are fixed), the server constructs short-cuts (from A to C direct) based on previous, more pedestrian movement. For Heylighen, such activity amounts to a form of self-organising, adaptive behaviour. The Principia Cybernetica Web does not simply produce a better search engine or more usable websites – it produces intelligence. Autonomous agents (artificial intelligences or life-forms) will also constitute part of this intelligence by suggesting and adding new links – or 'by making connections between concepts that did not previously exist'. When an autonomous agent researches a question or request similar to one it has previously encountered, it will be able to recall the information more easily. Heylighen claims that collective thoughts of the whole brain may stem from this kind of 'web on web' activity. According to Norman Johnson (Symbiotic Intelligence Project, Los Alamos), the global brain's intelligence – rather like the sum of all human intelligence – could as easily come from an assembly of more limited intelligences spread across different networks, each with its own area of expertise.

For Heylighen, the global brain is at the centre of the global superorganism which relegates people to 'playing the role of cells in the body'. As Brooks (2000) points out, this is 'not a very comfortable self-image for a species used to considering itself the pinnacle of creation'. Where Heylighen's science fictional vision of the Internet is dystopian (threatening existing power structures, creating an information underclass, disconnecting individual users who fail to answer requests for information from a system capable of knowing where its gaps and inadequacies lie), Johnson's is more utopian because agency and control are not so freely relinquished to the machine. For Johnson, the global brain is a collective intelligence – an extension of human society rather than its apotheosis. By extending amazon.com's practice of telling people which other books buyers have bought, the Internet makes collective knowledge available while constituting an intelligent agent with perhaps some claim to kinship with people, but not to their over all control. Even so, Brooks asks, 'how many people will relish the prospect of being assimilated in this way? Are we really doomed to become the Borg?' (2000: 27).

In *Darwin among the Machines*, George Dyson (1997: 215) also presents a view of the Net as a self-organising intelligent entity rather than as a mere environment or ecosystem for self-organising intelligent entities. He points out that 'for a long, long time . . . we have awaited the appearance of a higher intelligence from above or a larger intelligence from without' and that science fiction has sounded clear warnings that alien/artificial intelligence is likely to challenge our evolutionary status and its attendant privileges of power. It is presumptuous, according to Dyson, to think that artificial intelligence will be comprehensible and 'as we merge toward collective intelligence, our own language and intelligence may be relegated to a subsidiary role or left behind' (224). Here, the fears attendant on AI machines transfer to ALife systems and principally to aggregate structures of computers which constitute a 'larger' if not 'higher' intelligence.

Steve Grand is not a believer in Internet ontology, and his lack of faith stems in part from his critique of evolutionism as a hegemonic discourse in ALife:

> There is much talk at the moment of how intelligent systems might exist inside the Internet and serve useful purposes. Some people even speculate that the Internet could become an intelligent entity all by itself. Several hi-tech companies are currently banking their future on 'intelligent search agents' which can supposedly 'live' in the World Wide Web and find useful documents for people. But there is a real problem. The Web is a consistent world, but it is not *self*-consistent.
>
> (Grand 2000: 137)

In other words, neither the Internet or its agents are, in Grand's (1998a) view sufficiently autonomous to qualify as entities and even if they can be said to evolve, there is more to life – in a holistic synthetic view – than evolution. In Kelly's view, evolution 'as a tool' is good for three things: 'How to get somewhere you want but can't find the route to. How to get somewhere you can't imagine. How to open up entirely new places to get to' (1994: 342) These three things can be characterised as being practical, mystical and science fictional and to correspond, respectively, to genetic algorithms, emergence and the desire for new worlds. Two out of the three things that evolution as a tool is good for stem from the realms of belief and fantasy and the other one corresponds to an engineering practice which Steve Grand fears is type-casting and restricting ALife to reductionist and mechanistic criteria: 'In practical terms, this means that research into morphogenetic, chemical and neural mechanisms, communcication, perception and intelligence is being eclipsed by the overwhelming shadow of evolution'

(Grand 1998a: 20). Kevin Kelly also senses the limits of evolution (by natural selection) as a means to the future of artificial life: 'artificial evolution will not be able to make everything' (Kelly 1994: 387). Artificial life may not be reducible to evolution, but artificial life itself can, and will evolve either within (Grand) or as (Kelly) the computer ecosystem. To this end both Kelly and Grand offer a blueprint, a 'how-to' guide to the creation of life, in which they defer only nominally to God. Grand presents 'God's Lego Set' in the form of cybernetic building blocks which, once plugged together into networks, 'can generate the vital spark that makes these inanimate parts into a living whole' (2000: 133). Echoes of Frankenstein are here intended. Kelly (some six years earlier) presents 'The Nine Laws of God' in the form of 'bio-logic' principles which, once incorporated into network systems, signal a new epoch: 'When the Technos is enlivened by Bios we get artifacts that can adapt, learn, and evolve. When our technology adapts, learns, and evolves then we will have a neo-biological civilization' (Kelly 1994: 471). In Kelly's Neo-Biological Age, something can be created out of nothing; the focus is no longer on the chaos within order but on the regeneration of order from chaos. The non-vitalist vitalism of emergence is this biological age's answer to the physics of entropy – and the job of postmodern science and culture is done. The paradigms of nature – concealing and revealing life itself – seem to have survived (if not unchanged) the search and destroy missions of post-structuralist epistemologies and to be newly deployed in and through the artefacts of information and communication. These artefacts weld together 'engineered technology and unrestrained nature' (Kelly 1994: 471) producing a bioculture which is at once more and less than the sum of its parts, but identical to none. Bioculture is not the biological culture of the petri dish any more than it is the forms and processes of everyday (human) life. Bioculture is the culture of analogous, information or network systems, self-organised within what has been referred to as network society.

Artificial societies

In 1996, Joshua Epstein and Robert Axtell published a pioneering petri dish approach to growing artificial societies which inspired Gessler's *Artificial Culture* and connected with evolutionary and computational approaches at the Sante Fe Institute, of which Epstein is a member. A recurring theme in the work of the Institute is the emergence of recognisable patterns and structures from the interaction of 'simple "unknowing" agents'. A characteristic preoccupation might be with how the interaction of these simple agents produces structures such as markets which in turn shape those agents into action patterns which recreate the structure. How such structures can exist at a level above that of the component individuals and exert a partially autonomous force on them is considered to be a fundamental question in the social sciences. Epstein and Axtell (1996: 2) adopt a computational approach to the social sciences, employing an agent-based evolutionary perspective influenced by work on cellular automata, genetic algorithms, cybernetics, connectionism, AI and ALife. The authors state that there have been very few attempts to bring these perspectives to bear on social science, and argue that 'in this approach fundamental social structures and group behaviors emerge from the interaction of individuals operating in artificial environments under rules that place only bounded demands on each agent's information and computational capacity' (4). Artificial societies are here viewed as laboratories in which there is an attempt to grow social structures from the bottom up *in silico*, with the aim of discovering the basic local mechanisms which are sufficient to generate global structures and behaviours. Agents are characterised as people, and have internal states and designated behavioural rules. Internal states such as sex, metabolic rate and vision are fixed (or genetic), but others such as health, wealth and cultural identity can change in interaction with the environment. The environment is the 'Sugarscape', a landscape of unevenly distributed edible resource which agents, in the first instance, must consume and metabolise. Behavioural rules apply to agents and environments and might incorporate a simple movement rule (governing resource location) or a rule of interactions (mating, combat or trade). The full behavioural repertoire from which 'a society is born' develops to incorporate 'movement, resource gathering, sexual reproduction, combat, cultural transmission, trade, inheritance, credit, pollution, immune learning, and disease propagation' (14). From a neo-Darwinist scenario involving genetic replication, diversity and selection in a competitive environment of scarce resources, the authors wish to reproduce a complex society with a familiar history or 'social story'. This is another example, like Cog, of directed development with a desired outcome necessitating something more than the input of evolutionary chance or emergence. Epstein and Axtell retell a Western social story in the language of biological science where ideologies of human social relations are naturalised. Their project in this retrospective sense is inevitably sociobiological. Its premise is sociobiological. The story line of the 'entire history' of artificial civilisation which they wish to grow is detailed as follows:

> In the beginning, a small population of agents is randomly scattered about a landscape. Purposeful individual behavior leads the most capable or lucky agents to the most fertile zones of the landscape; these migrations produce spatially segregated agent pools. Though less fortunate agents die on the wayside, for the survivors life is good: food is plentiful, most live to ripe old ages, populations expand through sexual reproduction, and the transmission of cultural attributes eventually produces spatially distinct 'tribes'. But their splendid isolation proves unsustainable; populations grow beyond what local resources can support, forcing tribes to expand into previously uninhabited areas. There the tribes collide and interact perpetually, with penetrations, combat, and cultural assimilation producing complex social histories, with violent phases, peaceful epochs, and so on.
>
> (Epstein and Axtell 1996: 8)

Stage one in this 'proto-history' is sexual reproduction. Agents are sexed, genetically heterogenous, but with the same rules of engagement – select a mate and if he or she is fertile and of the opposite sex, a child is born. 'Men' are fertile for slightly longer than 'women' but there is no innate aggressor or childbearer. Selection favours agents with low metabolisms and good 'eyesight'. In a more complex environment, with more varied resources to compete for, gender differences might be expected to emerge according to sociobiological principles. The second stage in the historical development of the Sugarscape is tribe formation, or 'social speciation', and this has led Stefan Helmreich to identify the project as the product of 'a white cultural imagination', characteristic of the West and specific to North American histories of kinship constructed, as elsewhere, at the Sante Fe Institute: 'Like Albert Einstein wearing a headdress in the Institute's stairwell photograph, Sugarscape agents are dressed up as Indians, but they also stand as translucent representations of scientific and Euro-American logic' (Helmreich 1998a: 159). Despite some degree of slippage between genetic and cultural identity, 'Sugarscape tribes come to be "pure", and purely identified with biogenetic kin groups' (160). By incorporating spice as an additional resource an commodity, and by allowing trading behaviour to be subject to the influence of nurture as well as nature, Epstein and Axtell (1996) are able to develop a model for opposing economic theories. They claim to create markets of agents which correspond to the ideals of economic textbooks – 'infinitely lived agents having fixed preferences who trade for a long time' – as well as those of agents who do not – 'agents that are non-neoclassical in so far as they have finite lives and evolving preferences' (1996: 10). Where the former produces equilibrium, the latter does not – and therefore raises questions concerning the efficacy of free markets. The broad aim of the project amounts to a continuation of E.O. Wilson's 'new synthesis' in that it implements 'a more unified social science, one that embeds evolutionary processes in a computational environment that simulates demographies, the transmission of culture, conflict, economics, disease, the emergence of groups and agent coadaptation with an environment, all from the bottom up' (1996: 19). Within this desire for a unified, computational social science there are policy implications for human social behaviour, and notably in this instance, economic behaviour. For instance, the authors claim that they might be able to distinguish conditions which lead to the unlikely emergence of efficient markets. This might enable them to answer the apparently rhetorical question: 'Is it reasonable to base policy on the assumption that if central authorities "just get out of the way" then efficient markets will self-organise in Russia?' (1996: 35).

Epstein and Axtell question the decentralised autonomous agency ascribed to the market in Kelly's 'network economics', an autonomous agency which is arguably only less celebrated in Castells' theory of 'network society'. The context of Castells' theory is the information technology 'revolution' and the 'transformation' of global economics into an 'interdependent system working as a unit in real time' (Castells 2000: 2). The 'incorporation of valuable segments of economies throughout the world' serves to accentuate uneven development, 'this time not only between North and South, but between the dynamic segments and territories of societies everywhere, and those others that risk becoming irrelevant from the perspective of the system's logic' (2). The economic system is sufficiently unified to possess a primary logic which may still be, in Castells' account, a form of self-organisation. He commits himself to the 'informational paradigm', understood as 'a specific form of social organization in which information generation, processing, and transmission become the fundamental sources of productivity and power because of new technological conditions emerging in this historical period' (21). Very much by implication, the new technological conditions which are 'emerging' must (by definition) be bio-logical if they are to determine a society in which information 'generation, processing and transmission' become fundamental sources of productivity and power. Such implicit biological determinism is underlined only by Castells' retreat from technological determinism, which he characterises as 'probably a false problem, since technology *is* society, and society cannot be understood or represented without its technological tools' (5). The bio-logic of information produces capital and informationism is 'the new material, technological basis of economic activity and social organization' (14). Castells describes a 'new communication system, increasingly speaking a universal, digital language' which is simultaneously integrating and customising cultural products. He refers to the exponential growth of interactive computer networks which are at once shaped by and shaping life itself (2). This global, metaphorically organic network is neither the benevolent ecosystem nor the more malevolent entity of other, more science fictional visions, but its relentless 'instrumentality' tells a similar tale of alienation and anti-humanism. One characteristic of the network society is the revival of religious fundamentalism which Castells associates with the search for social meaning through identity. Identity, he argues, is becoming the main source of meaning in a destructuring, delegitimising and ephemeral society. Individuals attempt to gain meaning not from what they do, but from what they are, while on the other hand, 'global networks of instrumental exchanges selectively switch on and off individuals, groups, regions, and even countries, according to their relevance in fulfilling goals processed in the network, in a relentless flow of strategic decisions' (3). Dystopianism shadows this self-organising (selfish?) network agent and the shadow deepens where the individual subject or self is dislocated: 'there follows a fundamental split between abstract, universal instrumentalism, and historically rooted, particularistic identities. Our societies are increasingly structured around a bipolar opposition between the Net and the self' (3).

Castells' view of alienation, or the bipolar opposition between the Net and the self denaturalises the operations of capital in the synergy between economic and evolutionary systems. It also evokes a political theory which arguably fails to capture the cultural complexity of such systems, and thereby overemphasises the disparity between system and self. In contrast to the economic self-sacrifice of free labour which fed the Net, enabled it to evolve and generated a form of anarcho-communism which Richard Barbrook termed the 'gift economy', Castells reinvokes a Marxist model of exploitation and alienation. Andrew Leonard charts the demise of the gift economy in the face of the increasing commercialisation and economic maturation of the Net, arguing that what breaks down is the 'tacit agreement among the technogeeks that all of the Net's problems could be solved to the public benefit through altruistic and volunteer sacrifice' (Leonard 1997: 131). But Terranova (2000) seeks to avoid the backlash against the glamorisation of digital labour through its reinscription as capitalist exploitation. She does this by pointing out that free labour is 'a trait of the cultural economy at large' and immanent to the flows of network society (Terranova 2000: 33). Labour is then not wholly inscribed within the economic requirements of capital but also in relation to the expansion of the cultural industries, and as part of an ongoing process of economic experimentation which creates monetary value from the forms and processes of information and communication (2000: 38). There is then a willing exchange, a free exchange between the self and a socio-economic system which evolves more than it progresses, and which remains, to an extent, metaphorically organic: 'Whether or not the Net is truly organic, whether or not bots are truly autonomous new creatures, we will continue to think of them in such terms and accordingly shape and guide their development' (Leonard 1997: 181).

Alongside and complementing the shifting patterns of work in network society is an investment in technocultural forms of autonomy and agency which relies on a dialectic and not a division between the Net and the self. There is no clear opposition between 'global networks of instrumentality' and 'the anxious search for meaning and spirituality' (Castells 2000: 22). The search for meaning through identity – and perhaps even the search for spirituality – occurs neither outside or inside the Net but in a dialectic articulated in part through the reproduction (symbolic and material) of agency and autonomy. The transfer of agency and autonomy to the (id)entities of the Network, although apparently anti-humanist, is – in one sense – a process of externalisation which enables agency and autonomy to be renegotiated and reclaimed within the identities of the self. The posthuman self thus engages the forms and concepts of posthumanism. Erik Davis (1999) envisages a spiritual posthumanism emerging from the dialectic of the self and the Net, and this vision is based on an identification of religious myths attendant upon the supposedly secular industrial and information ages. His premis is that 'we are beset with a thirst for meaning and connection that centuries of sceptical philosophy, hard headed materialism, and an increasingly nihilist culture have yet to douse', and that this thirst for meaning and connection 'conjures up the whole tattered carnival of contemporary religion' (Davis 1999: 7). The contemporary (over)emphasis on the matter and meaning of information causes it to 'crackle with energy' and act as a strange attractor to all manner of myths and metaphysics. There is, he suggests, 'a host of images' looping information into a spiritual imaginary: 'redemptive, demonic, magical, transcendent, hypnotic, alive' (9). What resonates most clearly with the digital zeitgeist is a mystical form of Christianity which arose in the late antiquity and is anti-materialist and pro-self-knowledge. Gnosticism anticipates the 'libertarian drive toward freedom and self-divination' and the 'dualistic rejection of matter for the incorporeal possibilities of mind' (80). It constitutes the unconscious metaphysics of information society and centres on the autonomous disembodied subject: 'Today's techgnostics find themselves, consciously or not, surrounded by a complex set of ideas and images: transcendence through technology, a thirst for the ecstasy of information, a drive to engineer and perfect the incorporeal spark of the self' (101). ALife's quest to engineer intelligent agency through emergence would then epitomise such techgnosis, but there is a twist in this metaphysical tale. Gnosticism's Demiurge or false creator who imprisons his creations in a realm which is unnatural but which they transcend through self-knowledge thereby achieving true identity, offers an interesting analogy with ALife gods and the prospect of emergent complexity. It is with respect to the bioethics as well as the metaphysics of autonomous agency that Davis calls for the revivification of the social and spiritual imagination.

Artificial subjects

The network identities which come to matter in the non-homogenous discourses of ALife may generally be characterised as posthuman but offer to realise a vision of posthumanism which lacks, in Davis's terms, imagination. N. Katherine Hayles (1999a: 2) identifies four main features of posthuman identity: information or form is privileged over matter; consciousness is an epiphenomema or evolutionary side show rather than the seat of human identity; the body is the original prosthesis and, most importantly, human beings are synonymous with intelligent machines. The posthuman subject is no longer in a 'state of nature' (3), and self-ownership, agency desire or will are complicated by its essential non-unity as a natural/artificial entity. The posthuman has 'a distributed cognition located in disparate parts' (3) and 'is "post" not because it is necessarily unfree but because there is no a priori way to identify a self-will that can be clearly distinguished from an other-will' (4). Hayles states clearly that the posthuman is located not just in specific cyborgian identities, but constitutes a prevalent concept of subjectivity. It is an epistemology as well as an ontology of the self in the Information Age and as such, necessarily engages with historical constructions of humanism. The construction of the liberal humanist subject has, as Hayles points out, been widely critiqued in feminist, postcolonial and postmodern theories which locate its 'notorious universality' in practices of disembodiment which erase the markers of difference such as race, gender and ethnicity (5). Without regretting the loss of a concept 'so deeply entwined with projects of domination and oppression', Hayles does argue that specific characteristics associated with the humanist subject – notably agency and choice – may need to be reconsidered in a posthuman context. She identifies a 'critical juncture', an opportunity in the present posthuman context 'when interventions might be made to keep disembodiment from being rewritten, once again, into prevailing concepts of subjectivity' (5). By implication, the re-embodiment of subjectivity is one – for Hayles the primary – route to post-liberal humanism. It would seem that what emerges from the apparently anti-humanist practices of ALife – rather like what emerged from the apparently anti-humanist practices of cybernetics – is an attempt to extend rather than subvert liberal humanism. What Hayles says of Norbert Wiener – 'he was less interested in seeing humans as machines than he was in fashioning human and machine alike in the image of an autonomous, self-directed individual' – can also be said of ALife engineers from Langton to Brooks to Grand. Perhaps this goes some way towards explaining the popular appeal not just of their work, but of their ideas. There is, to my mind, a strong and *fundamentally affective* appeal to humanism in Grand's writing and thinking. While this is predominantly a liberal humanism it also served as an invitation to dialogue; precisely, for me, a critical juncture. This may or may not be realised in any kind of agreement over the terms of posthumanism, but our dialogue – more specifically than, as Hayles characterises it, the posthuman itself – is certainly a 'resource for rethinking the articulation of humans with intelligent machines' (287).

A dialogue over the forms and concepts of posthumanism may be effected between and within narratives, individuals and disciplines. In 'Anti-Boxology: Agent Design in Cultural Context', a PhD thesis stemming both from the Computer Science Department and the Program in Literary and Cultural Theory at Carnegie Mellon University, Phoebe Sengers (1998) effects a dialogue between engineering and what she terms humanism in order to generate a novel agent technology. This technology is based on a practical and theoretical critique of autonomous agency and implements a view of the artificial posthuman subject as being – as a being – embodied and socially situated in human-machine relations. By doing so, the project (which is discussed in detail in Chapter 7) contributes to the important consideration not just of the ontology and epistemology of posthuman subjectivity, but to the ethics of network identities reclaimed from maths to meaning, from computation to communication. Sengers' agent does not evolve and its lifelike existence is not separate from, but dependent on the imaginative investment, the desire for alife manifested by engineer and 'humanist' alike.

Chapter 6

The meaning of life part 2
Genomics

Artificial life as wetware

eXistenZ is an entirely new game system incorporating biological hardware and software, and bodies with 'bioports' which supply power via the spine. The hyperbolically organic – womb-like, breast-like – 'metaflesh game pod' connects to a bioport in the player's back by means of an 'umbi-chord'. Designed by game goddess Allegra Geller, *eXistenZ* is owned by the Antenna Corporation and marketed as the ultimate virtual reality experience. During a demonstration of the game, security is breached by a would-be assassin bearing a gun made of flesh and bone and loaded with teeth (handy for eluding metal detectors, fillings permitting). Allegra is shot in the shoulder then whisked away by security guard Ted Pikel who reluctantly agrees to having a bioport fitted and escaping into her game world. Cronenberg's narrative underlines and extends *Videodrome*'s play on the gendered dualities of body and machine. This machine is both an extension of her mind and an externalisation of her body – her womb, her 'baby'. The body–machine analogy is literalised in that the pod is an 'animal grown from fertilised amphibian eggs stuffed with synthetic DNA' (*eXistenZ*, 1997). This organ(ism) pulsates, bleeds, lives and dies. It becomes infected by a contaminated bioport fitted on Pikel by a malicious garage attendant bearing surgical tools of alarming proportions. He creates a new opening in Pikel's back and Allegra lubricates it with WD40. Visceral displays of penetration – later she enters his 'excited bioport' with a wet finger – take place in a context of obfuscated gender and sexual roles. Initially resistant to the bioport procedure – 'too freaky, makes my skin crawl' – Pikel feels vulnerable, 'disembodied' and psychotic inside the game world and has to pause it. Cronenberg highlights transgression in a narrative which crosses and re-crosses the line between fantasy and reality. These crossings are embodied in the form of agents and double agents who simultaneously signify geopolitical revisions of power and resistance. Fantasists and realists, US and Russian agents operate in relation to two competing corporations – Antenna and Cortical Systematics – and two competing games – *eXistenZ* and *transCendenZ*. Allegra and Ted, the agents of fantasy in *eXistenZ* culminate as agents of reality – 'death to *transCendenZ*' – in the film if not the game narrative. Notably, as biotechnological agents or characters inside the corporate game, they are semi-programmed. The plot establishes lines which must be spoken, actions which must be taken; but they have 'just enough free will to make it interesting' (*eXistenZ*, 1997).

eXistenZ seems to invert the story of virtual reality as a masculine fantasy of disembodiment and transcendence (Braidotti 1996; Springer 1996), introducing hyperbolically feminine and organic technology in the form of manufactured life-forms: 'using mutated animals and nervous systems as game pod parts is certainly feasible' (*eXistenZ*, 1997). This science-fictional vision of farming hints at the emerging industry of 'pharming' (transgenic and cloned animals mass producing biochemicals for the pharmaceutical industry) and is consonant with the genetic, or specifically, genomic imaginary and with the existence of artificial life wet-ware.[1] As wetware – flesh, blood, bone and other organic matter – artificial life is constituted through genetic manipulation and practices such as cloning and transgenesis.

In *Clones, Genes, and Immortality*, John Harris (1998) offers an ethical response to the reality of new life forms manufactured to order. The issue, he suggests, is not whether to manufacture life, but how to do so responsibly. It is too late for denial or retreat: 'there is no safe path' (1998: 6). His is one of many responses to a landmark event in genetics – the announcement in 1997 of a successfully cloned sheep named Dolly – which significantly raised the stakes in debates on the transformation and transcendence of natural kinds – animal and human. The public preoccupation with genetics and especially cloning is, for Harris, 'not unrelated to the human search for immortality'. This search is simultaneously vain 'and well rewarded by the understandings that genetics has lent to what has formerly been an essentially theological debate' (Harris: 1998: 7). The corporate and consumer culture of life is, then, imbricated with the ethical and religious dimensions of genetic science; in the terms and conditions of existence and transcendence. Manfred Davidmann's earthly concerns with 'Creating, Patenting and Marketing of New Forms of Life' (1996) centre on monopoly ownership, privitisation and the continued exploitation by the West of developing nations. Focusing on the creation of transgenic organisms (where genes are transferred between plant and/or animal species) in animal pharming and crop farming, he offers a critique of the GATT (General Agreement on Tariffs and Trade) agreement which aims to protect the intellectual property of patent holders for twenty years. Patenting may signal a transfer of ownership from developing countries where the bulk of the genetic raw materials are grown to the laboratories of Western companies which then market branded, standardised products. It signals the privatisation and monopoly control of hybrid crops: 'What is happening is that genetic material from anywhere is being patented, mainly in the US and resulting seed marketed; this means that farmers will have lost their rights to their own original stocks, and not be allowed under Gatt to market or use them' (Davidmann 1996: 7). Where non-hybrid seeds can be resown, hybrids cannot and must be repurchased each year (Haraway 1997; Franklin *et al.* 2000) and 'it would seem that multinationals are working on genetic modifications aimed at converting non-hybrid plants such as wheat into hybrids which would compel farmers to repurchase seed from the multinational each year' (Davidmann 1996: 8). Large corporations are able to enforce patent rights and, in 1996, it was estimated that the US would earn $61 billion a year in royalties from developing countries. Genetically modified and transgenic organisms raise moral and ethical as well as political and economic questions concerning the integrity of natural kinds – 'should we tinker with the basic building blocks of the planetary environment and of life itself?' – and of the natural order: 'Do we as human beings have the right to meddle with a set-up which took so long to produce us, when we have only existed for such a comparatively short time?' (3). Interference entails a negotiation of risk which vascillates freely between the public and private realms; between the global and the personal, the species and the self. The production and consumption of genetically modified organisms (in biotech industries and institutions) creates real and imaginary risks for public health, global biodiversity and the survival of the human species. It also risks the integrity, uniqueness and authenticity of the individual or self as much embroiled in the discourses of myth as of science.[2] This chapter is concerned with the imaginative as well as instrumental construction of wetware artificial life-forms. It is concerned as much with the impact of myth as of capital, and it establishes the grounds of a feminist analysis on this split level.

Sarah Franklin (2000: 197) points out that 'life itself has always been inextricable from its invocation as a story', whether or not it is '*the* story' of Genesis, *The Origin of Species* or *The Double Helix*. The power of stories about the origin and evolution of life lies in their global appeal to 'a universal essence of humanity'. Therefore, a significant part of what she calls the 'syntactic power' of life resides in its 'storiedness' and in imagined futures 'dense with the possibility of both salvation and catastrophe' (198). Paul Rabinow (1999) locates those imagined futures of salvation and damnation in the re-emergence and rearticulation of Christian myth and specifically the story of purgatory. Christian mythology underpins not only the discipline but the wider discourse of alife. In that heterogenous zone 'where genomics, bioethics, patents groups, venture capital, nations, and the state meet', the air is filled 'with talk of good and evil' (1999: 4). Within the domain of bioethics, Rabinow argues, the fate of humanity is articulated in spiritual rather than purely material terms (7). Drawing on Feher and Heller's concept of the spiritual as an imaginary construction concerned with the interpersonal dynamics of 'everything 'that is not natural' and with 'very real and potent effects' (11), Rabinow claims that the deterministic or synecdochical identification of DNA with the 'human person' is a spiritual identification. Moreover, drawing on Foucault's notion of humanism as 'a manner of resolving (in terms of morality, values, reconciliation) problems . . . that it cannot pose' (in Rabinow 1999: 7), he argues that this spiritual identification is also humanistic. To equate the person with DNA 'is to provide a solution to a problem that has not yet been adequately posed' (16). The problem with (geneticised) life is that it exceeds both natural and philosophical classification. It is diffuse, abject, unrecognisable, anxiety-provoking. It results in a situation which Rabinow calls purgatory and invites what in other quarters is called reductionism. Purgatorial themes and tropes expressed, crucially, 'by subjects who are (in the majority) forthrightly secular moderns' include 'a chronic sense that the future is at stake', a sense of 'redeeming past moral errors and avoiding future ones', recognition of ambiguities and 'a heightened sense of tension between this-worldly activities and (somehow) transcendent stakes and values' (18). Of importance to this chapter's emphasis on myth and narrative is that Rabinow identifies a distance between these secular subjects and the Enlightenment faith in reason: 'One finds an urgent and uneasy sense of hesitancy and caution over the consequences of a felt imperative to know or to put that knowledge into action' (18). Also significant is the fact that in purgatory, between heaven and hell, salvation is possible by worldly intervention.

Although the reductionist determinism of myth-makers such as Dawkins might be characterised as anti-humanist, it is helpful to consider the problem of genetic determinism as a spiritual and humanist one where 'spiritual' refers to the circulation and infusion of non-nature between and within nature's perceived boundaries, and humanism, after Foucault, bars understanding by offering simplified, ready-made answers to questions which have not yet been asked. These questions might be organised around notions of human dignity, integrity, authority and uniqueness in need of protection and revalidation as universal attributes. If humanism, in the context of geneticised life – the origin and evolution of informationalised life – is exposed as a solution to a problem not yet adequately posed, then an opportunity arises to pose more adequate questions which explore more fully the relationship between DNA and human identity, the limits of genetic influence and, indeed, the efficacy of the genetic manipulation of life (Rabinow 1999: 16). What is clear is that a strong current of anti-determinism runs counter to deterministic, and at least in Foucault's sense of the word, humanistic thinking within science, popular science and science journalism. As purgatory is the state (of mind) between heaven and hell, salvation and damnation, complexity occupies the (epistemological) space between genetic cause and (physical, behavioural, psychological) effect. What is also clear is that complexity, as a form of anti-determinism and anti-humanism, fails to replace universal attributes such as human dignity with an *ethics of difference* (Cilliers 1998; Varela 1999). Complexity as 'the world we live in' (Cohen and Stewart in Meek 2001a) is, for some, itself a universal attribute of networked bodies, technologies and societies which exceeds (human) knowledge: 'If the web of electricity cables, microwaves, rails, roads, airways, computers, fibre-optic links, retailers, distributors, sewage systems, phone lines, warning systems, farmers and manufacturers was simple enough for us to understand it, it would be too simple to exist' (Meek 2001a: 2). For others, such as Lewontin (2000), complexity means a rejection of the 'over-weening pride' of the natural sciences (as well, perhaps, as a rejection of the humanism of the social sciences) and the belief 'that everything about the material world is knowable and that eventually everything we want to know will be known' (xxi). Complexity, as Lewontin demonstrates, features in a recognition of the fallacy of genetic determinism by the US National Bioethics Advisory Commission which nevertheless 'continues to insist on the question of whether cloning violates an individual human being's "unique qualitative identity"' (280). For Harris (1998), complexity is the reason why Lenin (like Hitler in *The Boys from Brazil*) could not be resurrected through cloning: 'So many of the things that made Vladimir Ilyich what he was, cannot be reproduced even if his genome can. We cannot re-create pre-revolutionary Russia, we cannot simulate his environment and education, we cannot re-create his parents to bring him up and influence his development as profoundly as they undoubtedly did' (1998: 8). Complexity, as Harris's evocation of history, politics, education and family suggests, problematises not just the relationship between genetics and identity but the very notion of environment as opposed to genome. With complexity there are no singular or unified entities as well as no simple relationships between them. Matt Ridley (1999) apologises for creating an illusion of simplicity around the human genome for the sake of exposition and communication. The genome, like the world, 'is not like that. It is a world of greys, of nuances, of qualifiers, of "it depends"' (1999: 65). Short of rare genetic diseases, 'the impact of genes upon our lives is a gradual, partial, blended sort of thing' (66). Ridley introduces the concept of pleiotropy – 'a technical term for multiple effects of multiple genes' – in conjunction with that of pluralism. Pluralism combines multiple genetic, with multiple environmental causes with the effect of rendering the relation between cause and effect entirely non-linear and non-predictive. Complexity, for Ridley, offers a critique not only of genetic determinism (of genes 'for' this or that) – 'You had better get used to such indeterminacy. The more we delve into the genome the less fatalistic it will seem' – but also of social or environmental determinism. He is critical of science writers who reject one form of determinism only to establish another in its place: 'It is odd that so many writers who defend human dignity against the tyranny of our genes seem happy to accept the tyranny of our surroundings' (303). Social determinism then becomes another form of biological determinism, as illustrated in the argument that child abuse is caused not by any genetic predisposition but by the abuser's experience of abuse. There is no point, Ridley argues, in demanding rigorous proof for the genetic cause of behaviour while passively accepting social ones. He is no more contented with the opposition between genetic determinism and environmental choice. When environmental and genetic influences co-exist, determinism and its conventionally cited opposite, free will, are neither poles apart or the sole prerogative of genes and environment respectively: 'The crude distinction between genes as implacable programmers of a Calvinist predestination and the environment as the home of liberal free will is a fallacy' (303). Genetic determinism for Ridley is a reality, just not a fatalistic one. It says more about input that it can about output. Its effects are, in other words, not linear but circular: 'and in a system of circular feedbacks, hugely unpredicatable results can follow from simple deterministic processes' (311). This then, by the end of the twentieth century is a cybernetic biotechnological genome – the genome as a complex system comprised of the species-self.

The species-self

Ridley's (1999) book on the human genome is subtitled *The Autobiography of a Species in 23 Chapters*, and his use of complexity (or chaos) to trouble the myth of Calvinist predestination is ultimately defensive and strategic.[3] It allows him to revise and recuperate genetic determinism as a more subtle, nuanced, sensible and sensitive universal. The genome may not reduce individuals to their genes, but in authoring 'the' autobiography of a species, it reduces the species – all species – to one universal individual. The posthuman genome reveals the junk DNA inherited from our earliest ancestors but rendered redundant (like the instructions which tell us how to make fins or tails). Ultimately, it insists on unity not diversity – 'the unity of life is an empirical fact' – or on the notion of the species as its own, autonomous self (22). Predestination is, in a crucial sense, the myth that must be shattered in order to preserve both genetic determinism and the genome as a unifying law of humanity. Ridley makes repeated references to Calvinism which, like the mutated version of the Wolf-Hirschhorn gene (which causes Huntington's chorea) offers 'a prophecy of terrifying, cruel and inflexible truth' (56).

Confessions of a justified sinner

> be thou partaker of the afflictions of the gospel according to the power of God; who hath saved us, and called us with an holy calling, not according to our works, but according to his own purpose and grace, which was given to us in Christ Jesus before the world began.
>
> (2 Timothy 1:8–9, in Wain 1983)

Jean Calvin (1509–1564), a Genevan protestant reformer, denied the possibility of salvation by good deeds and argued that some souls were predestined to heaven and some to hell by God's will (human free will being anathema) before the act of Creation. The Atonement offered by Christ on the cross offered salvation only to the elect and did nothing to help those already damned. From here, 'it is easy to see how the doctrine of antinomianism, the view that those predestined to salvation are exempt from the moral law, could take root' (Wain 1983: 14). James Hogg's *The Private Memoirs and Confessions of a Justified Sinner* (1824) situates Calvinism in seventeenth-century Scotland and offers a metafictional comment on the memoir of a self-justified sinner by framing his narrative within that of the editor. Robert Wringham is proclaimed as a member of the elect by his adoptive reverend father and assured that he is 'a justified person, adopted among the number of God's children', and that 'no bypast transgression, nor any future act of my own, or of other men, could be instrumental in altering the decree' (Wain 1983: 124). With his redemption sealed, Wringham is overcome by a sense of exhaltation which places him above ordinary unelected humankind: 'I deemed myself as an eagle among the children of men, soaring on high, and looking down with pity and contempt on the grovelling creatures below' (Wain 1983: 125). No sooner does he express these sentiments when he encounters a stranger to whom he is attracted by an 'invisible power' or 'force of enchantment', and who he recognises – with astonishment – as 'the same being as myself!' (125). This same being or 'second self' (126) adopts the appearance of others at will and debates theology at length with Wringham, reflecting and extending his Calvinist credo towards antinomianism. Wringham's double offers him no name except that 'which may serve your turn' (136), coming up with Gil-Martin perhaps in connection with Wringham's earlier persecution of a boy named M'Gill. Gil-Martin declares that 'I have no parents save one, whom I do not acknowledge' (136) and draws Wringham towards fratricide by using the scripture: 'none of us knows what is pre-ordained, but whatever is pre-ordained we *must* do, and none of these things will be laid to our charge' (134). The damned are dispatched on the basis of the belief that they cannot, in any way, be saved or reformed. Ultimately, the appearance of Wringham's demonic self – much like Dorian Gray's – becomes hateful to him and the consequences of his (dis)association with evil return to destroy him. Try as he might, he cannot loosen his shadow from his side: 'Our beings are amalgamated, as it were, and consociated in one, and never shall I depart from this country until I can carry you in triumph with me' (187). Hogg's novel is part of the gothic tradition which abounds with doppelgangers, psychological and moralistic explorations of the merging distinction between good and evil acted out by the individual split protagonist. The alter-ego personifies not only evil but also death. Narratively, death and destruction are the consequences of projected, disavowed sins which the protagonist is always given an opportunity to own, and always fails to take responsibility for until it is too late. Ontologically, death is the alter-ego's non-being, its status as shadow, other, antithesis (of creation) which is traceable to Milton's Satan. Satan is the source of the Faustian tendency to aspire, to over-reach, which haunts genetic discourse. What Milton makes clear is that God set the stage for the human sin of over-reaching:

> God to remove his wayes from human sense,
> Plac'd Heav'n from Earth so farr, that earthly sight,
> If it presume, might err in things too high,
> And no advantage gain.
>
> (Wright 1980: 294)

Alter-egos are autonomous agents (of death) who, by embodying the projected sin of over-reaching aspiration, double it back on the protagonist who is ultimately self-regulated. The gothic novel offers an insight into the relation of the self to the self, and as a genre which stemmed from Romanticism and the literary exhaustion of nature, it centres on 'unnatural acts performed by improbable characters in unlikely places' (Kiely 1972: 4). The genre is precisely against nature and it, arguably, is the autonomous narrative agent, the myth-bearing double which shadows contemporary genetic stories of unnatural acts and improbable characters.

The first draft of the map of the human genome, in conjunction with other over-reaching developments in cloning and transgenesis (interesting that the name for the transfer of genes between species should also connote the transgression of genesis, the story of creation) has raised the spectre of the doppelganger; of entities beyond nature; of genetic predestination and even of a genetic elect. The story told, in a broad range of science writing, is of the race to map the genome – understood as that which will reveal the essence of human identity – between two main characters: Sir John Sulston of the Sanger Centre in Cambridge UK and head of the publicly funded international Human Genome Project, and Craig Venter of Celera Genomics, Maryland, US – pioneer of the privatised genome. Where Sulston aims to 'tell us what we are' (Porter 1999a: 4), Venter, the story goes, aims to sell us what we are. The two men were profiled in two successive issues of the *Guardian* in April 1999. In 'And How Tall Would you Like your Child Sir?', Henry Porter explores the myriad manipulations made probable by the forthcoming book of life and its authors: 'In the next 25 to 35 years we can expect to see genetic manipulation of human embryos as routine practice. We will be able to eliminate inherited diseases, but also to enhance the new human being with genes for height, energy, spatial ability and intelligence' (1999a: 2). The path to these probabilities is not presented as an easy one. Although approximately twenty organisms had, in 1999 already been sequenced, the largest was a worm called *C. Elegans* with 19,000 genes whereas the estimated number of human genes was approximately 80,000. Although, for Porter, the human organism is fundamentally computable – DNA being completely conflated with identity – this signalled 'goals of almost unimaginable computation' (2). What surprises Porter is that his concept of genetic determinism is not matched by the claims of any of the eight scientists he interviewed, including Sulston and Venter. They all insisted that nature and nurture share responsibility for human development in roughly equal proportions. They all evoked free will. Porter then observes that public opinion, or the genomic imaginary nevertheless veres towards a sense of predestination: 'this won't prevent the reflex opinion that a person is programmed with certain flaws' (3). He also observes that what scientists are saying is not, ideologically, necessarily consonant with what they are doing: 'it is ironic that while scientists generally make efforts to buttress our ideas of free will, their work silently makes the case for the supremacy of genetic control' (3). Porter, then, fails to be reassured, anticipating nightmare scenarios in the news reports of the first quarter of the twenty-first century: 'there will be reports about irreversible genetic pollution. We will hear how DNA snatchers are trying to recreate Elvis and Marilyn, and one morning we will wake up to see a living dinosaur on our television screens'. It will be a crude caricature, manufactured from the genes of several existing creatures, 'but man will be the creator' (3). Such wreckless aspiration will ultimately result, via pride, in a sense of (species) self-revulsion reminiscent, for Porter, of Hamlet's: 'Man delights not me' (3). On the printed page, above Porter's text, is a photographic portrait of Sulston standing in open ground outside the Sanger Centre, baggy jumpered, hands in pockets, sporting a full whitish beard and a benevolent expression. A veritable halo of white cloud enshrines his head, and he is described as 'a cautiously spoken man who wears sandals' (3). The dispute between Sulston and Venter which is manifested as one of money (public versus private finance, public versus private gain) and method, slips incontrovertibly into one of moral good versus evil. God and Satan are conjured up in what is billed as the ultimate story of scientific research, set to influence the future of humankind. The genomic imaginary is productive on many levels. In his article on Craig Venter entitled 'The Joker', Porter (1999b) evokes Sulston's dark shadow. Venter is apparently described by his opposites as 'a Marvel Comics villain – the Joker of Genetics, bent on world domination' (10). Even less invested critics refer to 'the industrial scale of his ambitions' and the shoddiness of his version of the genome. For Porter, he is a likeable demon. While dismantling Venter's demonic image with one hand – 'there are no trails of sulphur in his office' – he slowly builds it up again with the other, referring to Venter's quite unnatural intelligence (see also Ridley 1999; Rifkin 1998) and 'exceptional audacity' (Porter 1999b: 10). Venter's methods of locating and tagging genes, though 'brilliant', never received due recognition by the establishment, and 'after repeatedly being turned down for government funding, Venter fled to the private sector' (10). The outcast turned ambitious entrepreneur – Celera has laid claim, through patenting, to hundreds of genes – is finally demonised by his photograph emphasising a bald head, bushy eyebrows, raised chin and very piercing eyes. These portraits of good and evil, though perhaps a little less heavily toned, appear in other contexts where they consistently overwhelm serious-minded attempts to describe and analyse the different scientific methods of the public and private genome enterprises. Kevin Toolis (2000) generates a similar portrait of Sulston and Venter as bearded, sandaled humility versus the sharp-suited, sharp-witted ego. 'Darth Venter', according to many of his peers, is 'a ruthless competitor who sold his soul to American business' (Toolis 2000: 11). These characters drive the narrative which focuses on the conflict between good and evil, public and private property, capital and social gain and which enlivens the complex detail of genes, genomes and mapping methods. The sanctioned public method entails carefully dividing the genome into segments, analysing the sequence of nucleotides (A's, C's, T's and G's of which genes are comprised) in each segment and then painstakingly putting the jigsaw back together (17). Venter's 'shotgun method' blasts the whole genome into fragments which are reassembled by powerful sequencing machines. Though controversial – the human genome produces 60 million fragments, 2000–10,000 letters long and reassembly is said to leave gaps – this method sequenced the fruit fly genome in March 2000. The completion of the first draft of the human genome was announced jointly by the public and private projects and by both Bill Clinton and

Tony Blair on 26 June 2000. In 'The End of the Beginning' in the *New Scientist*, Coghlan and Boyce (2000) describe a tactical rather than moral or methodological reconciliation. Reciting the familiar mantra of data mass (humanity newly encoded in hundreds of potential telephone directory sized tomes), hiatuses and hermeneutic challenges, Coghlan and Boyce (2000: 4) state that 'staggering statistics and engineering feats failed to detract from the central drama of the genome project, the end of the beginning of which saw President Clinton flanked by Sulston's US equivalent Francis Collins on one shoulder, and Venter on the other'. Clinton's showbiz triumphalism – 'we are learning the language in which God created life' – is itself said to have outdone Blair's sensible optimism in a kind of meta-portrait of duality (Meek 2000: 3). With the public project due for completion in 2005 (having begun in 1990), Venter started the race by announcing that he would be finished by 2001, and where attempts to establish genuine collaboration failed, the race became close enough to produce a draw at an apparently arbitrary point in the sequencing process. On the day hailed as that 'when humankind learned, in a sense, what it is to be human' (Coghlan and Boyce 2000: 4) the estimated total number of genes in the genome was still in the 80,000–100,000 region. Since then, it has been revised down to less than half of that estimate – a development which significantly troubles such confident humanism. Publishing in *Nature* and *Science* respectively, Sulston and Venter's teams agreed, in February 2001, that the genome was not what it seemed. Venter attributes the error to the prevailing determinism of the genomic imaginary: 'if you think that we are hard-wired and that everything is deterministic, there should be a lot of genes, because we have a lot of different traits. So I think a lot of people were expecting that to be the case' (in Radford 2001: 6). Whereas the prospect of fewer genes loosens the (causal) relationship between DNA and individual identity – introducing far greater complexity and far more significance to small diversities – it strengthens the relationship between the individual's DNA and not only the human species but also the totality of life itself. According to Sulston, 'it is the unity of life, of nature being conservative, or the idea of the Blind Watchmaker – the notion of evolution as a constant reworking or random recombining of parts. You convert your Austin 7 into a Mercedes, but basically it is the same underneath' (in Radford 2001: 6). What makes the difference between the Mercedes and the Austin 7 is the variety and subtlety of genes which control other genes, doing sophisticated management jobs (6). Put another way, the human genome is 85 per cent identical to that of a dog. More ontologically reassuring is the argument that people are, genetically, 99.9 per cent identical to each other and that the variation between individuals within racial groups may be greater than that between them: 'Nobody knows the identities of the people who donated their DNA for the publicly funded team, but Dr Venter deliberately selected DNA from five individuals, one Asian-American, one African-American, one Hispanic American and two others and found no way of telling which was which' (6). The continuum of life stretches through both race and species and back to the beginning of evolutionary time. Meanwhile, and despite similar findings, the personalised feud about money, method and morality in genomic research reportedly continues in a less than godly fashion: 'Everybody keeps wanting to turn this into a pissing contest of whose is bigger and whose is better' (Venter in Radford 2001: 7).

The relatively minimalist genome itself offers to shatter the myth of predestination in order to preserve both genetic determinism and the genome as the unifying law of humanity. It does this by facilitating a split between the individual and the species such that the former comes to signify the diversity and malleability of human life and the emergence of the new autonomous 'GM' self, and the latter signifies the unity and immutability of life (itself) and the integrity of the true species-self. These split selves are doubles, doppelgangers, two-in-one; the emerging subject of genetic modification functioning as the (narrative) agent of projected desires and anticipated retribution. Put another way, the price the individual will pay for the vanities of self-modification is integrity and authenticity (subsuming all other risks). Genomics has a familiar moral code. The GM self becomes malleable with reference to both health and lifestyle, disease and destiny and becomes the centre of a supposed golden age of medicine in which treatment is personalised: tailored to suit individual lifestyles and genetic backgrounds. The key to genetic self-modification lies, according to Katherine Brown in SNPs – Single Nucleotide Polymorphisms. These are small variations which occur when a single letter in the genome is altered, and they account for most of the genetic differences between people (K. Brown 2000: 44). SNPs could 'help researchers figure out which bits of a person's genome code for green eyes, a knack for mathematics or an increased risk of diabetes, cancer or heart disease' (44), and in a privatised if not golden age of medicine, all of these bits could become subject to consumer choice. As Mariam Fraser (2001) has pointed out in her examination of the links between the brain and psychopharmacology, there is an ironic sense in which, after decades of criticism centred on biological determinism, biology once more becomes destiny, but this time it is the destiny of individual choice or enablement based on the mutability of (brain) matter (Fraser 2001: 72). Of course, in the genetic market as in any other, consumer power must be analysed in the context of corporate power; choices made looked at in the context of options available. Enablement is ultimately a factor of governmentality (Rose 2001). In the debate about genetic screening and selection, a quite panoptic vision of genetic identity counteracts a more voluntaristic one and genetics is portrayed more as a technology of control than as a technique of the self. Both Ridley and Harris reflect this vision and attempt to deflect it by recourse to a very naïve political voluntarism in which 'we' (as individuals) refrain from censoring state and corporate genomics for the sake of overall public good and/or scientific progress. Ridley argues that 'there is a danger that the hobgoblin of genetic insurance tests and genetic employment tests will scare us away from using genetic tests in the interests of good medicine' (Ridley 1999: 269). His own hobgoblin is state controlled genetic screening, data and policy. On the topic of intellectual property he is unequivocal – 'It is yours, not the government's, and you should remember that' (270) – and he is clearly optimistic about directions in governmentality. In as far as genetic screening re-opens the door to eugenics – 'increasingly, today, genetic screening allows parents to choose the genes of their children' – Ridley is happy to see selection as a facet of individual choice rather than government prescription. Genetic screening then becomes the euphemised term for sanctioned consumer eugenics which 'is about giving private individuals private choices on private criteria' (299). As long as it is not centrally controlled it is acceptable: 'Eugenics was about nationalising that decision to make people breed not for themselves but for the state' (299). Ridley finally evinces a Thatcherite genetic individualism, arguing that the difference between private and national eugenics is overlooked in the rush to define what 'we' should or should not allow in the new genetic world: 'who is "we"? We as individuals, or we as the collective interest of the state or the race?' (299). Harris (1998) is less certain about the effects of genetic screening as a means to consumer eugenics: 'while manipulation of the human gene pool to create a super race may be fanciful, fears about ideal individuals of various sorts may be more realistic' (Harris 1998: 13). But he is generally sanguine about the selection of adult phenotypes such as gender. Recognising that gender selection has a long history and multiple causes, some of which are ethically more problematic than others (sexism co-exists with, but is not equivalent to selection on the basis of sex-linked diseases), Harris concludes that there is, on balance, nothing wrong with it: 'Rather than trying to evaluate various reasons that might be given it seems better simply to recognise the legitimacy of parental choice' (192). Consumer choice, for Harris, is ultimately self-regulatory (a gender imbalance in one generation is likely to be corrected in the next) and it autonomously ensures diversity with respect to gender and 'other' phenotypes (brown eyes, curly hair and so on): 'it is likely that there will be a sufficient diversity of wishes to ensure that humanity continues to flourish in a myriad of ways, while at the same time reducing barriers to flourishing such as disappointment and disability' (194). Gender is subsumed within a list of phenotypic differences and politics is subsumed by an organic, autonomous, self-regulating market in genomics. In addition, disability – and whatever 'disappointment' may be – is not presented here as a legitimate aspect of human diversity but as a barrier to it. Harris's eugenics of choice slips in quietly. It is almost reassuring to note that if, in this free market of reproduction, people begin to look a bit, well, standardised, then 'we could of course review the question of the desirability of controls' (194). Like Ridley, Harris is optimistic about the balance of consumer and corporate power. Acknowledging the already iniquitous practices of insurance companies which discriminate against individuals who even take, let alone 'fail' an HIV test, and looking towards a future which consists of only two insurance groups (good risks and the uninsurable), compulsory screening and certificates of genetic hygiene, Harris remains confident in consumer clout. Quite simply, 'if we fear such a world we must make sure it does not become a reality' (278).

Lewontin (2000) has a rather different grasp of the relationship between genetic knowledge and power, depicting a struggle over biological information between individuals and institutions. For him, 'the genome is becoming an element in the relation between individuals and institutions, generally adding to the power of institutions over individuals' (Lewontin 2000: 166). Knowledge of the genome tends to concentrate existing relations of power (Nelkin 1992) and the choices of individuals take place within institutional structures such as government agencies, hospitals and the family. A personal account of embryo screening illustrates this point. Leah Wild (2000) describes herself as being 'genetically disadvantaged', having the condition referred to as a 'balanced translocation' where two fragments broke off from two of her chromosomes and swapped places, balancing each other out. Although the condition does not affect her health, her child could inherit the chromosomes in unbalanced form, resulting in a major disability and likely miscarriage. There is no cure, and 'despite the Human Genome Project, no prospect of a cure' (Wild 2000: 5). Genetic advances, according to Wild, realistically focus on 'sorting good embryos from bad' (5). Through IVF treatment, tests are made on eight-cell embryos. The testing process is called preimplantation genetic diagnosis (PGD). According to Wild, there have been no more than a handful of live births as a result of PGD for chromosomal translocations: 'This is the reality of creating a designer baby', and in this sense, 'I am hoping to have a designer baby' (5). The emphasis then, is not on curing but on eliminating genetic disadvantages (through 'genomicide') which lead to disability or disease, and Wild makes it clear that her choices are circumscribed not just by her condition – 'I either make this genetic selection, or risk further multiple miscarriages and accept infertility' – but by government authorities. The Human Fertilisation and Embryology Authority has to license each new PGD test, and forbids testing for 'social or psychological characteristics' and 'normal physical variations' (Wild 2000: 5). It permits testing only for serious medical conditions and those associated with disability. In the UK, PGD is not permitted for Downs' syndrome, presenting Wild with a scenario where – at 40 years of age – she may be successfully implanted with a screened embryo only to discover, 12 weeks into the pregnancy, that it has Downs' syndrome. The issue then, 'in this brave new world of human genome sequencing' is 'what should we screen for, when and why?' (5).

Many of the issues raised by genetic screening and selection are explored in Andrew Niccol's film *Gattaca* (1997), set in a future when screening precedes *in vitro* fertilisation and where selection, though voluntary, has a clear gold standard. Vincent is born to liberal-minded parents who resisted screening but subsequently changed their minds. He is genetically imperfect ('in-valid') with poor eyesight and a heart defect. He is uninsurable since 'at seconds old, the exact time and cause of my death was already known'. Despite legal controls on 'genoism', which outlaw discrimination, Vincent's employment prospects are bleak and he can work only as a cleaner at the space centre *Gattaca*. Employers evade the law, if necessary, by taking DNA samples from the saliva on sealed application forms, and however knowledgeable Vincent is, 'my real resume was in my cells'. Vincent's brother Anton was a designer baby, selected first for gender and then for other specified characteristics such as premature baldness and a propensity for violence. Anton qualifies as a member of the 'superior' class where class is no longer related to social status or to race. Genetics is now the only significant marker of difference and the only real source of discrimination, but in this near-future there is both a means and a market for resisting genetic predestination. Superior individuals who happen to fall from biological grace (through non-genetic influences or accidents) enter the black market and sell their genetic identities to the less fortunate. So, Jerome Morrow, genetically elect swimming star with a broken back resulting from a car accident, sells his biological property (blood, urine, hair) along with his name and identity to Vincent, who then passes his interview as navigator, first class, with a urine sample. At this point he becomes a 'de-generate', someone who crosses from one genetic class to another. Suited, scrubbed (the new Jerome has to remove all traces of his own identity through scrupulous daily grooming) and generally slicked-back, he catches the eye of a female colleague – Irene – who has him sequenced by submitting one of his acquired hairs for an instant DNA test.[4] Being herself one notch short of genetic perfection, she feels obliged to reciprocate – 'if you are still interested let me know' – and so romance blossoms. The arrangement between the original Jerome and his copy succeeds despite the intervention of Vincent's estranged brother turned degenerate detective. Vincent, as Jerome, makes it onto the space mission of his dreams and is last seen ascending to the heavens, while Jerome, whose 'accident' was actually an attempt to relieve himself of the 'burden of perfection', is last seen stepping into an incinerator. The original, if not the true self, pays the price for genetic self-modification here.

In *Gattaca*, genetic predetermination is a fact countered by strategies of resistance. Like many other science-fiction films it both absorbs and contests specific paradigms, contributing, in this case, to the genomic imaginary and its more dystopian aspect. Central to the genomic imaginary is what Rabinow (after Foucault) terms humanism, or what Haraway prefers to call fetishism. Fetishism occurs where the gene substitutes for, and is reified as life itself and Haraway argues that there are both economic and psychoanalytic forms of gene fetishism. In economic or commodity fetishism, the gene is perceived as a 'thing-in-itself' where 'things are mistakenly perceived as the generators of value, while people appear as and even become ungenerative things, mere appendages of machines, simply vehicles for replicators' (Haraway 1997: 135). Where commodity gene fetishism is as facet of the multinational biotechnology industries, psychoanalytic gene fetishism exists in the unconscious mind of what Haraway insists is not the universal subject, but the culturally specific scientific subject. Here, the gene serves as a phallus substitute, or the object through which its loss is disavowed. Disavowal involves an uncomfortable and 'odd balancing act of belief and knowledge' where 'belief in the self-sufficiency of genes as "master molecules", or as the material basis of life itself, or as the code of codes' persists despite 'the knowledge that genes are never alone, are always part of an interactional system' (145). Indeed, Haraway insists that all supposedly autonomous entities are defences and that gene fetishism defends the scientific subject 'from the too scary sight' of biological reality (146).

An extension of gene fetishism is genome fetishism and the reification of maps of the human subject. The genome is the sum of all the genetic information in an organism, and the Human Genome Project is a prime example of gene and genome fetishism. The Human Genome Project (HGP) is an example of a form of reification 'that transmutes material, contingent, human and non-human liveliness into maps of life itself' and then conflates the representations with the object (135). So the map is as absolute and unquestionable as nature or life itself, and is presented in the form of disembodied knowledge – Haraway's 'god-trick' or the view of everything from nowhere. What appears to be laid bare in the map of the human genome is the universal human subject; the genetic archetype or the model of our sameness. The human in the HGP 'has a particular kind of totality, or species being' (247). By facilitating the insertion, removal, transfer, synthesis and manipulation of genes, the HGP has significant long term and hotly debated implications for the nature of human nature. Haraway refers to the ethical debates surrounding the HGP, including complex questions of agency ('people are in the information loop, but their status is a bit iffy in the artificial life world' (247)) and representation or 'who, exactly, in the human genome project represents whom?' (247). Such questions are amplified in the oppositional Human Genome Diversity Project (HGDP) which aims to recognise genetic diversity by incorporating 700 groups of indigenous peoples on six continents, but which, according to Haraway, works with some distorted, impositional ideas of racial difference and has stumbled over 'the agency of the people who did not consider themselves a biodiversity resource' (250). The stakes, she argues 'are about what will count as human unity and diversity' – they are about kinship on the basis of sameness or difference.

If human diversity and difference is threatened by the HGP which may well be used to help eliminate defective and undesirable genes and move the species increasingly towards genetic sameness, then it is clear that diversity is eliminated in perhaps the ultimate example of gene fetishism – cloning. Cloning raises the prospect of producing carbon-copy humans or multiple identical selves. In an article published in 1974 in the journal *Nature*, molecular biologist Gunther Stent described cloning as 'a fantastic facet of genetic engineering which, though it seems taken straight from the pages of Aldous Huxley's *Brave New World*, is actually likely to become a practical reality before long' (in Kolata 1997: 77). Cloning could potentially transform the human race by enabling us to 'abandon the old-fashioned genetic roulette of sexual reproduction' and replace a diverse species with 'identical replicas of carefully chosen, ideal human genotypes' (78). Stent pointed to the ethical paradox in the quest for human perfection which drives much of the research in biotechnology (especially genetic engineering) and which is at the heart of utopian social and political theory. We may relish the idea of another Einstein or Monroe, but 'the idea of having hundreds or thousands of their replicas in town is a nightmare' (78). According to Stent, the basis of this contradiction 'is the generally shared belief in the uniqueness of the soul. Even though the soul is incorporeal, it is supposed to fit the body, hence it is not conceivable that a unique soul should inhabit each of thousands of identical bodies' (78).

Self-replication through cloning threatens to destroy the notion of the self in as far as it is tied to the concept of a unique, authentic soul or spirit. The ethical and public debates on cloning counter a reductionist physics of self with a reactionary metaphysics. There is as much talk of spirituality and of the battle between good and evil, as there is of materiality and politics. Cloning dominated the ethics agenda from the inception of the ethics movement in the 1970s, and public concern was also addressed in film and in fiction such as Ira Levin's *The Boys from Brazil* (1976). It is a distinct theme in 1990s films such as *Blade Runner*, *Jurassic Park*, *Gattaca* and *Alien Resurrection*, and the less mainstream science fantasy *The City of Lost Children*. In this, the directors Jeunet and Caro invent characters such as the Octopus (evil Siamese twins), the Cyclops (baddies with ear and eye enhancing technologies), Irvin (a disembodied brain in a fish tank), Krank (who cannot dream and therefore has no soul), the Inventor (the clone's original who has lost his memory and therefore his sense of responsibility), One (a freak show strong man and the film's hero) and Miette (an orphan). The clones are unnamed and assist Krank in his quest to become more human by stealing the dreams of young children. Krank, Irvin and the clones are the monstrous creations of the Inventor, and where Krank is overcome by the spirited Miette who refuses to relinquish her dream/soul, the other incomplete humans are conquered by One and the return of the Inventor's memory. The good, whole (mind plus body plus soul) beings prevail and the clones recognise their status and need no longer argue about which of them is the 'original' and which are the 'cheap copies'.

The City of Lost Children plays with the idea of the unique individual who is set against the freakish and monstrous creations of science and technology. The clones invoke familiar social and political fears of a soulless, mindless, unthinking mass as well as the mythical horrors of technoscientific hubris. Gina Kolata refers not only to the ancient stories of Prometheus ('who stole fire from the Gods') and the Fall, but to their retelling by Mary Shelley and H.G. Wells. The story of *Frankenstein* is often invoked in (particularly feminist) critiques of medical and reproductive science (Braidotti 1994; Franklin 1993; Haraway 1997; Kember 1998) and Wells's cautionary tale of 'a scientist who glories in creating monsters' (*The Island of Dr Moreau*) has twice been made into a film (1977, 1996).

The self as other

The myth of Narcissus (retold in Oscar Wilde's *The Picture of Dorian Gray* [1890]) is invoked in the vanity of cloning which inevitably produces the spectre of the doppelgänger, 'the mysterious double, a person who may look exactly like you but who is a stranger, doing deeds you would never have contemplated' (Kolata 1997: 72). The myths of Prometheus, Narcissus and the Fall, retold in literary and media texts, in popular science and in science journalism seem to me to both inform and transcend their incorporation into discourses of risk. Put another way, the discourses of risk take place in purgatory. At this point it is expedient to return to Rabinow's assertions about the spiritual concerns of secular moderns and the distance between these subjects and the traditional faith in reason. Again, there is, Rabinow argues, 'an urgent and uneasy sense of hesitancy and caution over the consequences of a felt imperative to know and to put that knowledge into action' (1999: 18). This sense of hesitancy and caution, while responsive to perceived risks and ethical decision-making in genomics, is ultimately mythical and specifically Faustian. In their introduction to *The Risk Society and Beyond* (2000), Barbara Adam and Joost Van Loon acknowledge that risk discourse is now focused on genetics and is as much a product of imagination as it is of calculation (Adam and Van Loon 2000: 12). Risk calculation and predictability aimed at establishing binaries and certainties like safety and danger (good and evil) give way to complexity when the 'foreign bodies' of Big Science enter the network society and the 'scientists and engineers have inescapably lost control over the effects of their creations' (6). Risk discourse in genomics is mythical precisely because it cannot rely on old seeming certainties. Moreover, the myths in question became myths precisely because the binaries around which they were constructed were never stable and needed to be constantly reworked and redefined (Lévi-Strauss). They were always already complex. In purgatory then, myths are instrumental; they are helping humanity work out where 'it' is going – they are helping people and populations work out an ethics. The most cited award, so far, appears to go to Goethe's *Faust* or Christopher Marlowe's *Dr Faustus*. Jeremy Rifkin's analysis of the *Biotech Century* (1998) concentrates on public debates about genetic risks and ethics, and 'everywhere' he hears 'talk about playing God and manipulating nature' (preface). For him, the biotech century 'comes to us in the form of a grand Faustian bargain'. The lure of scientific progress and 'a bright future full of hope' is balanced by 'the nagging question "At what price?"' (Rifkin 1998: xiv). With reference to cloning, specifically, Matt Ridley adds that 'we have drummed into our skulls with every science fiction film the Faustian sermon that to tamper with nature is to invite diabolic revenge' (Ridley 1999: 256). He also comments that people tend to be more cautious as voters (on cloning) than as consumers (of cloning). Temptation then, is the name of the game.

In Marlowe's *The Tragical History of Doctor Faustus* (1616), the role of temptation is played by Mephistopheles who becomes, at Faustus' bidding, an alter-ego in the mode of Gil-Martin:

> I do confess it, Faustus, and rejoice.
> 'Twas I that, when thou were't i' the way to heaven,
> Dammed up thy passage; when thou took'st the book
> To view the scriptures, then I turned the leaves
> And led thine eye
>
> (Steane 1982: 334)

With the good angel and the evil angel constantly whispering in his ears, Faustus waivers considerably and his moral status is never less than ambiguous: 'I do repent, and yet I do despair. Hell strives with grace for conquest in my breast' (Steane 1982: 329). At the beginning of the play, the chorus recites Faustus' biography from infancy 'born, of parents base of stock' to his qualification as a doctor 'excelling all' when, 'swol'n with cunning of a self-conceit, His waxen wings did mount above his reach, And melting, heavens conspired his overthrow' (265). Faustus then repeats, in Milton's terms, the fall of man. For Faustus, life as a doctor proved a little too dull and limiting. He is unable to respect his profession because he can neither confer immortality or resurrect the dead: 'Couldst thou make men to live eternally, Or being dead, raise them to life again, Then this profession were to be esteemed' (266). It is magic, not medicine which offers to make him 'a demi-god' (268). Faustus mortgages his own soul and is the author of his own contract with Mephistopheles with whom he shares and enjoys many misdemeanours. Mephistopheles does not allow Faustus to abdicate his responsibility – ''Twas thine own seeking, Faustus, thank thyself' (285) but the chorus' final summation of the case may not be as morally unambiguous as it seems:

> Faustus is gone. Regard his hellish fall,
> Whose fiendful fortune may exhort the wise
> Only to wonder at unlawful things,
> Whose deepness doth entice such forward wits,
> To practice more than heavenly power permits.
>
> (Steane 1982: 339)

In his introduction to the play, Steane highlights a sense of the fascination and attraction of over-reaching – the wonder at unlawful things – which shadows exhortations of fear and repulsion, and points to the stern, 'greyly negative' and limiting force of heavenly power (23).

Dorian Gray is similarly the author of his own fate, even if he himself is portrayed as the creation of his hedonistic friend (Wilde 1980: 117) and his portrait, literally, 'has a life of its own' (166) and 'a corruption of its own' (168). The portrait becomes evil, death and decay in order to preserve Dorian's life as beauty, art, style and form which, for Wilde are 'the nearest things to eternal verities there are' (Murray 1980: ix). Dorian perceives neither morality or humanity in terms of simple dualities: 'To him, man was a being with myriad lives and myriad sensations, a complex multiform creature that bore within himself strange legacies of thought and passion, and whose very flesh was tainted with the monstrous maladies of the dead' (187). His second murder is of his conscience – 'Yes, it had been conscience. He would destroy it' (253) – which results in an identity both good and evil, art and life, immortality and death.

When they entered they found, hanging upon the wall, a splendid portrait of their master as they had last seen him, in all the wonder of his exquisite youth and beauty. Lying on the floor was a dead man, in evening dress, with a knife in his heart. He was withered, wrinkled, and loathsome of visage. It was not till they examined the rings that they recognised who it was.

(Wilde 1980: 254)

The myths of Prometheus and Narcissus, told and retold through these narratives, are morally unclear. Good and evil are brought into play through the characterisation of the double or divided self, but the closure, the punishment and destruction of humanity or the (species) self for its over-reaching acts of vanity leaves room for uncertainty. In his article on cloning, Munawar A. Anees (1998) argues that it is being debated primarily within this mythical register. The 'new world of identity, rights, responsibilities' opened up by cloning as 'neo-Genesis', new birth, is, on this level at least, ancient. Cloning through the transfer of genetic material from the nucleus of the adult donor cell to the enucleated egg cell of the host, was the technique employed by Ian Wilmut and his colleagues at the Roslin Institute in Edinburgh to produce Dolly. As the first successfully cloned mammal, Dolly raises the spectre of human cloning which was laid to rest during the 1970s. For Anees (1998: 2), the 'incessant realism of this blossoming technology constantly conjurs up images of fear', the biggest of which is the prospect of '*homo xeroxiens*', a new (evolutionary) competitor for *homo sapiens* but 'in his own image, of his own doing'. The prospect of human cloning, and of techniques which may not be prohibitively expensive leads, on the one hand to plans for commercial exploitation by individual entrepreneurs such as Richard Seed, Rael and Severino Antinori and on the other hand to censorship attempts in the international community of the great and the good. While Richard Seed, referred to as an early but discredited prophet of human cloning, plays a minor role in tales of cloning (Ridley 1999; Rifkin, 1998; Lewontin 2000), Rael plays the role of religious eccentric with unspecified financial backing. Rael 'and a group of investors' founded Clonaid®, 'the first human cloning company' in 1997 with the clear aim of exploiting the techniques used in Edinburgh. He is also the head of the Raelian movement, an international religious organisation with some fairly cosmic views on life, namely that life on earth was genetically engineered by an extraterrestrial race called Elohim who also resurrected Jesus through cloning (http://www.clonaid.com). Cloning, according to Rael, 'will enable mankind to reach eternal life' provided that individual members are prepared to part with £200,000. Based in the Bahamas, the company aims to build a laboratory in a country where human cloning is not illegal and to offer its services to wealthy infertile and homosexual couples. Related services include Insuraclone® which banks cell samples for future cloning requirements at £50,000 and Clonapet.[5] On the comments page of Clonaid's website, a female advocate of cloning asks why people should fear a discovery which will eliminate the fear of death. Better, she argues, to fear death itself and a life too short to enable us 'to accomplish ourselves'. Cloning, for her, means eternal youth or an endless supply of new bodies. Moving slightly closer to scientific respectability is Severino Antinori, head of the International Associated Research Centre for Human Reproduction. Antinori is known for assisting the conception of children in post-menopausal women, and for guaranteeing the birth of millennium babies. A Catholic based in the centre of Rome, his work was described by the Pope as 'evil' (Meek 2001b: 3). Nevertheless, the success of his fertility clinics adds authority to his project of treating male infertility through cloning. The technique would involve removing the nucleus from the cell of an infertile man and injecting it into an egg with its chromosomes removed. Following Wilmut's procedure,

> a tiny pulse of electricity will be applied to fuse egg and transplanted nucleus into a single cell and, by a process which is still not understood, reprogram the transplanted genes to operate as if they were 'young' again, enabling the cell to divide and differentiate into all the cells of a human body.
>
> (Meek 2001b: 3)

The cell is transformed into an embryo, and although Antinori plans to produce many embryos he does not plan to copy the procedures used in animal cloning by transplanting the embryos into multiple surrogate mothers in order to maximise the chance of success. Careful screening and selection of the embryos, based on available techniques, should prevent this from being necessary, creating a success rate equivalent to conventional IVF treatment. The result of adopting this solution to 'a human problem' would be 'ordinary children who grow up to be unique individuals' (Antinori in Meek 2001b). Ruling out science-fictional scenarios of resurrection and seeking to protect the ordinary right of men to genetic immortality, Antinori attempts to counter the problem of identity which lies at the heart of the censorship of cloning by the UN, the EU, the Council of Europe, the World Health Organisation and 'a catalogue of other bodies' (3) including the UK's Human Fertilisation and Embryology Authority (HFEA).

The HFEA was set up in 1991 to monitor standards in UK fertility clinics. It has licensing and policy responsibilities, is responsive to national debate and consists of 21 members – medical and non-medical 'experts' appointed by UK health ministers. Its policy on cloning preceded a consultation document, produced in January 1998 in conjunction with the Human Genetics Advisory Commission (HGAC). This attracted approximately 200 responses from groups and individuals and 'widespread support for the views initially expressed by the HFEA and the HGAC that human "reproductive cloning" (i.e. the *deliberate* creation of a cloned human being) should not take place' (http://www.hfea. gov.uk). The HFEA had already stated that it would not issue licences for reproductive cloning. Following this 'consultation', a report entitled 'Cloning Issues in Reproduction, Science and Medicine' was published in December 1998 (HFEA and HGAC 1998). The report validates existing safeguards, advises government to consider legislation banning reproductive cloning and makes a distinction between this and *in vitro* research 'using cell nucleus replacement technology with a therapeutic aim' (http://www.dti.gov.uk/hgac/papers). Here the distinction between reproductive cloning aimed at producing a whole being and therapeutic cloning aimed at producing human biological materials is drawn. The joint HFEA and HGAC report recognises the social as well as the scientific impact of Dolly as the first vertebrate cloned from the cell of an adult animal. This raised concerns both nationally and internationally about the development of the technology 'particularly in the context of the cloning of human beings'. In its construction of two types of cloning, it makes repeated reference to the production of 'genetically identical human beings', and frequent reference to individuality and 'human dignity'. The motivations for producing Dolly were primarily commercial even if the impact of her production is primarily ethical. The Roslin Institute and partners PPL Theraputics PLC aimed to improve the production of livestock including transgenic animals functioning as biochemical factories for the pharmaceutical industry. Although two sheep were cloned in 1996, Dolly was the first to be produced from an adult sheep rather than from an embryo. This fact must then have proved significant in creating fears about human identity. In elaborating the distinction between two types of cloning, the report states that reproductive cloning, the unsanctioned variety, produces a whole being from a single, possibly adult cell, by asexual reproduction. It therefore raises issues concerning gender and sexuality, making gender-free procreation possible alongside that of homosexual couples. Therapeutic cloning, on the other hand, does not carry this kind of political baggage, not because the basic technique of nuclear transfer is necessarily different, but because adult beings are never produced by the implantation of the embryo in a uterus. Therapeutic cloning produces human materials for research (and commodification) leading to ethically unproblematic treatments such as the replacement of damaged or diseased tissues or organs without risk of rejection. So, for example, skin tissue might be cloned to treat patients suffering from serious burns. The report tackles the issue of twins as 'a natural form of cloning' which occurs through sexual reproduction, and therefore raises questions concerning the efficacy of artificial as opposed to natural intervention. Embryo splitting is an older technique than nuclear transfer and involves the artificial division of a single embryo in a process which replicates the natural process which can give rise to twins. What the phenomena of natural occurring twins shows is that 'genetically identical individuals are far from being identical people: they may differ from one another physically, psychologically, in personality and in life experience' (HFEA and HGAC 1998: 10). The link between DNA and identity is thus loosened. Interestingly, the report suggests that twins can be cultural but not biological clones; that is they are made but not born identical – the challenge to uniqueness in human identity is then profoundly unnatural. Difficulties experienced by twins trying to establish their own identity 'usually arise when the children have been treated as an indistinguishable and inseparable pair' (10). The experience of natural identical twins suggests, then, 'that a unique genetic identity is not essential for a human being to feel, and be, individual'. Therefore, the report asks, 'what is meant by the assertion that individuals have the right to their own genetic identity?' As part of a consultation process, the question invites a response and is not entirely rhetorical. With the humanist fallacy of genetic determinism exposed, there are no easy answers but it is interesting that the report then moves directly to a number of imaginary scenarios which suggest that the refrain of unique genetic identity invokes a sacred essence (or soul) to be protected at all costs from the Faustian desire for powers of resurrection and immortality.[6] Three scenarios are outlined as sources of concern, and they all involve making copies of an individual: parents might want to 'replace' an aborted foetus, dead baby or child killed in an accident; parents might want to clone an organ donor for a sick child and 'an individual might seek to use cloning technology in an attempt, as that individual might see it, to cheat death' (10). The report concludes that whether or not reproductive cloning is desirable, 'there is considerable doubt about whether it would even be possible' using the techniques used to produce Dolly (11). Risks of wastage and malformation underline, and in a sense secure, ethical uncertainties since Dolly was the only normal lamb born from 276 attempts. Only 29 embryos were implanted and all but the one resulted 'in defective pregnancies or grossly malformed births' (11). Concerns raised here about ageing – Dolly's DNA may be the same age as the original sheep and therefore she may have a shortened life-span – have not subsequently been confirmed (although she is said to suffer from arthritis). The question that remains, then, is whether safety concerns in general make reproductive cloning ethically unacceptable (11).

In his review of the 1997 US National Bioethics Advisory Commission report on human cloning, Richard Lewontin (2000: 277) points to the pervasive error of conflating DNA with a person which persists despite the commission's awareness of the error of genetic determinism (280). The report addressed four main ethical issues: 'individuality and autonomy, family integrity, treating children as objects, and safety' (278) and an intersection of religious and secular concerns. Lewontin criticises the report as an 'incoherent' attempt to rationalise deep-seated religious myths or 'a deep cultural prejudice' centred on 'the uncontrollable power of creation' (277). This attempt at rationalisation is manifested in an overemphasis on safety issues at the expense of far more complex and ambiguous ethical and religious ones: 'The reliance of the commission on purely technical matters of safety for their recommendation seems a neat way of finessing the political problems raised by the ethical, but especially the religious, issues'. In short, 'if it is unsafe, we really don't have to struggle over all the rest' (2000: 299). The joint UK HFEA and HGAC (1998) report leaves open the possible elision of ethics by safety concerns and leaves hanging the religious connotations of genetic identity. The government response simply reasserts a position taken two years earlier in 1997 that 'the deliberate cloning of human beings' is 'ethically unacceptable' (http://www.doh.gov.uk/cloning),[7] agrees to look further into the risks and benefits of therapeutic cloning and 'accepts the report's conclusion that the protection of genetic identity, so far as it relates to the issues raised in the report, does not appear to raise any new ethical issues at this time' (3). The contradiction highlighted in the report's views on genetic identity is simply glossed over. Finally, the UK government signals its allegiance to the Council of Europe's 'Additional Protocol to the Convention for the Protection of Human Rights and Dignity of the Human Being with Regard to the Application of Biology and Medicine, on the Prohibition of Cloning Human Beings'; and to the UNESCO Universal Declaration on the Human Genome and Human Rights. The Council of Europe document states 'that the instrumentalisation of human beings through the deliberate creation of genetically identical human beings is contrary to human dignity and thus constitutes a misuse of biology and medicine' (http://www1.umn.edu). Genetic identity refers to the 'nuclear gene set' of an individual. UNESCO's declaration is based on three main principles: that the genome is part of the 'heritage of humanity'; that the dignity and rights of the individual must be respected regardless of his or her genetic characteristics; and a 'rejection of genetic determinism' (http://www.unesco.org). Article one of the declaration states that the genome 'underlines the fundamental unity of all members of the human family' as well as the 'dignity and diversity' of individuals, and article two states that it is a matter of dignity that individuals should not be reduced to their genetic characteristics. The incoherence of these declarations stems from their simultaneous rejection and reinstatement of genetic determinism not as a biological but as a philosophical universal where determinism is conflated with humanism and the gene with the soul. Once the uniqueness and dignity of the individual and the species are universalised, made sacrosanct and separated, to an extent, from the operations of the gene and genome, then the soul and the spirit of mankind (gendered) is invoked and the power of genomics is necessarily mythical. So banning human cloning quite simply reflects and protects 'our' humanity – 'it is the right thing to do' (Anees 1998). Humanity is protected not just from scientific but from ethical and moral uncertainty which risks not just our minds and bodies, ecologies and economies but our very essence. Unanswerable questions of benefits and costs, safety and danger give way to those of good and evil, salvation and damnation as people bear witness to the manipulation and replication of human identity.

The Faustian fears and desires of individual citizens are perhaps more freely expressed than those of appointed groups who represent the national and international community and have the future of humanity on their shoulders. In just a brief sample taken from a web-based bioethics forum ('Cloning Special Report', *New Scientist*, 12 March 2001), a variety of viewpoints on cloning can be heard; many of which have the myth of unlawful aspiration as their touchstone. Andrew's premise is that 'all humans have a compelling urge to pursue knowledge and push the boundaries or even break them' and there are many references to God's law and the natural order – whether or not these are to be upheld. John thinks 'we may be able to clone the human body but the spirit, which comes from God, will never be cloned', while there are votes both for and against the instrumental use of human clones in medicine and pharmacology. An anonymous contributor who invokes God's law sees destruction or damnation around the corner – 'Go ahead and try and see what happens. This world is getting way ahead of itself' – while Debbie foresees a mass identity crisis: 'cloning humans would also cause mass hysteria with doubles and triples of everyone walking around'. The prospect of resurrecting extinct or endangered animal species is advocated by two (out of thirteen) contributors, two others see no difference between cloning and 'natural' forms of reproduction and three subsume ethical concerns within the promise of medical progress. The bioethics forum illustrates the attractions and repulsions at the heart of the international (not global) ban on human cloning. As well as shedding some light on 'the unique problem of identity' (Lewontin 2000: 280) raised by cloning in a secular–sacred debate which conflates the gene and the soul, the forum touches on some key issues regarding life and death; not whether but where the promise of human immortality resides – in salvation of the soul or in progress of the gene. Both Harris (1998) and Rifkin (1998) relate cloning to the search for immortality: 'With human cloning, one's genetic information can be replicated endlessly into the future, creating a kind of pseudo-immortality' (Rifkin 1998: 218). In the genomic imaginary this is the immortality not just of genes but of identities. In the genomic imaginary, the clone is the double (or triple) self as other, who, like Satan, Gil-Martin, Mephistopheles and Dorian Gray's portrait is antithesis and embodies evil and death as the split, disavowed and abject part of the (good, immortal) subject. The clone is ultimately the monstrous Hyde to yet another aspiring doctor who seeks to manufacture himself anew. Stevenson's novel (*Dr Jekyll and Mr Hyde*), like Shelley's (*Frankenstein*) is the result of a nightmare. Like Hogg's (*Confessions of a Justified Sinner*) it has Calvinist connotations of good and evil which Stevenson troubles through realising 'man's dual nature' (Calder 1979: 11). Hyde is representable only through his monstrosity (34) and the somatic symptoms of trauma experienced by those who see him (77). Through Hyde, Dr Jekyll realises that 'man is not truly one, but truly two' or rather many: 'I say two, because the state of my own knowledge does not pass beyond that point. Others will follow, others will outstrip me on the same lines; and I hazard the guess that man will ultimately be known for a mere polity of multifarious, incongruous and independent denizens' (81). Still, the difficulty remains of how to contain these dualities or multiplicities within the self: 'If each, I told myself, could but be housed in separate identities, life would be relieved of all that was unbearable' (82). Melanie Klein's psychoanalytic concept of splitting recognises the intolerability of containing conflicting desires, emotions or 'identities',[8] and Stevenson's story of moral dualisms raises the question of exactly what, in the (gothic) myth of cloning is being rehoused in order to relieve life of all that is unbearable. In the myth of cloning, death is the ultimate evil which shadows and haunts life and which is expunged through the creation of another self. The clone removes death, takes it away – albeit temporarily. As a wetware form of artificial life then, the clone (rather than cloning) represents death-in-life (undeath) or epitomises the idea that alien life-forms are humanity's doppelgängers (and vice versa). The gothic and science-fictional imagination coincide within genomics where cloned organisms (as doppelgängers) co-exist and literally coincide with transgenic organisms (as aliens). In *Alien Resurrection*, Ellen Ripley is brought back from the fiery furnace into which she had plunged with a creature bursting from her chest (Kember 1996). She is cloned from the DNA in a finger nail, but she is cloned as a transgenic organism – a human with alien genes. The creature she was carrying is re-gestated and removed for farming and experimentation, and Ripley herself is reborn as an adult. As a hybrid she has unexpected powers and also as a hybrid her cloned sibling is part-human and gives birth to Ripley's offspring. The foolish instrumentalism of the biotech industry is represented by United Systems Military, which is determined to farm aliens at the expense of all safety, moral and ethical considerations. The paternalistic as well as proprietorial gendering of biotechnology (Franklin 2000) is symbolised in the fact that the ship's computer is no longer 'mother' but 'father'. The complexities of cloned and transgenic kinship are brought home when Ripley ends the suffering of failed transgenic experiments, becomes familiar with a robot of the same sex and has to deal with the conflicting interests of her human and alien kin. Through Ripley's relations and relationships, the narrative emphasises the concept that humanity is not the sole preserve of humans: 'I should have known. No human being is that humane', she says as she places her hand inside the robot's wound.

The other species

The technology of cloning is increasingly inseparable from that of transgenesis (and xenotransplantation) where the miscegenation of human and other species threatens 'the "sanctity of life"' and 'categories authorised by nature' (Haraway 1997: 60) with a whole new level of difference which itself promises to open up and close down on familiar questions of identity. In their introduction to *Global Nature, Global Culture*, Franklin, Lury and Stacey (2000) place feminism at the centre of debates on the redifferentiation of both 'kinds' and 'types' in the global imaginary. They explore concepts of globalisation as an evolutionary development of capitalism, an extension of modernity not synonymous with postmodernity (but sharing 'detraditionalising characteristics': Franklin *et al.* 2000: 2), and as an aspiring project of cultural and economic homogenisation (5). The authors investigate 'the constitutive power of the global – as a fantasy, as a set of practices, and as a context', focusing on the multiple ways in which the global produces 'de- and renaturalised identities, subjects, properties and worlds' (5). They enact a concept of feminism as 'an analytical tradition concerned with the production of difference, not only of male and female, but of *kind and type*' (6), extending Haraway's grammatical concept of gender and her analysis of the shift from kind to brand. Haraway argues 'that while classifications of kind and type have been denaturalised through proprietary marking, as is the case with the patenting of transgenic organisms, the conventional brand or trademark has, in the same process, been naturalised through an attachment to the reproduction of new life forms' (Franklin *et al.* 2000: 9). Franklin *et al.* (2000) argue that the increasing isomorphism of nature and culture is not adequately described through concepts such as implosion, inversion, assistance or elision but that it creates epistemological (and ontological) instability and an ongoing process of re-differentiation which is largely manifest in a tendency to re-inscribe nature and re-establish its authority (10). Nature is 'pivotal to the new universalisms of global culture' within which both nature and culture appear to be 'self-generating' (10) and the global 'provides the model for a context which is isomorphic with itself'. In the terminology of artificial life, renaturalisation is a facet of the universalisation of two main properties – autonomy and self-organisation. Whereas Franklin *et al.*'s analysis 'seeks to trouble this process of auto-reproduction' by describing it as being partial, contested and uneven it does not seek to offer a model of change or of intervention in the ongoing processes of redifferentiation as Haraway does through figuration. Figurations are visual or verbal images which embody transformations in knowledge, power and subjectivity (Kember 1998). Haraway's original figuration, the cyborg, is joined by a new trinity of Onco-Mouse, FemaleMan and Modest Witness in order to produce 'diffractions' or 'interference patterns' (Haraway 1997: 14) which 'can make a difference in how meanings are made and lived' (16). Haraway rejects the contemporary critical practice of reflexivity in favour of diffraction because reflexivity is predominantly an act of displacing (in order to replace) the self (same) whereas what is necessary 'is to make a difference in material and semiotic apparatuses, to diffract the rays of technoscience so that we get more promising interference patterns on the recording films of our lives and bodies' (16). What Haraway seeks is a sense of self which is not (just) reflexive, but which (literally) makes a difference – one which diffracts to produce or allow the production of something/someone else. She selects a prominent transgenic organism, and the first ever patented animal, OncoMouse™, and attempts to refigure it within the associated realms of science studies, feminism and biotechnology. OncoMouse is a commodity created and owned by a biotech company called Du Pont, and used in cancer research. The mouse carries a transplanted human tumour-producing gene (oncogene) which produces breast cancer and is, Haraway (1997: 79) argues, 'truly our kin', suffering on behalf of 'her sisters' so that 'we' can live and can 'inhabit the multibillion dollar narrative of the search for the "cure for cancer"' (79). Haraway refigures the instrumental production of transgenic organisms by claiming kinship with them. 'Who are my kin', she asks, 'in this odd world of promising monsters . . .?' (1997: 52). Her refiguration is simultaneously an attempt at redifferentiation which celebrates difference as that which is against nature. It is precisely an attempt to discover kinship in a world of 'promising' monsters. Kinship, in this context, works not towards the assimilation and exploitation but towards the recognition of other species. It breaks the taboo of species miscegenation and opens out/denaturalises both humanity and humanism. The unity and diversity of identity is renegotiated through the terms of politics, ethics, epistemology and ontology having been appropriated from the terms of corporate and consumer genomics.

The story of corporate and consumer genomics is focused on pharming patented and cloned transgenic organisms whose kinship with humans is at once avowed and disavowed through a concern with suffering as well as safety. Lewontin (2000) outlines some of the key developments since Dolly, highlighting an announcement in 1998 that a successful embryonic culture had been developed from cow cells containing human DNA. Visions of the horned Minos were conjured up but 'nothing can be said about its humanity' since the culture did not progress beyond the early embryonic stage. What can be said is that its potential economic value is vast 'because it makes possible the production of large quantities of all kinds of human proteins that can be used in disease therapy' (2000: 302). The majority of UK research on cloning has focused on replicating transgenic animals to mass produce human proteins and organs for human transplantation. Xenotransplantation involves the transfer of cells, tissues and organs from one species to another and 'the current situation society now faces focuses mainly on organs from pigs meant for humans' (Greek and Greek 2000: 8). Mass produced pigs are likely to be cloned and in March 2000, PPL Therapeutics cloned Millie, Christa, Alexis, Carrel and Dotcom. Both *Science* and *Nature* reported this development and its associated risks, the editor of *Nature* making 'a compelling case for a moratorium on clinical trials' while the risks are evaluated. These may revolve around the transfer and mutation of viruses. According to *Nature*, the momentum towards clinical trials of xenotransplantation is also 'unstoppable' due to multimillion dollar investments by biotech companies (Greek and Greek 2000: 8). The pigs were cloned using the same techniques of nuclear transfer used to make Dolly and her transgenic successor Polly, but with new steps designed to lead to new patent applications and verify PPL's ownership of the genetic material. Cloned cells (taken from the parent pig) were sent to a lab for testing before the pigs were born: 'That is so we couldn't be accused of taking the cells from the pigs themselves. That means we were absolutely certain they were clones' (Ron James in Radford 2000: 3). Patent licensing criteria centres on the distinction between invention and discovery where invention is defined by novelty and industrial applicability: 'If you find something in nature, then finding some way to separate it and to make it into something useful can be an invention' (spokesman for UK patent office in Clark and Meek 2000: 31). A British intellectual property company, BTG, patented Factor Nine – a gene involved in blood clotting – and licensed it to PPL Therapeutics which is competing to produce an artificial version. Patenting protects the company's research investment by preventing it from being copied. BTG's patent attorney argues that actual ownership of genetic material is, however, a fallacy: 'What we own is the right to exploit it' (Clark and Meek 2000: 31). The patenting of cloned and transgenic organisms raises the prospect of a limited number of heavily marketed brand name genotypes – Dollys, Millies and ANDi's – who, as a result of market forces, eliminate the competition (Rifkin 1998: 112) and acquire an autonomous agency as types if not kinds. ANDi is in this sense analogous to Nike (Lury 1999) – it is the brand name that has a life of its own. ANDi (the name stands for 'inserted DNA' backwards) is a transgenic monkey made in October 2000 by a US Primate Research Centre. He carries a jellyfish gene linked to fluorescence and is part of a pharming project which has raised alarms about an increase in animal and especially primate experimentation. The jellyfish gene was a first stage experiment which might lead to minimising Alzheimer's or any other human disease. According to Sue Mayer of GeneWatch UK, 'Experimentation on primates is particularly problematic because they are closer to us, because we know they are much more likely to suffer in similar ways to us' (in Meek 2001c: 1). While broadly in support of animal protection movements, Haraway remains critical of the extent to which they become reactionary appeals to the integrity of natural kinds 'and the natural *telos* or self-defining purpose of all life forms' (Haraway 1997: 60). For her, such references to natural differences and given categories signify a fear of mixing and of racial impurity: 'In the appeal to intrinsic natures, I hear a mystification of kind and purity akin to the doctrines of white racial hegemony and US national integrity and purpose that so permeate North American culture and history' (61). Harris (1998) outlines an 'instinctive hostility' to transgenic practices which is centred on 'the idea of usurping God's prerogative', disrupting the course of nature and 'the horror and perhaps taboo attached to crossing species boundaries' (Harris 1998: 178). In response, he points out that 'if the practice of medicine has a coherent aim it must be seen, if anything, as the comprehensive attempt to frustrate the course of nature' (178), while the fear and taboo attached to the contamination of species 'deserves about as much respect as objections to miscegenation' (181). The only valid objection to transgenesis for him is the prospect of physical and psychological suffering, and here the validity of the objection is graded according to the organism's proximity to the human species: 'Clearly there will be an important difference in people's likely reactions here depending on whether or not the hybrid is or is not part human and an important moral difference depending on whether or not the hybrid is or could become a person' (179). Transgenic practices thus enact an ethics of kinship based on sameness not difference, and Harris's defence of the significance of human suffering over and above the suffering of other species reproduces the racial puritanism which he criticises elsewhere. The pharming of transgenic animals for human medicine is made possible by reinforcing the boundaries of natural *human* kind – not by destroying them.

In John Frankenheimer's film *The Island of Dr Moreau* (1996) a white UN Peace Envoy named Edward Douglas is involved in a plane crash en route to Jakarta. He is taken to an island formerly colonised by various nations and currently ruled by a Nobel Prize winning scientist referred to as 'the father'. The scientist, Dr Moreau, was exiled from the US by animal rights activists due to his 'obsession' with animal research which now involves transgenic animals, fused with human genes. Moreau's relationship with these 'children' is one of paternalism and propriety, and, through Douglas's eyes, their suffering is witnessed. He sees and hears monstrosity and distress and his judgement has both secular and sacred origins: 'Has it ever occurred to you that you have totally lost your mind? This is just Satanic'. Moreau's aim is to perfect the purification of the human species by eliminating the 'devil' of human destructiveness which is ultimately 'nothing more than a tiresome collection of genes'. His desire to create a 'divine creature, pure, harmonious, incapable of malice' produces 'necessary by-products' whose suffering is justified by the quest for human perfection which, Douglas discovers, is to be realised through him/his genes. In this highly racialised and gendered story of transgenics and eugenics, it is the sanctity of the compassionate colonial subject which is highlighted along with the suffering of his nearest kin; the almost human daughter of Moreau who was destined to receive Douglas's biological blessing and become perfection. The film's ambiguity is highlighted by the distinction between Douglas and Dr Moreau's henchman, Montgomery, and by the suggestion that when he leaves the island, Douglas returns to a world full of 'unstable combinations' or destructive hybrid monsters.

Species hybridisation is celebrated in Eduardo Kac's transgenic art which aims to establish a critical forum and create a space of ambiguity between corporate transgenesis and its opponents. Like ANDi, Kac's transgenic rabbit – Alba – has the jellyfish gene associated with a green fluorescent protein. Unlike ANDi, in whom the gene has not produced the protein, Alba is green. Also unlike ANDi, Alba's purpose is not purely instrumental and her kinship with humans is not ultimately denied (as being insufficiently close). Eduardo Kac's transgenesis represents, for him, a paradigm shift away from individual self-expression in art and science towards an inclusive biotechnological language, spoken by more people and open to the possibility of viable inter-species difference. For Kac, 'biotechnology right now is a language only a few people speak' and 'it is imperative that those who are not experts . . . learn the elements of this language' (in Schueller 2001: 35).[9] His use of a transgenic animal is a provocative invitation to debate the desirability of a future with more than the three current transgenic humans, treated for Severe Combined Immune Deficiency and living 'regular, fulfilling lives' (36). Alba herself stands for nothing but difference; she offers neither salvation or damnation and cannot be exploited either by the proponents or exponents of corporate medicine. She 'will not cure cancer' and she does not suffer: 'GFP, green fluorescent protein, is a standard marker for genetic research because it does not cause any morphological or behavioural transformations' (36). Because she cannot cure and does not suffer on 'our' behalf, Alba merely exists, and her mere existence is what facilitates an ethical debate clear of the sidelines of human progress. Kac appropriates this technology of manipulation because he does not think that it, or any other technology of manipulation should belong only in one set of hands: 'Computers were developed by the military for the military, but today they are everywhere. The idea that a technology is the exclusive domain of a single profession is not acceptable' (37). This is reminiscent of Haraway's refusal to leave 'in the hands of hostile social formations, tools that we need to reinvent our lives' (Haraway 1991b: 8). Although Kac will not work with humans, he claims responsibility for Alba as 'part of my family' (in Schueller 2001: 37) which he seeks to extend, through transgenesis.

The story of artificial life as software, hardware and wetware is unfinished and at least to this extent contestable. The story is being told partly through genomics, presented here as a related set of practices and discourses (encompassing science studies, fiction, policy and journalism) which together constitute an imaginary. Genomics, as part of alife, is overwhelmingly informed by Christian mythology which frames and to an extent constrains the construction of posthumans (pre)destined to salvation or damnation. But the Faustian act of over-reaching is morally ambiguous – the distinction between good and evil never quite absolute – so where the autonomous agents, the doppelgängers of cloning embody the projected sin of creationist ambition, they also embody self-regulation and the survival of the soul-as-gene. Similarly, the unholy monsters of transgenesis both threaten and secure the survival of the human self-as-other-species. It is these ambiguities, functioning as the internal contradictions of genomics, which invite debate and, optimally, bioethical dialogue on the future of posthumanism. Such dialogue is premised on dispersing the spectre of the science wars from all aspects of artificial life.

Chapter 7

Evolving feminism in Alife environments

Alife-as-we-know-it

This chapter will draw together and develop elements of the feminist critique of alife presented both here and elsewhere. Specifically, it will highlight the constitutive discourses of alife – biology, computer science and cognitive psychology – in order to elaborate the concept of autonomous agency which has been associated with the description and fabrication of alife systems in hardware, software and wetware. These informatic systems are regularly, routinely anthropomorphised and yet denied both meaning and morphology. Their autonomy is not least that mythical autonomy of the technical realm from the contaminating spheres of the social, political and psychological (Keller 1992). My aim then, is to contaminate the notion of autonomy as an ontology, epistemology and ethics of the self in technoscientific culture. In so far as this self can be, and has been designated 'posthuman', to what extent does its autonomy signify an era of posthumanism? What is the role of (post)humanism in expanded alife environments, and what is the relationship between feminist and humanist, political and ethical projects? Neither feminism or humanism can be treated as given. Rather, both will be interrogated with respect to their preferred adversaries – biologism, anti-humanism – and indeed with respect to the problem of polarisation which limits debate on all levels to a question of either/or: universalism or relativism; essentialism or constructivism; nature or culture. My argument is that since alife – by definition – exemplifies the redundancy of all polarities which lead to nature versus culture, a feminism evolved in alife environments (physical, social, historical, political) must adopt a new stance in relation to this aspect of the new biology. Biology is not reducible to biological essentialism and is arguably insufficiently homogenous to sustain its own claims to universality or counterclaims of social constructivism which rely on the stable division of nature and culture. In its new stance, feminism must relinquish its corner in recognition of the fact that biology, or at least some biology, already has. To extend the metaphor just briefly; this does not mean that the fight is over. There is a great deal of ground to contest even when there are no clearly demarcated territories, and the gendered and racial legacy of territorialism must surely be no great loss. Biology no longer occupies a territory that can consistently and reliably be named 'nature', and feminism does not preside over a pure, abstract extrapolation of nurture called 'culture'. So how should feminists contest the material and metaphoric grounds of human and machine identities, human and machine relations? This question exceeds the boundaries of this book and to an extent those of current feminist thinking on science and technology within which it is situated. This book and the thinking within which it is situated are both part of the problem and potentially part of the solution. To a large extent, the preceding chapters have rehearsed a constructivist argument developed in legitimate opposition not to biology and evolutionary theory per se, but in opposition to attempts to employ them in naturalising and discriminating discourses and practices. To a large extent, ALife appears to rehearse these discourses and practices in its attempt to create virtual worlds and artificial organisms governed purely by Darwinian principles. And yet, fears concerning the association between Darwinism and Social Darwinism may be grounded more in history than in the present. As demonstrated in Chapter 2, ALife is in-formed by 'a' biology which is far from unified and in fact highly contested internally. Perhaps the strongest critique of sociobiology comes from within the discipline (Rose *et al.* 1984; Rose and Appignanesi 1986). Similarly, as demonstrated in Chapter 3, ALife's foundation upon the hierarchical Platonic distinction of form versus matter is internally contested and ultimately productive. What is produced here is contestation over the terms of embodiment and disembodiment which has served as an invitation to dialogue, and perhaps even – in Donna Haraway's words – to 'diffraction' (Haraway 1997, 2000). Diffraction is an optical metaphor for a slight or sudden change in direction. ALife's foundational commitment to disembodiment – to kinds of behaviour and not kinds of stuff (Langton 1996 [1989]) – is diffracted by other ALife theorists (Emmeche 1994) and practitioners (Brooks and Flynn 1989; Tenhaaf 1997). My own dialogue with the field is partial in the sense of being invested, located, embodied and incomplete. It arose – organically – from the research process and was nurtured by a growing awareness that diffraction – making a difference – is dependent on, if not synonymous with the internal state of the system. It requires proximity and is not free from contamination. If diffraction is, for Haraway (2000), a contaminated 'critical consciousness', I will present dialogue as a critical methodology which entails what, after Isabelle Stengers (1997), might be called 'risk'. What feminism risks, in a dialogue with alife, is not complicity – complicity as illustrated in Sadie Plant's (1995, 1996) work bypasses dialogue and is tacit, given – but the complacency of a secure, well-rehearsed oppositional stance. Feminism risks its anti-biologism (and technophobia) by entering a more dynamic relationship based on contest and consent, on centrifugal and centrepetal forces (Bakhtin in Holquist 1981).[1] It risks entering a dynamic dialogic relationship which, like the strange attractor in chaos theory, is infinite, complex and will not reach equilibrium or closure.[2] There are no clear solutions to the problems raised in a close encounter between feminism and alife – even a very particular 'cyber' feminism and an expanded concept of alife – but the encounter itself, as this chapter hopes to show, can make a difference. My project, as has already been suggested, is by no means pristine and is, in a sense, doubly contaminated – not just by the sciences it moves (slightly) closer towards but by the position it started from. I do not mean to trace a linear trajectory from ignorance to enlightenment, not least because accusations of scientific 'illiteracy' fly all too often in the face of feminists from scientists entrenched in the science wars and invested primarily in preventing any form of outside intervention.[3] What I do mean to trace is the problem of entrenchment and a possible means of averting – in the case of alife – the stalemate of science wars. Dialogue does not entail a one-way movement but a two-way dynamic, and a step into 'no-man's' land for feminism is predicated on the apparent break in biology's ranks. The unfortunate macho metaphors of combat which suggest themselves in this context signal all too clearly the problem which needs to be addressed. This book is inevitably part of the problem, and where I started from is where I am now but with some hopefully significant differences which developed during a process which I am calling dialogue (not emergence). These differences do not centre so much on greater degrees of assent or dissent between the values of feminism and alife, but on the kinds of complexity opened up through internal contestation and made available to diffraction.

Since I am rejecting a teleology of feminist engagement with alife, what follows at least indicates the need for a genealogy attendant on the gaps and inconsistencies in both clearly non-homogenous discourses. My sustained interest and investment in discourses of alife is strongly informed by debates within cyberfeminism. Cyberfeminism may be defined in relation to its origins in feminist theory and practice of the late 1980s and early 1990s (Kennedy 2000: 285) which engaged with the emergent technologies of the information revolution. It was in part a response to the anarchic politics of cyberpunk (Squires 1996), characterised by both Andrew Ross (1991) and Rosi Braidotti (1996) as a realm of middle class adolescent male fantasies centred on a rebellion against the parent culture and a disdain for physicality, or the merely mortal. Computer hacking and science fictional depictions of transcendence as 'getting out of the meat' (Springer 1996) are associated with the then 'new' technologies of the Internet and virtual reality (VR), and with the notion of cyberspace as, in Woolley's (1992: 122) terms, the 'new frontier'. The new final frontier, rather like the old one, was swiftly colonised by cowboys and so cyberfeminism was in part a kind of Calamity Jane for the new media, creating anarchy more specifically within patriarchal culture and strategically employing anachronistic or essentialist images of women. There were the Riot Girls (Braidotti 1996: 14) and VNS Matrix (1994) whose computer game heroine called Gen sabotages Big Daddy Mainframe and does for Circuit Boy ('a fetishised replicant of the perfect human HeMan') by bonding with DNA sluts and consuming plenty of G-slime. Where parody and humour may have mitigated against the troubling aspects of essentialism here, the same cannot be said for Plant's (1995) analogy of weaving, women and cybernetics in which a supposedly feminised technology is described as being autonomous, self-organised and imminently apocalyptic. Plant's (1995) work enjoyed popular appeal despite (or because of) its technologically determined apocalypticism and biological essentialism. This appeal was arguably part of a widespread millennium fever, the rather quiet passing of which may be said to have signalled the failure and obsolescence of the cyberfeminist project. But this argument presupposes a degree of homogeneity within cyberfeminism which did not, and indeed does not exist. Where the dystopian spirit of cyberpunk sci-fi inspired some cyberfeminists to anticipate a sudden end to patriarchy, others remained grounded in the less fictional realms of science and technology studies and emerged with the more measured if somewhat utopian concepts of change exemplified in Haraway's (1991) figure of the cyborg and Braidotti's (1994: 36) figure of the nomadic subject – or 'cyborg with an unconscious'. Haraway's (1991a) hugely influential cyborg manifesto directed cyberfeminism towards the impacted fields of science and technology, and specifically identified biotechnology as a branch of technoscience where some of the most important political, ethical, social and economic issues of the day converged. Haraway's concern with cybernetics, reproductive medicine, immunology and genetics is developed in subsequent work (Haraway 1997) and is contained within her conviction that biology is perhaps *the* hegemonic discourse of the late twentieth- and early twenty-first centuries. In *How Like a Leaf* Haraway (2000: 26) validates the argument that biology, 'woven in and through information technologies and systems . . . is one of the great "representing machines"' of the century, superseding film – or literature in the nineteenth century. From health and food industries to environmentalism, management and intellectual property law, 'there is almost nothing you can do these days', she argues, 'that does not require literacy in biology' (26). Here, Haraway indicates the need for a renewed and enhanced (cyber)feminist engagement with biology which, in *Modest_Witness*, she effects through a sustained examination of the meaning of kinship in a biotechnological age. Kinship is the central theme in a book which, via the email address of its title 'is situated as a node that leads to the Internet, which is synechdochic for the wealth of connections that constitute a specific, finite, material-semiotic universe called technoscience' (3). The Internet is seen to embody the primary concerns of technoscience (including identity, democracy, access to knowledge, globalisation and wealth), and technoscience 'extravagantly exceeds the distinction between science and technology as well as those between nature and society, subjects and objects, and the natural and artificial that structured the imaginary time called modernity' (3). Haraway's refusal of binarism and her insistence on the potent connectedness of apparently discrete disciplines, objects and entities is captured in both the form and content of her argument. Her book is a network of debates on science, biotechnology and feminism which draws as much on art and literature as it does on science and policy. Among the principal connections she insists on is that between the subject and object of technoscientific knowledge, and it is helpful to consider her views on their kinship in relation to actor-network theory which seeks 'to enfranchise the world of objects' (Adam 1998: 63). Haraway's relation to actor-network theory is more clearly examined by Alison Adam (1998: 60) who defines it as a 'growing interest in looking at the process of creating scientific and technical knowledge in terms of a network of actors or actants, where power may be located throughout the network rather than in the hands of individuals'. In so far as actor-network theory affords agency to machines and other entities previously denied such status (including animals and other 'others') it may be seen as a useful tool for feminism. Despite pointing out the limitations of this approach for feminism – including the lack of address to gender, race, class and the body in knowledge-production – Adam suggests that it 'strikes a significant chord' with Haraway's feminist cyborg manifesto which is concerned with 'transgressing boundaries between machine, human and animal' (1998: 173).

The cyborg manifesto attacked the binaries and boundaries of mainstream patriarchal technoscience and sought to refigure them in favour of those who they discriminate against – including women (Kember 1998). *Modest_Witness* extends this project by asserting that the subjects and objects of technoscience are not only equal but also related (see Chapter 6). So Haraway's cyborg is joined by a trilogy of (re)figurations all of whom are kin. Where the modest witness is a figure in science studies, the FemaleMan is 'the chief figure in the narrative field of feminism' and OncoMouse 'is a figure in the story field of biotechnology and genetic engineering' (Haraway 1997: 22). Her 'tendentious' point is this: 'that the apparatuses of cultural production going by the names of science studies, antiracist feminism, and technoscience have a common circulatory system. In short, my figures share bodily fluids' – fluids which are mixed where they meet in 'the computing machine of my email address, named Second Millennium' (22). Modest witness (@second millennium) is 'the sender and receiver of messages in my email address' (23). S/he is Haraway's subject position and the link to her central methodology and epistemology: situated knowledge. The modest witness is situated inside 'the net of stories, agencies, and instruments that constitute technoscience' (3) and s/he 'is about telling the truth, giving reliable testimony' while 'eschewing the addictive narcotic of transcendental foundations' in order to 'enable compelling belief and collective action' (22). S/he works to refigure the subjects, objects and 'communicative commerce' of technoscience, and Haraway declares herself to be 'consumed' by the project of refiguration because she believes that it is central to both technoscience and feminism (23). She derives the term 'modest witness' from a story about the development of the air-pump (Shapin and Schaffer in Haraway 1997: 23) in which the modesty of the witness is dependent on his invisibility. This self-invisibility 'is the specifically modern, European, masculine, scientific form of the virtue of modesty' (23) and the one which stakes its claim to truth and objectivity on the basis of disembodiment. This witness appears to be free from his 'biasing embodiment' and 'so he is endowed with the remarkable power to establish the facts' (24). Haraway seeks to 'queer' rather than oppose the myth of disembodiment and to enable 'a more corporeal, inflected', self-aware and accountable kind of modest witness to emerge within the worlds of technoscience (24). It is crucial for Haraway that her modest witness is implicated, that s/he does not seek to stand clear and maintain the dubious distinction between theory and practice, politics and technology, which ultimately reinforces the tradition of invisibility: 'I remain a child of the Scientific Revolution, the Enlightenment, and technoscience. My modest witness cannot even be simply oppositional' (3). Rejecting oppositional science studies – and particularly those of Bruno Latour (1987) – she argues that 'the point is to make a difference in the world, to cast our lot for some ways of life and not others. To do that, one must be in the action, be finite and dirty, not transcendent and clean' (Haraway 1997: 36). Preferring Sandra Harding's (1992) case for strong objectivity because it 'insists that both the objects and the subjects of knowledge-making practices must be located' (Haraway 1997: 37), corrects a common misunderstanding about the meaning of location. This is 'not a listing of adjectives or assigning of labels such as race, sex and class' (37). It is not as self-evident or transparent as that. Rather, it is the 'fraught play of foreground and background, text and context, that constitute critical inquiry' (37) and it is partial in as far as it is incomplete and favours some worlds over others. Situated knowledges are then founded on this sense of location, and they are methodologically and epistemologically employed by Haraway (1997) in the figure of the modest witness.

Alison Adam works with the body of writing on feminist epistemology and science and technology within which Haraway has become central. In *Artificial Knowing* Adam (1998) is concerned primarily with the inscription of gender in artificial intelligence. Through a focus on embodiment concerned with the role of the body in knowledge production, she argues that 'AI systems, in taking a traditionally gendered approach to knowledge which reflects the style of mainstream epistemology, incorporate a view of the world which tacitly reflects a norm of masculinity, both in terms of the knower and the known' (Adam 1998: 8). What this leaves out is 'other types of knowing subject and knowledge, particularly that which relates to women's ways of knowing' (8). Adam shares a concern with Haraway, Hayles and Turkle in the field of ALife which, she argues, demonstrates at best physical rather than cultural forms of embodiment and is tied to sociobiology. In 'Embodiment and Situatedness: The Artificial Life Alternative', Adam (1998) discusses the relation between the gendered body and knowledge formation (in which masculinist knowledge is perceived as being disembodied and feminist knowledge strives for re-embodiment without recourse to essentialism) prior to making the important distinction between physical and social situatedness. She argues that the social constructivist position in science studies, including actor-network theory, is concerned only with social situatedness 'as it seems to shy away from dealing with messy bodies, maintaining a masculine, transcendental (albeit not necessarily rationalist) position' (Adam 1998: 129). On the other hand, research on AI which attempts to be situated, 'looks at the problem almost exclusively from the physical stance' (129). Haraway (1997: 302) notes that Helmreich also correctly distinguishes between materialised entities in ALife research and the concept of embodiment which incorporates located and accountable lived experience. Situated robotics lies partially within what Adam (1998) refers to as 'a wider AI discipline' of artificial life which has produced, among other entities, the physically but not (yet) culturally embodied Cog. She claims that the primary goal in emergence-based artificial life is evolution in the purely biological sense. Transferred to the human realm either in hardware or software, it becomes 'sociobiology in computational clothing' (Adam 1998: 150). Since research on the development of artificial human societies and cultures is taking place (see Chapter 5), and ALife has entered popular culture through film, computer games (Chapter 4) and online user-oriented ecosystems it does seem timely to respond to Adam's assessment that 'there is no room for passion, love and emotion in the knowledge created in A-Life worlds' (155). Adam maintains that ALife's attachment to sociobiological models is premised on 'an essentialist view of human nature and women's nature; where cultural ways of knowing are to be explained and subsumed in deterministic biological models' (155). Adam's early warnings about the dangers inherent in ALife environments constitute a valuable contribution to further feminist critiques. They effectively highlight sociobiology and particularly the idea of a *subsuming* biological hegemony for future analysis. Any criticism of Adam's argument would focus on the contradiction in her simultaneous critique and replication of the social constructionist position. Adam shares with many feminist epistemologists a desire for embodiment which stands clean and clear of essentialism. This is only possible in a concept of embodiment which privileges social over physical located-ness and thus distances the 'messy bodies'. Alternatively, it requires an address to biology which realises the malleability and indeterminacy of sexed and other forms. Anti-essentialism has, for many of us, become too much of a mantra and there is a need to turn to, for example, Braidotti's (1994) outline of 'essentialism with a difference' or at least Spivak's recognition that 'you are committed to these concepts [universalism, essentialism], whether you acknowledge it or not' (in Kirby 1997: 155).

In *Life on the Screen*, Sherry Turkle (1997) effectively locates the consensual culture of alife through key concepts such as emergence and connectionism. These articulate 'something that many people want to hear', and in particular, 'the non determinism of emergent systems has a special resonance in our time of widespread disaffection with instrumental reason' (Turkle 1997: 143). ALife's emphasis on biology rather than physics, bottom-up rather than top-down processing, holism rather than reductionism is ideological rather than purely methodological. It fits, alongside chaos and complexity theory within a zeitgeist identified by Turkle, Emmeche and others as postmodernism and is consonant with 'a general turn to "softer" epistemologies that emphasise contextual method-ologies' (144). ALife's 'constituent agents' express feelings of fragmentation (theorised within psychoanalysis) while simultaneously decentring and recentring the myth of the autonomous, unitary identity: 'Today, the recentralisation of emergent discourse in AI is most apparent in how computer agents such as those designed to sort electronic mail or scour the Internet for news are being discussed in popular culture' (145). Whereas the intelligence of such agents emerges from the function of a distributed system, there is 'a tendency to anthropomorphise a single agent on whose intelligence the users of the program will come to depend'.

Analogies are then created between this 'superagent' and a butler or personal assistant (145). For Turkle, such attempts to re-engineer the autonomous agents of liberal humanism are as unstable as concomitant attempts to reinforce positivism in a postmodern technoscientific context. But this viewpoint tends to belie the fact that connectionism and emergent AI/ALife can highlight the status of biology and evolution as grand narratives or universals. Connectionism, as a metaphor in social and cultural theory serves to naturalise relationships and processes of change (Plant 1996). Conversely, Haraway's emphasis on the connection between entities and her concern with the processes and direction of technocultural change is resolutely political and a facet of what she calls 'nature-culture' (Haraway 2000: 105). Criticisms of Plant's cyberfeminism (Adam 1998; Kember 1996) are based on her descriptive employment of ALife paradigms such as connectionism and emergence, whereas Haraway's engagement with the forms and concepts of biotechnology is effected partly through parody or strategic refiguration which aims to make a difference/diffraction rather than reflect the supposedly already diffracted status quo. N. Katherine Hayles's attendance to the erasure of embodiment within cybernetic and alife discourses and their consequent recentring of autonomous agency (Chapter 5) precedes an act of 'rememory' (Hayles 1999a: 13) which may be strategically linked with both Haraway's refiguration and Helmreich's call for a feminist, queer and postcolonial intervention into how life comes to matter (Chapter 3). This intervention is predicated on the latent critical potential within the material and symbolic configuration of 'life-as-it-could-be', currently constrained but not determined by the terms and conditions of 'life-as-we-know-it'. While being differently located within and across the disciplines of science and technology studies, anthropology and literature these three critical strategies share two principal elements: a commitment to change (not evident in Turkle's engagement with ALife) and a dynamic, dialogic tension between a purely internalist and an interventionist methodology (not demonstrated in Adam's critique of AI and ALife). Haraway's identification as a scientist and feminist cultural theorist influences but by no means guarantees her commitment to contamination and rejection of oppositional science studies, while Hayles and Helmreich are similarly emplaced through narrative and ethnographic immersion within already contested ALife spheres. My own contribution to a political and methodological engagement with alife which will continue to expand the parameters of cyberfeminism, is to clarify and highlight the dialogic and subjective elements of the relationship between feminism and alife through an immersion in alife cultures or environments.

Natureculture

> HAMM: Nature has forgotten us
> CLOV: There's no more nature
> HAMM: No more nature! You exaggerate
>
> (Beckett 1964 [1958]: 16)

The encounter between feminism and ALife is grounded on the ever shifting terrain of what counts as nature and what is categorised as culture. In recent debates, particularly within Australian feminism, this inevitably epistemological problem of definition has been opened further to the presence of matter and ontology. In these debates, epistemological perspectives are seen to foreclose the question of ontology which is then offered as the means to revise a hierarchical distinction underlying the project of modernity. Jacinta Kerin (1999) outlines work by Elizabeth A. Wilson, among others, which is critical of social constructionist arguments that maintain the distinction between the realms of 'ideality' and materiality: 'As long as a radical break between matter and ideality is assumed, such that the former is inaccessible while the latter is the only possible site of investigation, the nature/culture distinction is re-installed and nature operates as the subordinate term' (Kerin 1999: 91). Wilson (1998: 13) extracts her methodology from a feminist critique of traditional and cognitive psychology which are rethought through the neurological facets of connectionism. In what must, in part, be an implicit rejection of the post structuralist project of critical psychology (Henriques *et al.* 1998), Wilson expresses her impatience with arguments which 'deliver tired rearticulations of antiessentialist, antibiological, antiscientific axioms, and thus promote a kind of interpretive eugenics that breeds out the bastard children of any liason with biological or scientific systems' (Wilson 1998: 4). Her position is shared with Vicky Kirby (1997), who approaches 'the problematic of corporeality' in a way which is as critical of a postmodernism which regards 'the apparent evidence of nature as the actual representation of culture' as it is of an empiricism 'that perceives data as the raw and unmediated nature of the world'. Her aim then is 'to pose the nature/culture (body/mind) question in a way that cannot be resolved by taking sides' (Kirby 1997: 2). In the context of VR for example, forged on the split between the virtual and the real, embodiment and disembodiment, Kirby notes with interest that 'the nature of corporeal substance <u>as such</u> is not the matter of contestation: the debate concerns the body's ethical valuation as either expendible or necessary' (Kirby 1997: 136). The VR debate recites a familiar problematic of the body as a self-evident 'residual "something" that technology is articulated against' even while it seems to displace it (137). According to Kirby, even Allucquere Roseanne Stone's critique of the 'Cartesian trick' in VR is reproduced in her reminder that 'virtual community originates in, and must return to, the physical' (Stone in Kirby 1997: 138). Stone's sense of an inevitable return to the body in VR 'implies that at some point we successfully took leave of it' (139). Moreover, her insistence that virtual subjects have bodies attached 'understands personhood as a divided composite of mind <u>and</u> body, where the body is the separable outer envelope of the *cogito*' (139). Wilson's revision of the body in feminism is also predicated on an unwitting reinstatement of Cartesianism effected by a widespread insistence on the significance of an abiological body. Where this proves to be most limiting is in any engagement between feminism and the natural sciences. ALife, in this context, can be understood as an unnatural science which seeks a new kind of literalised synergy between technology and biology. For Wilson, a politically engaging feminist critique of biological and behavioural sciences is hampered by a 'naturalised antiessentialism' within feminism. After all, she asks, 'how can a critical habit nurtured on antibiologism produce anything but the most cursory and negating critique of biology?' (Wilson 1998: 16). At the heart of the problem is the focus on gender as defined against biological sex. As well as positing a malleable constructed cultural identity against a supposedly immutable, given and natural one this distinction serves to reinforce scientific orthodoxy by sanctioning a realm of neutral (unsexed) knowledges where feminism appears to have no place or purchase. In other words, gender-led feminist critiques of science have focused on areas such as reproductive medicine where women function as the objects of knowledge, leaving areas such as cognition and neurology relatively uncontested (Wilson 1998: 19, 34). Londa Schiebinger (1999) seems to confirm this argument in her conventional investigation into the extent to which feminism has changed science. Where gender-based feminisms have made inroads in primatology and biology – in terms both of access and epistemology – there have been less significant advances in physics and maths which do not lend themselves as readily to gender analysis (Schiebinger 1999: 159). Indeed, a feminist critique of alife is most readily validated in relation to alife's biologism which is often regarded internally as a 'softening' (Turkle 1997) or feminising (Grand 2000) of AI's physics-based epistemology. An intervention is then implicated, to an extent, in the essentialism it seeks to challenge at least until it incorporates those aspects of alife – including computation, neuroscience, cognition (see Chapters 4 and 5) – which are harder to validate within the framework of a feminist analysis. Inevitably, at these points, the question of gender is deferred. Regarding the threat of biological essentialism as a constitutive effect of gender theory, Wilson (1998: 55) calls for a deferral of gender as a restrictive category (50) prior to the establishment of a concept of gender which is not premised on the exclusion and pathologisation of biology (51) or inert sexual matter: 'the first premise of this book is that if our critical habits and procedures can be redirected so that biology and neurology are not the natural enemies of politics – that is, if we defer gender theory from the start – then we will find a greater critical productivity in biology than theories of gender would lead us to believe' (62). Wilson finds this critical productivity in connectionism as a neurological process of self-organisation which, like biology per se should be subject to denaturalisation without recourse to antinaturalism: 'My point is not that biology requires no critique, not that it is given, self-obvious material, but rather that in the regime of gender theory biology can only ever figure as poisonously foundational, originary, and normative'. As a consequence 'the critical habits assembled in this regime are limited to supposing that any critique of the foundational, the originary, or the normative comes from a place other than biology, and should be mobilised against the biological' (62).

Wilson turns to the work of Judith Butler (1989, 1993) for a more nuanced feminist account of the relation between sex and gender, but finds that Butler subsumes sex within discourse with the consequent re-negating of biology. There is no sex her, only 'sex' (Wilson 1998: 59). Kerin's critique also maintains 'that the problem of matter cannot be adequately approached merely by analysing the discursive models through which it is interpreted' (Kerin 1999: 91). Consequently the question of ontology has to be raised in order to recognise that matter exists in multiple forms. Feminist work such as Butler's tends to homogenise both natural science and its objects, and breaking with this tendency entails thinking 'in terms of modes of being, and the effects that different manifestations of matter have on their constitution as objects of human knowledge' (Kerin 1999: 91). A return to ontology then, also involves the difficulty of rethinking it outside of the terms of realism or 'as other than a pre-given ground in which certain naturalised truths can be located' (92). For Kerin the problem with Butler's work and that of other feminist epistemologists is that they assume that this is an impossible task and so foreclose the realm of ontology from critical inquiry. Even those philosophers of science who acknowledge the productive force of natural phenomena within scientific theories do so, according to Kerin, inadequately. She cites Evelyn Fox Keller, and N. Katherine Hayles's formulation of 'constrained constructivism' (Kerin 1999: 94). Wilson (1998) and Kirby (1997) focus some of their critique on Donna Haraway; the former deferring to the latter's analysis of the cyborg as a hybrid figuration which seeks to rework gendered structures of domination – nature/culture, object/subject – in traditional technoscientific cultures. For Kirby (1997: 147), the cyborg is actually 'the most recent of Cartesian recuperations because it is formulated against a pre-cyborgian realm of unity and identity'. What this critique of Haraway (1991a) perhaps overlooks is that the cyborg is forged 'within the belly of the monster', meaning that it is made both within and against the persistently dominating humanist and positivist structures of postmodernist culture. In other words, it is a parody. Kirby goes on to suggest that Haraway's defence of biology in the face of suspicious anti-essentialist feminists is inadequate because she attempts to recuperate biology within discourse: 'Biology is not the body itself, but a discourse' (in Kirby 1997: 147). Here Kirby seems to conflate biology with the body in a way that neither Haraway nor her interrogators nor certain biologists themselves (Rose 1997) are willing to do. Where Kirby's arguments for an ontology are strong, they are at times made against critiques of epistemology which are more partial, less careful. So a more generous reading of Haraway's argument would recognise it as an attempt to prevent the body from being subsumed within discourse by de-naturalising biology. For Kerin, two of the three philosophers of science (the third being Wilson) who offer a more adequate account of the productive force of matter include physicists Isabelle Stengers and Karen Barad. While both Wilson (1998) and Barad (2000) retrieve ontology 'in a manoeuvre that is deemed necessary to feminist inquiry' (Kerin 1999: 94), the extent of Stenger's aversion to epistemology distances her from this recuperative project. In his foreword to Stenger's (1997) *Power and Invention*, Bruno Latour (1997) warns feminists, ecologists and leftists who think she is an ally to expect some hard lessons derived directly from her hard science. The ingredients of domination in politics are the same as those in the laboratory 'that is, the inability to allow the people one deals with any chance to redefine the situation in their own terms' (Latour 1997: xviii). By extension, standpoint politics (Harding and Hintikka 1983; Harding 1992; Haraway 1991a) – 'where the outcome of the analysis is entirely determined from the start by the position of the speaker' – reflects scientific forms of domination. 'What should be said', Latour (1997: xviii) asks, 'of those thousands of radical tracts where the things to be studied – science, art, institutions, medicine – have no chance to say anything other than that they have been marked by the domination of white male capitalists?' Where this point seems to play along the reaction–inversion axis of classic dualism (where the denaturalising position is assumed to have become naturalised and the marginal voice become dominant), Latour goes on to make a more interesting point about risk:

> Like most critical thinking, they [the radical tracts] reproduce exactly at the outcome what was expected from the beginning, and if they have to be rejected, it is not because they are political, and not because they are not scientific enough, but simply because the writer incurred no risk in being kicked out of his or her standpoint in writing them.
>
> (Latour 1997: xix)

In seeming to incur risk, Kerin argues that Butler resists it. Or, in rethinking the role of matter in discourse (as a response to criticism of her earlier project *Gender Trouble*) as 'a process of materialisation' she effects not an ontology but an expanded epistemology: 'She wishes to open up the symbolic to alternative articulations of what is valuable' (Kerin 1999: 95). Since Butler's assumption of ontology as a means to the restoration of positivism 'is one which no feminist "ontologist" is advocating we return' Kerin raises the question of alternative conceptions which work against the historical grain. It might be added that no such possibility is afforded to epistemologists. Illustrations from genetics are offered 'because this discipline exemplifies a field in which scientific methodology has had to transform in the face of its object of inquiry' (100). So, for example, Evelyn Fox Keller's (1983) account of Barbara McClintock's 'feeling for the organism' is offered here as an ontological turn in molecular biology. Kerin (1999: 100) uses her examples to argue that 'feminists will not be able to intervene effectively in these areas of science unless we risk familiarity with the specifics of molecular genetics'. The question then remains as to whether, as she suggests, an ontology rather than an epistemology is necessary to this laudable goal. Haraway's critical and creative familiarity with biology (highlighted by her as a hegemonic discourse); her handling of the complex interplay between science, philosophy and politics and above all her figuration-based renegotiation of the distinction between nature and culture, subject and object seems to me to offer one prominent example of the limitation of the ontological approach. Not constrained, as Butler is, within a post-structuralist psychoanalytic framework

(Kerin 1999: 97), Haraway, herself a scientist, places great emphasis on her rejection of both relativism and universalism, constructionism and essentialism. Her epistemological (and also methodological) formulation of situated knowledge and partial truths is one which seeks to find a between ground and therefore offers a *means* of recognising matter without resorting to realism. An alternative means is not offered within the 'return to ontology' perspectives and arguably could not be offered without an epistemology (or some sense of how). It would seem that the arguments of Wilson, Kirby and Kerin are, to an extent, weakened by a 'flip-flop' (Kirby) between ontology and epistemology which Haraway, for example, does not enact. Her figurations are, in fact, not purely epistemological. They are a response to the different modes of being emerging from the mutually inflected fields of biotechnology, feminism and science studies. OncoMouse, FemaleMan and Modest Witness (and indeed the older relation – the cyborg) are kin, and Haraway's work on kinship is precisely a speculation on modalities of existence which offers passage points 'through which the terms of (and relations between)' nature and culture, subjects and objects emerge (Kerin 1999: 101). In what for her is the imploded discursive realm of 'naturecultures', Haraway encounters heterogenous entities and fleshy bodies which have agency enough to pose questions within human discourse (and in fact to resist the level of anthropomorphism familiar within sciences such as AI and ALife):

> Within this context I have written about cyborgs on the one hand and animals on the other, specifically about primates. And these primates raise the question of human-nature relationships differently than cyborgs do. In particular, evolutionary history emerges in sharp ways, issues of biological reductionism and the lived body, the fleshy body and who we are related to. Our kin among the other organisms is raised in potent ways in the primate story, much more so than in the cyborg story. The cyborg story raises questions about our kin among the machines – our kin within the domain of communication – while the primate story raises questions about our kin in the domain of other organisms and raises questions of the nature-culture interface that has been articulated in the human sciences.
>
> (Haraway 2000: 106)

It is clear from this that primatology, biology, technology are, for Haraway, discourses or more specifically narratives, stories, But these stories are not conflated with the organisms and entities they tell and are told by. Her intervention in the 'gapped reality' of nature and culture 'has to do with the idea of worldliness, an act of faith in worldliness where the fleshy body and human histories are always and everywhere enmeshed in the tissue of interrelationship where all the relations aren't human'. And so, 'my fundamental epistemological starting points are from this enmeshment where the categorical separation of nature and culture is already a kind of violence' (Haraway 2000: 160). How can matter which is unmeshed with discourse speak in a way which refuses this kind of violence? The enmeshment of matter and discourse is the starting point of Haraway's epistemology. In her interview with Thyrza Nichols Goodeve, Haraway (2000) is invited to embrace more of an ontology. With reference to Haraway's story of the immune system, Goodeve asks 'where is the practice of "science", of the facts of the immune system that do not respond to interpretation?' Haraway (2000: 107) answers with what seems to be a conventional constructionist argument: 'There's no place to be in the world outside of stories'. Once more she is pushed to acknowledge 'a kind of biologism' in her writing. She responds with a story about *Mixotricha paradoxa*, a microscopic entity which is 'more than a metaphor'. An organism that lives symbiotically with five other kinds of entities in the gut of a termite, *Mixotricha* is 'a fabulous metaphor that is a real thing for interrogating our notions of one and many' (Haraway 2000: 83). A boundary creature which is similar to, but different from, for example, a transgenic one, *Mixotricha* in-forms and is informed by technoscientific discourse. Is this constructionism? Only, Goodeve suggests, in a reductionist reading (Haraway 2000: 85). For Haraway, being accused of having an anti-materialist, constructionist view of science constitutes 'a kind of literal-mindedness. And that's why figures are so important to me, because figures are immediately complex and non-literal' not to mention instances of real pleasure in 'language *and* bodies' (85). Without wishing to iron out the inconsistencies which co-exist with the insights in Haraway's creative projects, it does seem important to underline the limitations of a reductionist equation between epistemology and ('strong') constructionism. Wilson herself appears at times to become 'trapped' in discourse, conflating biology with its objects and relying, strangely, on a structuralist semiotic model of connectionism. As a result, she replays the dualism of cognitivism/connectionism, top-down/bottom-up, centralised/distributed and AI/ALife systems. Interestingly, ALife constitutes the absent reference in Wilson's argument in favour of connectionism as an embodiment of psychology. Within ALife, connectionism tends to function more as an abstracted informational metaphor of self-organisation which is realised in biotechnological systems like mobots, robots and knobots. It is perhaps premature then, for 'ontologists' to claim to be able to resolve the tension between biological determinism and social constructionism. Nevertheless, a valid and important point which emerges from both the epistemological and ontological perspectives is the sense of necessary risk-taking involved in a feminist response to the changing matter of life (human or other) in the contemporary field of biotechnology.

Risk

In the previous chapter, I outlined a notion of risk which, particularly in the context of biotechnology, exceeds the boundaries of calculation, prediction and prevention, incorporating both imagination and complexity. In *Power and Invention*, Isabelle Stengers (1997) actually advocates a 'cosmopolitics' of risk-taking as a methodological and ontological approach to a world of complex objects. It is this approach, which, for her, distinguishes good from bad science. She opposes the consensual viewpoint that sorting good from bad entails a critique of scientific epistemologies and representations and a focus on human limitations and misunderstandings:

> To be sure, our society, language, mind and brain could be cause for some misunderstanding, but the main partner to be interrogated for sources of uncertainties is the complexity of the world, which does not wait outside and does not remain equal to itself. Against epistemology and against social construction, Stengers directs our attention to the ways in which the world is agitating itself and puzzling us.
>
> (Latour 1997: x)

Latour refers to Stengers' work with Ilya Prigogine on chaos theory which offered an ontological touchstone and constituted the first phase of her work which developed from the philosophy of science to a more general philosophy (Prigogine and Stengers 1985). At the heart of Stengers' cosmopolitics is a philosophy in which scientific realism and social constructionism are not opposed by synonymous (Latour 1997: xiv). Put another way, Stengers works with a notion of risky constructionism:

> There are constructions where neither the world nor the word, neither the cosmos nor the scientists take any risk. These are badly constructed propositions and should be weeded out of science and society . . . On the other hand, there exist propositions where the world and the scientists are both at risk. Those are well constructed, that is, reality constructing, reality making, and they should be included in science and society; that is, they are CC [cosmopolitically correct], no matter how politically incorrect they may appear to be.
>
> (Latour 1997: xiv)

Stengers advocates a philosophy in which the object, the thing, the world is recognised as having something to say for itself. It is about embracing the risk which is therefore posed to science and to the scientist. The equation, Latour (1997: xix) points out, is simple if hard to actually realise: 'no risk, no good construction, no invention, thus no good science and no good politics either'. Risk is simultaneously cosmopolitical and personal: 'In the obscure fights of the science wars, one can safely predict, she will be seen as a traitor to all the camps' (xix). In the context of complexity, Stengers seeks to distinguish between the complexity of the world and the complexity which is articulated within scientific and other discourses about the world, where it functions indirectly as a defence of rationalism. Complexity here recuperates science's loss: 'it allows one both to defend science against the charge of "reductionism" and at the same time to envisage science's conquest of what until now had escaped it' (Latour 1997: 3).

The risky question concerns the status of complexity as, or in relation to science: 'Can we use what present themselves today as "complex objects" to underline the general problems they raise, rather than the particular models of solution they determine?' (5). Can complexity function not as an epistemology but as part of a social scientific ontology, where, for example, in chaos theory the strange attractor remains a question mark in, and not a model of knowledge? (Latour 1997: 7; Kember 1991, 1996). The question of complexity is distinguishable both from that of emergence (which for Stengers entails physical but not conceptual genesis) and from that of complication (which in opposition to simplicity, is often a means to guarantee scientificity). Furthermore, the proper response to it 'is not theoretical but practical' requiring 'the enculturation of science' through risk. Risk is ultimately related to the practice of dialogue; a dialogue in which science and the world speak to each other (and which works to counter critiques of science as a means to 'othering' nature) but also a dialogue in which 'the value of communications between philosophical and scientific interrogations cease being suppressed by compartmentalism or destroyed by a confrontational attitude' (Prigogine and Stengers 1997: 53). A respect for nature should be extended to other intellectual approaches (57) leading not perhaps to interdisciplinarity as much as to 'interbreeding' which provides 'novel communications that enable us to deal with the unprecedented demands of our era' (57). Transgenesis, as Haraway (1997) indicates is a fabulous metaphor as well as the 'shiny news' (55) and complex 'stuff' of our time. It directs our attention towards 'undreamed of objects' (58) which demand a response from technoscientific societies within which 'new alliances' may well need to be re-forged if the response is to be at all adequate.

One such alliance, for Patricia Adair Gowaty, is that between feminism and evolutionary biology. As a Darwinian feminist, Gowaty (1997) has her own tale of risk to tell, and her edited anthology on this subject stemmed from a highly contentious and confrontational conference. Here, Gowaty rediscovered the impediments to 'cross-germination'; the first being the over-wrought debate about genetic determinism and the second being a widespread anti-scientism. Before the conference, Gowaty (1997: 1) 'thought the second impediment had nothing to do with me or mine and that the first was solved fifteen years ago'. Apparently not. Instead of finding herself facilitating a 'new dialogue' at the frontiers of both feminism and biology, Gowaty served as an interpreter and apologist at the contested '*old* boundaries between them' (2). These old boundaries mark the terrains of nature and culture which, though successfully crossed, Gowaty herself does little to disturb. This is partly because she identifies 'human nature' (and human natural behaviour) as the shared object, the common ground of investigation. The how and why of human behaviour may then be explained by the same evolutionary biological methods as the how and why of other animal behaviour, disregarding the complex interplay between the biological and the social factors which distinguish post-Pleistocene, post-industrial human life. Gowaty subsumes (feminist) politics within (evolutionary) biology in her attempt to establish a common scientific methodology and epistemology of behaviour. More usefully, she marks two sets of debates within evolutionary biology which might facilitate ways of rethinking the nature/culture, sex/gender dichotomy structuring the opposition between and within the disciplines. The first is the internal debate over universal selection pressures and their ability or inability to determine universal traits. Gowaty questions both feminist and biological claims for the existence of universal, fixed gender-specific traits, arguing for the (universal) existence of significant variation and diversity between groups and individuals. The possible or probable existence of universal selection pressures which work differentially on women and men in their respective biosocial environments does not translate directly into universal traits:

> The reasons for this include stochastic variations (that could lead some individuals to one way to skin a cat and to another way for other individuals), countertactics (that could favour some cats escaping via one hatch and other cats via other hatches), and the dialectical forces of interaction on traits in one gender and countertraits in another.
> (Gowaty 1997: 7)

Arguments in favour of the existence of innate differences in men and women are difficult to sustain, Gowaty argues, beyond a few examples such as menstruation, childbearing and (more questionably) lactation. On the other hand, interaction between selection pressures and individual adaptive responses accounts for 'the existence of men who – perhaps facultatively – exhibit traits usually associated with women and vice versa' (7). So, for the non-essentialist feminist evolutionary biologist like Gowaty, there are few essential differences between men and women but strong differences in the selection pressures operating on them: 'that is, differences in the environments that men and women inhabit and that our social behaviour and organisations create' (7). The second related debate within biology, stemming from the first, is over the relationship between statistical means and variance. This, for Gowaty, is in effect a political debate and is comparable with concerns about normalisation and resistance in social science: 'When we experience conditioning or social pressures to match our own characteristics to culturally sanctioned ideals, (means), one of the most liberating and important concepts that we can have is the idea of variation and variance' (15). Evolutionary biology is not, then, at odds with feminism provided that it accommodates variation in the lives of women. At this point 'we edge closer to a truer approximation of the nature of nature and to the fulfilment of our dreams for a more egalitarian world' (16). But what is this 'nature' which can sustain the full range of biocultural variation? Gowaty's assertion of evolutionary biology in the face of anti-biologism, her desire to apprehend feminism and the social sciences within scientific methods and epistemologies ultimately causes her to reassert the old boundaries which she has, in places, been at pains to cross.

Russell Gray (a psychologist), more like Haraway, seeks to challenge evolutionary biology from within, outlining developmental systems theory in order to more fully contest the terms of the nature/culture debate. Gray (1997: 387) argues 'that conceiving of evolution as the differential reproduction of developmental processes profoundly changes the nature of evolutionary explanation and thus of evolutionary accounts of sex/gender'. Developmental processes cut across the distinction between innate (genetic) and acquired (environmental) behaviours which structure the nature/culture split. Supposedly innate behaviour is actually dependent on acquired or experiential inputs and 'these inputs are not just secondary and supportive but "positive, informative and constructive"' (389). Not all behaviour is learned, and learning may also be dependent on internal factors. It is contingent, for example, on the sensory system and on genotype or species (in ethology). Developmental analysis expands and coalesces the categories of 'innate' and 'acquired' behaviour, rendering them meaningless: '*All* phenotypes are the *joint* product of internal and external factors' (390). However, the likelihood that all biologists are now interactionists and not determinists may not be enough to banish the spectre of modern science's foundational dichotomy. Gray points out that euphemisms for determinism such as 'predisposition' are still in play. The solution to this mere dilution of the problem comes in the form of a developmental systems approach which is founded on the interaction of genetic and environmental influence, and on contingent not determined development. In response to challenges from biologists as to the necessity of this approach – given the probable contingency of all genetic 'determinism' – Gray (1997: 397) argues that the central issue is 'whether the genome has a privileged role in development that generates some asymmetry between itself and all other developmental resources'. Without denying that there are distinctions among developmental processes, what the proponents of this theory reject is the hierarchical separation of two kinds of process – internal and external, innate and acquired. There is no autonomy, and no dominance of the gene over the cell, the organism over the environment. The biopolitical relationship which is subsequently outlined is one of interdependence and mutuality. Evolutionary biology is not to be conflated with the biopolitics of autonomy and domination since its central Darwinian principle – natural selection – is itself not genetically determined: 'Natural selection requires heritable differences in fitness-related attributes. The exact mechanisms of inheritance are not important' (399). In other words, just because a trait evolves does not mean that it is genetically rather than environmentally determined. In deconstructing the nature/culture divide and the attendant science wars (as exemplified in debates on evolutionary psychology), it is important for Gray to note that 'the traffic can go both ways'. If there is no autonomous realm of the biological or the social 'then not only should we be exploring the causal connections between the two, but we should also be exploring *potential methodological links*. Social scientists who have some enthusiasm for the critique of biological determinism may be less enthusiastic about this' (404). The methodological link which is implicated here is the already existing, the nascent dialogue which must itself be nature/nurtured given the ease with which any argument can be recuperated and subsumed within oppositional terms. It is important that, especially if there is currently much lip-service being paid to interactionism, feminism and science do not, as Gowaty learned, get ahead of themselves. Getting beyond the sex/gender divide is dependent on a dialogue at, or indeed beyond, the limits of both science and feminism which recognises 'the ways in which experiential inputs shape physiology and anatomy, and the ways in which physiological and anatomical differences shape behaviour and experience' (Gray 1997: 406).

Pengi and the Expressivator

In the previous chapter, I examined some examples of risk-taking ALife art (Prophet and Selley 1995; Tenhaaf 1997), and earlier in this chapter I mentioned the undoubtedly risky (if not downright dangerous) practices of essentialist cyberfeminism. There is, of course, a much greater legacy of dialogue between art and science which is currently so well established as to be institutionalised in, for example, the Wellcome Institute's annual Sci-Art competition. The historical relationship between science and post-structuralist social science/feminism/ cultural theory is much more problematic as the debates on sociobiology and now evolutionary psychology clearly illustrate. Andrew Ross wrote of the *Science Wars* in 1996, and there is a degree of consensus that these are by no means resolved (see, for example, Rose and Rose 2000). So, in order to ground this discussion, I now turn to examples from the intersection of computer science and cultural theory which specifically articulate a concept of risk and dialogue and which work in particular with the technical and discursive implementation of autonomous agency. It is through this kind of work that feminists might engage (computer and biological) scientists in a dialogue about subjectivity which defers the question of sex and gender. One advantage of doing so is that feminism may attend to the details of computer science which, as is internally acknowledged, is by no means a neutral field (and not just because it currently incorporates biological principles). Another advantage relates to the fact that there seem to be as few true biological essentialists as there are genetic determinists in science and a focus on the question of autonomy and agency offers a more expansive route to the discussion of gender and other aspects of identity within the framework of (post)humanism. AI researchers in particular are aware of critiques of their discipline which characterise it as being predominantly instrumentalist, rationalist, masculinist and anti-humanist. There is also a tendency for them to conflate non-scientific disciplines, the humanities, with humanism and so the conflict is set up internally, inviting a response.

Phoebe Sengers' project (introduced in Chapter 5) aims to establish an 'alliance' between AI and cultural studies of science by studying AI as a culturally situated activity and by using the results to generate novel technology (Sengers 1998: iii). Sengers is critical of the process of atomisation (or what she calls 'boxology') in autonomous agent design, regarding it as something of a catch-22. Atomisation is effected through the modularity of agent design; literally the way in which behaviours are broken down into fairly isolated modules. While this may be a necessary strategy for building comprehensible code 'it is fatal for creating agents that have the overall coherence we have come to associate with living beings' (1998: iii). Atomised agents lay claim to autonomy of both parts and whole (they do not effectively interact on any level) and to agency but are, for Sengers, essentially 'autistic' (1998: 20). Agency functions as a productive, profitable anthropomorphic metaphor in a field which internally contests the proper or most effective model of agent design. Here, Sengers characterises the relation between AI and ALife (otherwise known as 'alternative' or 'nouvelle' AI) as the often fraught contest between cognitive and connectionist models of mind and between autonomous disembodied agent designs on the one hand and situated embodied agent designs on the other. She presents a strong argument that these contests are not merely scientific, technical or methodological but that they are epistemological and ethical disagreements about subjectivity and the role of the body in the formation of knowledge and intelligence (6). Technical disputes about design are 'one specific area where the implications of ideas that are rooted in background culture are worked out' (9). Sengers explores some of those background ideas about the relationship between technology and subjectivity through the concept of schizophrenia. This is at once a technical problem concerning the difficulty of integrating agent behaviours and a philosophy of the subject in industrial and post-industrial societies (Jameson 1984; Deleuze and Guattari 1988). In both contexts, the 'agent' is atomised, mechanised, alienated and incoherent. Working through the technical aspect of AI's schizophrenia is one means of acknowledging the economic, social and psychological dimensions of the same problem which is not, in practice, obscured but highlighted in her decision to recreate the illusion of coherence in her agent design. Schizophrenia in AI is not a solvable problem for her, but an inevitable outcome of the design process – at least in classical AI. By implication, Sengers, unlike Grand, does not see ALife as the means to resolving the problems of AI. Rather than turning to biology she looks more closely at psychology in order to generate not real, but believable agents. For her, these exhibit more fluid, less fragmented behaviours and are therefore more lifelike, if not alive. What she does draw from ALife is a belief in the efficacy of situated agents, although she is keen to extend ALife's notion of the environment from a purely physical to a sociocultural one (Sengers 1998: 92). The sociocultural context within which agents are located is partly that of objects and other agents and partly that of the designer and user. Sengers thus begins to regard agents not as autonomous entities but as forms of communication between the designer and user. They become literal expressions of subjectivity: 'An agent is a representation . . . a mirror of their creator's understanding of what it means to be at once mechanical and human, intelligent, alive, what cultural theorists would call a subject' (94). An agent is not a 'pristine testing-ground' for theories of mind and matter – not a creation but a construction of life. This agent is a risky construction in so far as it is designed to bridge the perceived gulf between AI engineering and its 'humanist' detractors. What Sengers calls the 'standard' critiques of AI 'while not always easy for researchers to hear, could potentially help AI researchers develop better technical practices'. The problem is that these critiques are often unintelligible to computer scientists, leading to a 'ghettoisation' of both camps: 'technical practices continue on their own course without the benefit of insight humanists could afford, and humanist's concerns about AI have little effect on how AI is actually done' (1). This is an argument for dialogue premised on the acquisition, the learning of new languages – be they technical and/or theoretical. It would not be enough here to defer to interdisciplinarity,[4] since this term does not convey either the problem or potential solution of the science wars. These are characterised by a 'siege mentality' (11) and frequently by levels of hostility which attract media attention (Gross and Levitt 1998 [1994]; Sokal 1996; Thornhill and Palmer 2000) and encourage further entrenchment and position-taking. The principal inspiration for Sengers is the work of Philip E. Agre, and particularly Agre's formulation of 'critical technical practices'.

Agre (1997) seeks to embody anti-Cartesian philosophy in agent design. He uses technology as a form of epistemology which opposed the Cartesian tradition of mind over body in classical AI. This strategy is at the heart of his critical technical practice which, although common to artists is risky in the context of computer science where it is difficult to gain recognition for the problem, let alone any possible solution. The problem of disembodiment in AI is compounded by the division between computer science and critical theory. Believing his own discipline to uphold 'false' conceptions of human experience, Agre also maintains that 'critical analysis quickly becomes lost unless it is organised and guided by an affirmative moral purpose' (1997: xii). His own moral purpose is to offer one possible route to a corrective (critical) practice. What is stressed here, is that philosophy serves as more than a supplement or means of supplanting technical methods. The point is 'to expand technical practice in such a way that the relevance of philosophical critique becomes evident as a technical matter' (Agre 1997: 2). Agre's internalist revision is developed through a history of increasing anthropomorphism in computer science since the Second World War. The developing process of automation, for example, lent itself to anthropomorphic descriptions based on the autonomy and intentionality of machines. Moreover, Claude Shannon's Information Theory 'soon provided mathematical ways to talk about communication' in computers (Agre 1997: 1). The increasingly 'suggestive confluence of metaphor, mathematics and machinery' led the way for a 'counter-revolution against behaviourism and a restoration of scientific status to human mental life' (1). The new anti-behaviourist cognitive psychology (in which the emphasis shifts from external to internal programming) sought to realise the other side of the human–machine analogy in a computational theory of mind. Once digital technologies were developed, 'a powerful dynamic of mutual reinforcement took hold between the technology of computation and a Cartesian view of human nature, with computational processes inside computers corresponding to thought processes inside minds' (2). The rationale for the artificial intelligence project is founded in this dynamic. Agre's aim is 'to reorient research in AI away from *cognition* – abstract processes in the head – and toward *activity* – concrete undertakings in the world' (3). His aim is to re-embody computer models of human knowledge and experience by engineering 'situated, embodied agents living in the physical world' (4). In AI, Agre points out, the word 'agent' is both common and productively ambiguous, providing useful ways to talk about entities ranging from robots, insects and cats to people without 'reducing all of them to a bloodless technical order' (4). By stating that an agent is situated, he is emphasising that its actions do not make sense outside of the specific situation in the physical and social world in which it is located (4). An embodied agent does not simply have, but, after Merleau-Ponty (1962 [1945]), '*is* a body', it '*exists as* a body' (Agre 1997: 4). As a physical being it then has limited experience and finite capabilities. The point is not to assert the truth of these ideas – which for Agre is unlikely to be contested – but to demonstrate their impact on foundational ideas in both computer science and psychology.

The broader aim of Agre's critical and constructive project is 'to ease critical dialogue between technology and the humanities and social sciences' (1997: 4). Agre's agent is not an autonomous thinker in an implicitly hostile world of problems requiring intelligent solutions, but is one among other 'doers situated in a usually benign environment' (Sengers 1998: 32). Pengi is a computer program designed to play a commercial video game called Pengo. The game consists of a geometrical arrangement of cartoon figures – a penguin, some bees and a number of ice cubes. The aim is for the penguin to kill all of the bees using the ice cubes and to avoid being stung or hit in return. Agre describes the adversarial nature of the game as 'unfortunate', but claims that it does not detract from the principal technical-discursive innovation which is to implement successful agent activity without recourse to a cognitive map or plan. Problem-solving, 'mentalist' AI programmes rely on a planner: 'that is, a complicated set of machinery for building and maintaining world models and for constructing and executing plans' (Agre 1997: 265). On the other hand, Pengi's architecture is simple, more improvisational and more interactive. Instead of executing a plan derived from a whole world map, it visualises possible movements in a more localised contingent and biologically realistic strategy (Agre 1997: 269). Its fitness within the environment is not based on its autonomy and it is one partial model of an alternative praxis. Agre problematises (cognitive) autonomy as one of the generative metaphors within his own discipline and Sengers develops this problematic in her own agent architecture which she names the Expressivator. Where Agre derives his architectural innovation from a theory of representation, Sengers derives hers from structuralist models of communication, such that an agent's behaviours are recast as signs within a sign-system. The design of the Expressivator is based on the management of units of meaning rather than units of information, and it is its communicative 'nature' which situates it in a world of embodied agents rather than intelligent machines. By engineering links or transitions between otherwise atomised units of behaviour, Sengers seeks to present a coherent, co-dependent believable agent to potential users both in and beyond the art and entertainment industries. Believable agents come to life like characters in narrative rather than like organisms in evolution – they are intelligible not intelligent. Senger's AI thereby denaturalises the role of the agent in ALife and retraces the almost imperceptible but fundamental shift from metaphor to analogy in the 'creation' of biological technology.

The Expressivator is an agent only in so far as it is afforded agency by both designer and user. This then is a strategic (if not quite parodic) anthropomorphism. Animate intelligence, as the designers of Cog have come to realise, is in the eye of the beholder. It is more a facet of interpretation than observation and it is in this way that Sengers marks her critique of Enlightenment epistemology in computer science. The interesting question posed by both projects is whether computationalism (as an approach to the study of intelligent life) adapts successfully to the realisation that its object is unfixed (life as a moveable feast). Is it possible, in other words, to compute a dynamic nonlinear intersubjective and embodied process in a meaningful and/or useful way (bearing in mind that Sengers and Agre integrate narrative and representation – respectively – with computational methods and that they make appropriately modest claims for their designs while even those more conventional AI approaches – such as Pattie Maes's have reported difficulties in 'scaling up')? In so far as this is a goal in contemporary or alternative AI (ALife), it is one which is widely deferred. The outstanding risk – actually for both perspectives – concerns the survival of computation in its dialogue with critical philosophical discourses. Currently this risk is managed by the marginalisation of dialogue within small (but growing) communities of theorists, practitioners and engineers. But the generation and degeneration of metaphors is a leaky amorphous process (Keller 1995, 2000) and if the human–machine metaphor is literalised and subsequently de-literalised then what remains of it, and are there only remains, traces? There are certainly traces (as any trawl through the mass media would demonstrate),[5] but if we are not proposing to look at humans as machines, machines as humans then what does the conjunction of human and machine currently mean – what is its purpose? While the professional, technical, economic and ideological efficacy of biologisation has been explored in Chapters 3, 4 and 5, what I'm arguing here, in the context of anthropomorphic autonomous agency in particular, is that the purpose of the human–machine conjunction is becoming increasingly anti-instrumental. It is a material and metaphoric tool through which a desire for humanism is being rearticulated. This rearticulation is necessarily in productive tension with liberal humanism since it incorporates what the liberal tradition excludes as other than human – technology. Technology, in its incorporation in to human life, no longer clearly signifies the anti-humanism celebrated in postmodern praxis. It is beginning to signify at least the desire for a post-liberal humanism articulated within more dialogic, more risky (less deconstructive) discourses.

Alife-as-it-could-be

The concept of dialogism employed here as a feminist, scientific method entailing risk, is derived from Bakhtin's work on language and the novel. Critical of earlier approaches which resolve into dichotomies analogous with universalism/relativism (such as abstract formal versus ideological approaches to the novel or Saussurian structuralist divisions between langue and parole), Bakhtin argues that as language is characterised by heteroglossia – plurality not just of dialects but of historically located voices and experiences – so the novel is characterised by dialogism or the interaction between multiple oppositional viewpoints. As Michael Holquist (1981) states, Bakhtin's concept of language is based on a 'sense of opposition and struggle at the heart of existence, a ceaseless battle between centrifugal forces that seek to keep things apart, and centripetal forces that strive to make things cohere' (Holquist 1981: xviii). The battle does not resolve itself into the kind of binary opposition which characterises structuralist semiotics and aspects of the social sciences: 'That opposition leads from human speech to computer language; it conduces, in other words, to machines'. Bakhtin's sense of a contest between wider forces 'leads in the opposite direction and stresses the fragility and ineluctably historical nature of language' (xviii). Bakhtin's decidedly situated view of language as heteroglossia is modelled on 'two actual people talking to eachother in a specific dialogue at a particular time and in a particular place' (Holquist 1981: xx). Far from the abstract binary (zeros and ones) informational view of language as the communication of a message between sender and receiver, Bakhtin's model is polyphonous, polysemic, 'noisy', nonlinear. Dialogue brings into play the boundary tensions between languages, 'creates a feeling for these boundaries, compels one to sense physically the plastic forms of different languages' (Bakhtin in Holquist 1981: 364). What is evoked here as the structure and imaginary of a literary genre is suggestive in other contexts,[6] and without transferring as such might inform the analysis of specialist languages, of disciplines rather than dialects in tension. These too can be evoked in their global-historical and local dimensions and characterised with reference to heteroglossia. The plurality of voices structures and strains the disciplines of science and feminism and may also be realised, in part, through the kind of conversations or specific dialogues (between two people at a given time and place) which Bakhtin describes.

Autopoiesis and autonomy

Within ALife, one of the axes along which centrifugal forces outweigh the unifying centripetal tendency is that which concerns the relationship between organism/agent and its environment. This relationship centres on the question of autonomy or the degree to which the agent is self-determining rather than determined; merely responsive to its surroundings. In a talk entitled 'The "Whole Iguana" Mk 2' (6 June 2001) in which Steve Grand presented his work in progress on Lucy, the robot orang-utan, to delegates at the University of Sussex, the subject of autonomy was disputed. In the context of his attempt to engineer a self-organising entity, Grand reflected critically on the internal, cognitive capacities of his norns which he described, in conclusion, as 'very behaviourist'. In contrast, he hopes that Lucy – still in the process of acquiring a richer neural and sensory system – will have cognitive abilities lacking in norns, notably the ability to imagine, as distinguished from the provision of a plan. Grand's servo model of imaginative autonomy in which the agent is able to connect a current to a desired state (in thought and/or in action) is not based in nature, and what Grand outlines here is not biology-as-we-know-it ('biology isn't like this . . .') but biology-as-it-could-be (' . . . but it could be'). His infidelity to nature and to the evolutionary process were the grounds on which Lucy's viability was challenged by one of the delegates. How, he asked, could Grand sustain his claims to be able to engineer agency as complex as that of an orang-utan when it is still not possible to accurately model the intelligent behaviour of insects? Grand's defence of his risky project was articulated more perhaps on philosophical than on purely biological or technical grounds. It seemed to me not only that he spoke more as a 'strong' rather than 'weak' ALife engineer but that he spoke (in Agre and Sengers' terms) more as a 'humanist' than strictly as a scientist/engineer when he restated his investment and greater interest in imaginative self-directed artificial agents exhibiting complex behaviours in specific situations. Grand's insistence on an unconventionally holistic view of the organism *in situ* mitigates against an unproblematic autonomy and outlines a form of co-evolutionary interdependence between the organism/agent and its environment. In other words, the agent both acts and is acted upon, making it a 'self-organising, self-maintaining system in flux' (Interview, August 2000). Here, in a dialogue covering the terms of embodiment, situatedness, form and matter Grand refers to the concept of autopoiesis introduced by Maturana and Varela in 1971. Autopoiesis, for Grand, is predicated on the constitution of all agents (natural and artificial) as form not matter. In answer to my protestations regarding the importance of the body in definitions of (a)life, Grand maintains that:

> I'm not denying the importance of the body; I'm denying the very existence of matter . . . I totally agree that the localisation in space and time (the configuration and the history) are the only things that matter . . . in non-linear systems like you and me . . . But we can simply treat the matter as a placeholder – the markers for the configuration of ourselves and our minds, that reflect our previous history as an organism. The matter is *not* the organism. Nor is it special in any way that a physicist would claim. Atoms are form. They are self-organising, self-maintaining systems in flux. People are self-organising, self-maintaining systems in flux too (this idea is called autopoiesis). To focus too hard on matter is to give in to the physicists! Think of atoms as mere markers, pegging out ever-changing relationships. It is these relationships that make us. They extend into the body and outward into

society. What counts is the rules that govern the changing relationships, not the stuff of which we are made. Denying the very existence of that stuff liberates us from physics and makes our minds and societies and other relationships as important and as *real* as electrons and protons.

(Grand interview, August 2000)

Grand's philosophy of (a)life is strongly demarcated against the materialism of physics as the hegemonic scientific discipline. It is also against the rationalist, masculinist epistemology of physics-based AI that he poses his own brand of 'androgynous' and risky thinking which is, above all, antagonistic towards polemical forms of polarisation. Inevitably, this androgynous and risky thinking, as a counter-strategy, occasionally falls into its own trap: 'I have to overemphasise my case . . . it's then very easy to find myself having to defend the extreme view (that I don't believe) against the converse extreme view'. So 'I sometimes write that "matter doesn't exist – there are only relationships of cause and effect"', but 'I know perfectly well that relationships have to be *between* things, so there have to be some things as well as the relationships'. Grand also knows perfectly well that such a strategy can lead, and has led him in to battles akin to the Darwinist's 'arms race'.

According to Margaret Boden (2000), Maturana and Varela's concept of autopoiesis effectively does for evolutionary biology and cognitive science what Grand's concept of ALife does for AI and physics. Insisting 'in contrast to the dominant cognitivist, symbol-processing views of the 70s' on the embeddedness of the cognitive agent, 'natural or artificial', the term is nevertheless emblematic of a particular view of the agent–environment relation 'where its self-constituting and autonomous aspects are put at the centre of the stage' (Varela 2001: 5). Autopoiesis (from Greek, meaning self-producing) defines the minimal criteria for life and distinguishes the living from the non-living. The definition amounts to self-produced coherence, or self-organisation and maintenance as a discrete, unified entity. According to Varela (2001: 6), this is neither a reductionist nor a holist, a mechanistic or a vitalist definition of life. One of the central questions raised through autopoiesis is that of the relationship between autopoietic entities and their environment, and the paradox outlined by Varela is that this relationship is both autonomous and coupled, dependent but discrete. This paradox is maintained by the agency of the organism which acts upon and transforms its immediate environment, distinguishable for Varela, as the organism's 'world'. What Varela is doing, by weighting the balance of this 'dialogic coupling' slightly in favour of the organism, is looking at the meaning of life 'from the point of view of the living system' whereupon its environment gains a 'surplus of signification':

Thus a bacteria swimming in a sucrose gradient is conveniently analysed in terms of the local effects of sucrose on membrane permeability, medium viscosity, hydromechanics of flagellar beat, and so on. But on the other hand the sucrose gradient and flagellar beat are interesting to analyse only because the entire bacteria points to such items as relevant: their specific significance as components of feeding behaviour is only possible by the presence and perspective of the bacteria as a totality. Remove the bacteria as a unit, and all correlations between gradients and hydrodynamic properties become environmental chemical laws, evident to us as observers but devoid of any special significance.

(Varela 2001: 7)

So, this version of autonomy is literally that of the living from the non-living. It does nothing to undermine the dialogic coupling of organism and environment and says nothing about the relation between organisms. Varela insists on the existence of this 'truly dialectical relationship' in which the already embedded organism, looked at from the viewpoint of itself, acts upon its world in order to maintain itself – and he goes on to argue that this action is always already cognitive. Autopoietic identity is the only precondition for cognition, contra to 'the force of many years of dominance of computationalism, and the consequent tendency to identify the cognitive self with some computer program or high level computational description' (Varela 2001: 11). Here, Varela aligns his thinking with the ALife engineers, like Rodney Brooks, of situated and autonomous robots. Or, from the other perspective it is clear that ALife engineers who reject top-down AI 'command and control' programming and the design of internal plans based on symbolic representations of the environment are employing an autopoietic biological method instead. Cognition then is not about programming (either internal or external) and 'living beings in their worlds of meaning stand in relation to eachother through *mutual specification* or *co-determination*' (14). Since, as Celia Lury (1998) points out, the theory of autopoiesis holds that 'not only are living systems cognitive systems but . . . living is a process of cognition' it is 'sometimes described as an epistemological constructivism' (Lury 1998: 139). Such a description underplays Varela's paradox of agency and dependency and his distinction between world and environment. If the cognitive process of the living system 'consists in the creation of a field of behaviour through its actual conduct in its closed domain of interactionism and not in the apprehension or the description of an independent universe' (Lury 1998: 140), then this process is indicative of what Hayles (1996b) terms constrained constructivism.

Maturana and Varela's (1980 [1971]) primary concern is with the phenomenology of living systems, and their phenomenology encompasses epistemological and ontological concerns. In terms of epistemology, or 'the validity of the statements made about biological systems', they recognise that 'evolutionary and genetic notions have been the most successful' (1980 [1971]: 115). Yet, it is Darwinian evolutionary theory which they counter on the basis that although it provides an effective explanation for historical change, it does not adequately define the domain of biological phenomenology or the nature of living systems. For them, any universal theory is 'doomed to failure if it does not provide us with a mechanism to account for the phenomenology of the individual' (115).

Maturana and Varela's phenomenological reassessment of the evolutionary emphasis on species rather than individual is the basis for both a posthumanist ethics and a rejection of the information paradigm.[7]

Referring to Maturana and Varela's (1980 [1971]) defining text on autopoiesis, Margaret Boden (2000: 127) argues that it is closely aligned with a particular reading of metabolism but that the main differences result from the greater emphasis Maturana and Varela place on the self-maintenance of unity. This emphasis leads them to prioritise processes over the components of self-maintenance and to stress the importance of form over matter which Steve Grand and other ALifers find amenable to their work. It also leads them to decentralise the role of reproduction and evolution in biological definitions of life. They argue that reproduction, or genetic self-replication 'assumes the pre-existence of an identifiable unity' and that evolution, of course, presupposes the existence of reproduction (Boden 2000: 134). Outlining the conflict between autopoietic and evolutionary biology, Boden (2000: 128) suggests that 'part of the reason for regarding the maintenance of the bodily boundary as fundamental to life is that . . . one can conceive of living things that are self-bounded but not subject to evolution'. If autopoiesis is, as Boden says, 'biology as it could be' – a subversive alternative – then, specifically, the marginalisation of reproduction and evolution – bearing in mind that this is a difference in emphasis, not a wholesale negation of these processes – could surely offer much to a feminist, queer, antiracist formulation of alife as it could be? ALife artist Nell Tenhaaf (see Chapter 5) certainly finds it productive. Boden attempts to undermine the relation between autopoiesis and ALife by conflating autopoiesis with metabolism. Artificial life-forms, she points out, do not metabolise, even – as in the case of *Creatures* – where there is an attempt to simulate metabolism (Boden 2000: 122). Even robots, though bounded and in some sense unified, are engineered not autopoietic life (132).

Boden points out that the theory of autopoiesis is formulated not in relation to natural or artificial bodies as such, but in relation to the self-organisation of the cell – the formation of the cell membrane which bounds and constitutes the cell as 'an autonomous vital entity, distinguishable from its environment' (Boden 2000: 124). Attempting to explain this phenomenon is universally acknowledged to be one of the foundational problems of biology and, for Maturana and Varela it is '*the* philosophically and scientifically fundamental problem' (Boden 2000: 124). Nevertheless they do speculate on the nature of more complex forms of self-organisation including minds and societies and their critical intervention in cognitive science goes beyond the significance of the cell.[8] It may well be that they conflate cognition with adaptation in their assertion that all life is cognitive (no distinction between human and other forms, no distinction between minds and bodies), but this is a philosophically significant assertion which rejects the assimilation of Cartesian thinking into the biology of informationalism: 'Where every constituent of the system is an essential participatory element of it, talk of information processing is out of place' (Boden 2000: 136). That autopoiesis promotes a subversive view of the subject-object of biology and cognitive science is evidenced by Boden's attempt to de-limit and contain it with reference to metabolism: 'It would be better, then, to use the term "cognition" more strictly, so as to avoid the implication that autopoiesis necessarily involves cognition . . . If that restriction is made, the concept of autopoiesis in the physical space comes ever closer to the biologist's concept of metabolism' (Boden 2000: 140). What autopoiesis does, in effect, is displace the rational, cognitive, information processing disembodied and autonomous self from the centre of biology and psychology, replacing it with something more like the cell in the 'organismic dialectic of the self' (Varela 2001: 14). Lacking teleology or identity, this self-maintaining homeostatic unity is the basis of Maturana and Varela's anti-evolutionary, posthumanist bioethics already strategically employed in a dialogue with forms of social Darwinism.[9]

Both Hayles (1999a) and Helmreich (1998a) have offered useful accounts of autopoiesis in relation to ALife; underlining the association between autopoietic entities and their embodied, therefore historical emplacement within the environment (Helmreich 1998a: 222), and highlighting the possibility of non-terrestrial, non-carbon-based and formalised autopoiesis (223). Varela has made the link between autopoiesis and autonomous agent research (Hayles 1999a: 223). Where Helmreich (1998a: 224) criticises the conflation between autonomy and autopoiesis, Hayles begins to articulate a critique of autonomy within the new forms of liberal posthumanism (see Chapter 5). She does not relate this back to autonomous agent research or to the biopolitics of autopoiesis. Autopoiesis is a biopolitics of autonomy within which autonomy is delimited by the co-dependence of the autopoietic entity on its environment, and by its historical emplacement:

> This autopoietic process of self-production or self-organisation unfolds over history (both evolutionary and individual) and cannot be understood except through that history. Moreover, that history must be told from the point of view of the system in question. The processes that maintain an organism's identity cannot make sense except from within that identity, because that identity is fundamentally organised by its particular history of being situated in a domain of interactions.
>
> (Helmreich 1998a: 222)

Through Helmreich's somewhat tautological reframing of the individual in history and history within the individual, it is interesting to relate autopoietic biology (of essentially cellular organisms which may also be autonomous agents or indeed, societies of autonomous agents) to critical psychology and its mode of historicising and thereby delimiting the autonomy of the individual. Critical psychology problematises the status of the autonomous individual in traditional psychology where it operates as a bounded, pre-formed self at the heart of a dualism between the individual and society. Psychology derives this notion of an autonomous individual from its historical coincidence with the rise and hegemony of Darwinism in the late nineteenth and twentieth centuries. Darwin's construction of 'man' as the rational, moral, individualist animal able to affect control over the environment, becomes naturalised in psychology which then adopts a regulatory, governmental function in which self-reliance, rationalism, entrepreneurial individualism, self-transformation and autonomy are promoted as the biosocial ideal and defined against a pathologised 'other' self. This 'other', as Blackman and Walkerdine (2001) illustrate, refers to both the innately inferior gendered and racialised individual and to the social body of the irrational, vulnerable masses. By historicising the foundational concept of autonomy in psychology, critical psychology undermines the scientific authority of a discipline which claims that what it means to be human, 'what we are really like' is discoverable (Blackman and Walkerdine 2001: 6). Its strategies are articulated within a post-structuralist framework where identity is seen to be constituted in language and discourse and within this framework, critical psychology opposes Darwinian biological determinism with a semiotic form of cultural determinism:

> The sense of being human embedded within the human sciences, especially psychology, is that the possession of language and reason enables individuals to develop morality and a sense of responsibility. This is what makes us human and distinguishes us from the animals. Language represents our mental and cognitive processes imposing a structure on an otherwise chaotic world. Semiotic approaches view language and the structures of language as creating the very possibility of representing the world to us in a particular way. In order to understand what it means to be human, we cannot look inwards but instead need to focus upon the historical and cultural processes that make our sense of self possible.
> (Blackman and Walkerdine 2001: 20)

Critical psychology focuses on the embodiment of history in the individual as a reaction against a scenario in which the individual imposes himself on history. This reaction is anti-biological and anti-humanist. What then of the point of view with which autopoietic biology maintains the tension between the bounded, autonomous but not pre-formed individual and the biosocial environment?

Embodying Alife

Phoebe Sengers' work exists in what she calls a 'productive tension' between the two poles of liberal humanism – with its 'rigid notions of the human which often leave quite a few people living in the human-non-human boundary area' – and postmodern theories of subjectivity which 'seem to celebrate an abandonment of human-ism in a celebratory unity with technology' (Interview, May 2001). Her internal critique of AI is not formed by an a priori feminist agenda but by a conviction that the kind of subjectivity represented in AI programs – rational, problem-solving, efficient – omits important aspects of human experience such as 'emotions, expressiveness, sociality' (Interview, 2001). Her own computational subjects, or representations of subjectivity, are not, she suggests, particularly embodied themselves: 'It's the embodied subjects of the people BUILDING or USING the computational subjects that's interesting' (Interview, 2001). Sengers offers a productive method of countering the impasse in which scientists are confronted with often unconscious baggage which, by definition, is not readily available for them to review. This method is 'to work your way into the science until the attitudes can be discussed "scientifically", i.e., in a language that scientists understand. And this can also be a great way to create new scientific work with different attitudes, that then can help shape our societal self-understanding in different ways' (Interview, 2001). In a collaborative project with Simon Penny (artist, theorist) and Jeffrey Smith (roboticist) entitled 'Traces: Embodied Immersive Interaction with Semi-Autonomous Avatars' (Penny *et al.* 2000), Sengers explores the relationship between human users and their computer representations. Traces is an artwork which enables users to experience full body interaction with semi-autonomous avatars (avatars are generally understood to be computer programs which represent users directly in the computer environment, and can be similar to bots or agents). It does not constitute a navigable world, but a space within which the bodily movement of the users leaves behind volumetric and spatial-acoustic residues or traces. The project was designed to explore and critique four main issues in human–computer interaction: embodied interaction with computational systems; rapid and transparent learning of the computer interface by untrained users; immersive bodily interaction with software agents and the notion of interactivity itself (Penny *et al.* 2000: 2). In part, the project is an extension of Simon Penny's critique of VR as a transcendent, disembodying medium (Penny 1994, 1996). Traces is not based on VR technology with head-mounted displays, but on the 'inherently more embodying' technology of the CAVE. CAVE is a spatially immersive display consisting of a cubic room measuring three meters a side. By using multiple projectors, the CAVE displays stereo graphics on three walls and also on the floor. The project organisers wanted to enhance the embodying qualities of this technology by building a sensing system capable of modelling the entire body of the user. In order to achieve this, they built 'an infrared multi-camera machine-vision system' capable of constructing a volumetric model of the user's body in real time. They also developed customised three dimensional vision tools, graphic techniques and an agent behaviour environment (Penny *et al.* 2000: 3).

The artistic goal of the project is to facilitate a user experience which combines dancing and sculpture. By moving around in the CAVE environment the user 'dances' a 'sculpture' of entities into existence (4). Artificial entities are, in other words, generated by bodily movement and the environment offers not a detached panoptic spectacle but a spatio-temporal experience of embodiment. Crucially, the entities or traces generated by the user's movements become more active, more autonomous over time. At this stage, they fly around the space interacting both with the user and with each other, and exhibiting more complex behaviours (akin to Craig Reynold's 'Boids') of their own. A major goal of the project then, is to develop a range of techniques for generating semi-autonomous avatars in immersive spaces (5). These avatars, unlike those which characterise most virual environments, are not constructed as direct representations of the user (5). These are not 'soul-less bodies for which the user acts as mind' and they do not enact the user's instructions (19). Though co-dependent – they 'have their own behaviours and intentionality, but are intimately tied to the user's actions' – they do become increasingly independent: 'In Traces, user body movements spawn avatars that gradually become more and more autonomous' (6). Is this then, a developmental or evolutionary scenario which posits autonomy as the ultimate goal of (non)human life? It might seem so, but it is important to consider that in 'Traces' as distinct from 'Anti-Boxology', the authors/artists/engineers start from the non-autonomous end of the avatar–agent continuum. They treat as their problematic technical-discursive constructions not of autonomy but of non-autonomy in which 'the avatar, fundamentally, is the user' (19). This notion of 'avatar-equals-user' is now considered to be both technologically and philosophically inadequate (5) and so Sengers and colleagues use the concept of autonomy strategically in order to create new possibilities for user–avatar, human–machine relationships somewhere on the scale between self and other, sameness and difference. Semi-autonomous avatars 'can be thought of as on a range of autonomy, from the traditional fully passive avatar to the traditional fully active agent' (21). By constructing a scenario in which the user interacts physically with a series of avatars of increasing complexity and autonomy, these 'humanist'/engineers begin to fill out, flesh out the continuous scale of autonomy (21). By designating three stages of development/evolution from the Passive Trace to the Active Trace and finally to the Behaving Trace, they seek to incorporate the full range of current human-machine relationships and to validate that which moves beyond the dichotomy of passivity and activity characterising the familiar relationship between master and slave. Behaving Traces, initially formed like water droplets shaken off by the user, do in fact 'have articulated bodies that consist of a sequence of spheres, each of which follows the sphere before it' (26). Their embodiment, though, is primarily symbolic in that they come to constitute 'a kind of half-alien self' for the user who experiences their 'persuasive psychological effect' (28). The 'trace' functions as a new metaphor and the basis for a new technology of subjectivity which is not based on human identity and on the concomitant techniques of the anthropomorphising, narcissistic or omnipotent self. Traces are not embodied agents like or for humans, but they embody relationships between humans and machines which are a-like; like and not like each other. The project is about a non-fetishistic and ungodly desire for a(lien)life.

Towards situating Alife

The alignment between alife and postmodernism (Chapter 3) and alife and posthumanism (Chapter 5) has been introduced. Helmreich (1998) explains that the technologies of simulation and synthesis which characterise alife place it in an ambivalent relationship to notions of the postmodern, and may more easily align it with a sense of 'amodernity': 'Artificial life holds fast to modernist ideas about the unity of science and the world. But the practice of simulation pulls it along a postmodern trajectory that threatens/ promises to undermine these foundations'. So, as alife vacillates between modernism and postmodernism it undermines the stability of the category of the modern. From a perspective of amodernity – which refers to the embeddedness of science in culture – alife is 'simultaneously an empirical science, an engineering practice, a set of technologies for telling old and new stories about gender and reproduction, a tool for stabilising and undermining existing economic ideologies, and a religious epistemology' (236). Alife cannot be said to shatter foundational dualisms of nature and culture if these dualisms were never firmly in place. Postmodern science does not then overturn or supersede modern science but is, in a Derridean sense, a 'dissension' within it (Derrida 1978). Ihab Hassan describes the inflection of posthumanism within humanism in a similar way. Posthumanism is at once a 'dubious neologism' and a 'potential in our culture'. It is important to understand, he suggests, 'that five hundred years of humanism may be coming to an end, as humanism transforms itself into something that we must helplessly call posthumanism' (Hassan in Badmington 2000: 2).

Badmington (2000) amplifies the point by tracing the internal challenge to humanism through post-structuralist theory and popular culture. Popular culture, he reminds us, issues a consistent defence to this challenge lacking in theories which optimally celebrate the demise or deconstruction of humanism from within: 'Humanism never manages to constitute itself; it forever rewrites itself as posthumanism' (Badmington 2000: 9). What exactly is at stake in posthumanism remains, for him, totally open to question. What matters is that 'thought keeps moving in the name of a beyond' (10). If Alife is not to be regarded as a decisive break, a scientific, philosophical, cultural and epistemological intervention as if from the outside, then how else might it be situated within? What does it mean to situate alife as a possible dissension within (a)modern, humanist technoscientific culture? What constitutes its historical specificity? I have explored the conditions of possibility for alife in relation to the (molecular) biological hegemony of the late twentieth and early twenty-first centuries, and in relation to the failure of AI (in 2001). Alife may attempt to realise the scientific goals of molecular biology and AI – by engineering intelligent life forms – but it is not historically equivalent to these disciplines.

Paul Edwards (1996) has situated AI and cognitive psychology in a historical period encapsulated by cold war politics and US global power. He offers a revisionist history, a genealogy of AI which works to counter existing computer historiography. AI thus informs and is informed by the arms race, centralised power, instrumental rationality and masculine desires for mastery and control. Stemming from this historical moment, ALife is not entirely consistent with it. I have not offered a genealogy of alife, but have sought to develop a sense of the critical and creative dialogic potential which is facilitated by alife's heteroglossia. But, as with other cultural reflections on the subject, the outlines of a possible genealogy are inevitably drawn. The transition from what Steve Grand (2000) calls 'command and control' (AI) to 'nudge and cajole' (ALife) technologies is still a means to the end of efficient, useful, intelligent machines whose efficiency, usefulness and intelligence is now predicated more on biological than psychological, connectionist than cognitive processes. The processes tend to be characterised as feminine rather than masculine because they are more distributed, more co-operative, more chaotic, more based on the corporeal than the conceptual and more about growing and nurturing than programming life. Emergent biotechnological processes are key to the opposition of synthetic to analytic methods – holistic to reductionist ideologies; alife's feminism is a kind of essentialist eco-feminism – a nostalgic 1960s earth mother to the industrialist-militarist patriarchal cyborg figure of the 1980s. Biology is the 'softer', more feminine epistemology than that of the physics it displaces, even though physics reasserts itself through molecular biology and code-based informationalism. In its reconstruction of AI, alife becomes its own set of practices and philosophies which, though not unified, tend to be organised around autopoietic self-organisation/autonomy, evolution/emergence, self-replication/information processing. These conjunctions are culturally productive and have global-historical as well as local significance (not least through the proliferation of metaphors). Primarily, they correlate with the end of the cold war, the displacement (which, with the inauguration of US President George W. Bush and the events of 11 September 2001 is likely to be temporary) of the arms race by ideologies of greater co-operation, decentralisation and globalisation. The global bioculture is that which is evolving, emergent, self-producing and informational. It constitutes the (not) closed, distributed world of an era in which the individual and the species-self is becoming other, becoming a(lien)life. Through biotechnology, posthumanity is autopoietically self-producing and creating the conditions for posthumanism.[10]

The transition between cyborg politics and the politics of a(lien)life was marked in Haraway's cyborg figuration – her parody of the macho robot turned transgenic organism. Edwards's epilogue marks the destruction of the Berlin Wall, the 'symbol of the division of East and West for almost thirty years' in 1989, the same year that Langton gave his seminal paper on ALife. Along with East Germany, other communist governments throughout Eastern Europe also collapsed, and with the demise of the Soviet Union in 1991 'the central ideological conflict of the twentieth century finally vanished into history' (Edwards 1996: 353). Yet, the end of the cold war segues into the realisation of the global market economy – the 'ultimate achievement of world closure?' (353). The global market economy is linked to the decentralisation of power from the (struggling) US to the rather more fraught economies of Japan and East Asia, and with the development of the Internet. This, I argued earlier, retains its evolutionary status as an emergent global brain in spite of and, to an extent, by means of its commercial development (Chapter 5). So if, as Edwards (1996: 354) puts it, 'the closed world has not disappeared but merely been transformed in the post-cold war world, what has become of its cyborg subjects?' They have, he recognises, begun to transmute into artificial life-forms and may best be represented not by HAL 9000, but by Commander Data (Edwards 1996: 356). Edwards illustrates the ideological transition between AI and ALife by comparing *Terminator 2* (1991) with *The Terminator* (1984). But where this introduces, then terminates the humanised (father) machine in a normalised, liberal humanist family structure, a more useful comparison for this project is between Kubrick's re-released *2001. A Space Odyssey* (2001) and the Spielberg/Kubrick collaboration *AI* (2001).

AI, the eagerly awaited collaboration between two of the most prominent film directors of the twentieth century, retains its allegiance to the third and marginalised collaborator who wrote the short story on which it is based. In 'Like Human, Like Machine', Brian Aldiss (2001) reflects on his story *Supertoys Last All Summer Long* (1969) which was distinguished by two main narrative strands: a contemporary belief in the analogy between brains and computers and a meditation on the love between mother and child. Aldiss describes himself as being, at the time, excited by the abilities of early computers and suffused with a sense of possibility: 'I even shared a then common belief that the human brain worked like a computer, and that dreams were probably the computer downloading at the end of the day' (Aldiss 2001: 40). Moreover, it was not difficult 'particularly within the limits of a short story' to imagine an android boy programmed 'to believe himself to be a real boy, and to love his adopted human mother' (40). Even then, the story was focused more on a general preoccupation with 'love and the inability to love than the progress of computer science' (40). What succeeded as a short story arguably failed as a film more preoccupied with the encounter between Freud and the fairy tale, the Oedipal and Pinocchio, than with the current status of Artificial Intelligence. Where Aldiss questions his early faith in the informational paradigm, *AI* fails to explore the differences between HAL and the boy David, between disembodied and embodied consciousness, without becoming subsumed within a sentimental psyche. In one notable film review, Peter Bradshaw says of the 'unwholesome' romantic finale that it 'tells me more about Steven Spielberg than I ever wished to know' (*Guardian*, 21 September 2001: 13). For him, the film's true serendipity lies in its particular premonition of the events of 11 September offered in an apocalyptic vision of New York City and 'the World Trade Centre reduced to twin bungalows in the sea' (12). There are promising ethical and political elements which explore relations of servitude, exploitation, (ir)responsibility and cruelty between humans and machines and the formation of a sub-class, community or ghetto of artificial life forms; but these elements remain undeveloped and do not constitute the *E.T.* for grown-ups which was initially anticipated (Lyman 2001). If what remains is the premonition of a political landscape rather than a roundly humanised HAL – the after-effects of cold war politics and technology as a scene of destruction rather than reconstruction – then my preferred vehicle is not Spielberg's *AI*, but Anderson's *O Superman*. Performed in October 2001 at the London Festival Hall, this popular 1980s song derived from Massenet's aria, his plea for help ('O souverain, o juge, o pere'), was once a cold war cyborg anthem and is now, more clearly, a lament:

> 'Cause when love is gone, there's always justice.
> And when justice is gone, there's always force.
> And when force is gone, there's always Mom . . .
> So hold me, Mom, in your long arms . . .
> In your automatic arms. Your electronic arms . . .
>
> (Laurie Anderson, *Big Science*, 1982)

Chapter 8
Beyond the science wars

I have argued for a dialogic relation between cyberfeminism and artificial life in the context of emerging biocultures. This argument runs counter to the specific foundational anti-biologism in feminist theory and to the more general mutual hostility between science and the humanities famously outlined by C.P. Snow in his lecture on 'The Two Cultures' in 1959 and reconstituted in the science wars of the 1990s. Snow's defence of the benefits of scientific and technological progress against the then 'forces of conservatism' – a powerful literary intellectual elite acting as the arbiters of cultural value – was answered with interest by F.R. Leavis whose vigorous attack on Snow can be placed on a continuum of cultural combat:

> The 'Leavis–Snow controversy' can obviously be seen as a re-enactment of a familiar clash in English cultural history – the Romantic versus the Utilitarian, Coleridge versus Bentham, Arnold versus Huxley, and other less celebrated examples. And in this kind of cultural civil war, each fresh engagement is freighted with the weight of past defeats, past atrocities; for this reason there is always more at stake than the ostensible cause of the current dispute.
>
> (Collini 1998: xxxv)

The ostensible cause of the current dispute is the publication in 1994 of a scathing attack on an anti-scientific 'academic left' by two scientists, Paul Gross and Norman Levitt (1998 [1994]: 7). Claiming that Snow 'excoriated' traditional humanists for their ignorance of scientific principles, Gross and Levitt (1998 [1994]: 7) argue that 'ignorance is now conjoined with a startling eagerness to judge and condemn in the scientific realm'. Past atrocities and defeats weigh in with this recent development (in which science is condemned by the scientifically 'illiterate') in order to produce a combustible combination of professional pride and defensiveness which is situated in, if not confined to the academy. The context of the two cultures is mirrored and developed in that of the science wars; a widespread concern with the global socio-economic and cultural ramifications of a technoscientific 'revolution' alongside more local concerns about the expansion of higher education and the maintenance of academic standards. Collini points out that Snow did not respond directly to Leavis's 1962 attack until 1970, when 'he made clear that he felt Leavis had broken the ground-rules of debate – had misquoted him, had attributed to him opinions he did not hold, had made statements which were demonstrably untrue' (Collini 1998: xl). However, at this time, the debate had become 'inextricably entangled with the question of the expansion of higher education in Britain' (xl). Further developments over the past few decades raise the values and standards stakes through a combination of increasing specialisation and interdisciplinarity: 'in place of the old apparently confident empires, the map shows more smaller states with networks of alliance and communication between them criss-crossing in complex and sometimes surprising ways' (Collini 1998: xliv). It is just such an alliance and the epistemic diversity thereby made permissible that Gross and Levitt react against, perceiving it as an attack on the unity and integrity of scientific knowledge. Their constitution of the academic left is acknowledged to be somewhat loose and defined primarily by a vague sense of hostility and envy based on the apparent failure of socialist politics, increasing fragmentation and ideological impotence (Gross and Levitt 1998 [1994]: 26). What is most vexing for Gross and Levitt is that this aggressive attitude and the philosophical relativism which accompanies it has been rewarded by a gradual shift in power and status both inside and outside the US academy. One of their chief adversaries is Andrew Ross (1996) who argues that 'conservatives in science' who are 'seeking explanations for their loss of standing in the public eye and for the decline in funding from the public purse' have scapegoated the 'usual suspects – pinkos, feminists, and multiculturalists of all stripes' (Ross 1996: 7). Ross (1996) and Reid and Traweek (2000) locate the emergence of the science wars at the end of the cold war contract between science and the military, 'the epochal 1993 congressional decision to pull federal funding for the superconducting supercollider project' (Ross 1996: 7) and the downsizing of 'big science' in a decentralising economy. Hostilities were heightened by Alan Sokal's now infamous hoax (a science studies article published in 1996 by Ross as editor of *Social Text* and subsequently revealed as a parody) which, according to Gross and Levitt (1998 [1994]: xi), 'brought into the open a widespread reaction, years in the making, against the sesquipedalian posturings of postmodern theory and the futility of the identity politics that so often travels with it'. Ross (1996: 11) presents a scenario in which the internal reflections of the 'academic left' are inherent in the diversity of aims and methods and, at least to an extent, constitute the 'outcome of scientific self-scrutiny'. Certainly Gross and Levitt's (1998 [1994]: 3) initial thesis erased the input of certain 'working natural scientists', who subsequently emerged 'with some of the most hostile criticism' of it (ix) and were duly characterised as 'anxious apologists' for 'antiscientific, pseudosociological fads' (x). This dismissive, at times defamatory tone betrays fear and distrust in the face of boundary and collaborative discourse, creative and critical borrowings across the old divide. A particularly scathing attack is addressed to the allegedly uninformed misuse of scientific paradigms in postmodern theory (78) and to the excesses of feminist epistemology (4, 107). Containment is sought through the naturalisation of 'weak' constructivism – which allows that 'science is, in some sense, a cultural construct' (43) – as opposed to 'strong' constructivism in which 'science is but one discursive community among the many that now exist' and has no privileged access to truth (45). Where Gross and Levitt might have found the grounds for a dialogue on the limitations of relativism in the work of Sandra Harding or Donna Haraway, they avoid this challenge by producing the barest caricatures which fail to engage and function to dismiss their work. It would seem that this dismissiveness was already provoked by the fact that, in the US higher education system, it was inevitable that Gross and Levitt would regularly encounter students of Harding and Haraway . . . 'in our classes!' (251). Where Roger Hart (1996) enters the fray with an (arguably necessary) refutation of Gross and Levitt's refutation of Harding, Haraway and others – pointing out, for example, that their misreading of Harding is based on two sentences and that of Haraway on one, decontextualised interview – Hayles (herself subjected to an extensive and personalised dismissal) adopts a more strategic and less combative approach. Her motive in articulating a conceptual canon in the cultural and social studies of science is to undermine the purifying distinction of scientific content from the context of its production; the idea that the substance of science remains unaffected by the cultural institutions in which it is situated (Hayles 1996b: 227). The notion of a pristine core of science (the basis of 'weak' constructivism) is challenged by her three candidates for canonisation: 'it matters what questions one asks and how one asks them' (228); 'there is not one but several scientific methods' (230) and (after Maturana and Varela) 'everything said is said by an observer' (231). For Hayles, the inevitability of perspective, the presence of the 'enculturated' observer does not necessarily resolve universalism into relativism, objectivism into subjectivism. Rather than universal, 'scientific knowledge can, in the best case scenario, be reliable and consistent' (Hayles 1996b: 231). Following Haraway's formulation of situated knowledge, Hayles distinguishes consistency from universality on the grounds that it does not depend on absolute truth claims: 'It occupies the more modest (and humanly possible) position of providing constructs which provide reliable knowledge over the range of perspectives for which the constructs hold good' (231). This concept of a constrained constructivism is not then, at odds with a pragmatic science (233) and may be used strategically against the 'us-versus-them mentality' in order to forge alliances across disciplinary boundaries (233). Hayles asks 'what factors are necessary to make alliances with (some) scientists possible?' and argues against the familiar rehearsal of rhetorics of resistance, practised in a vacuum and providing fuel for arguments such as those of Gross and Levitt. Perhaps a greater sense of security, generosity and indeed risk is necessary to allow cultural and social theorists to acknowledge the reliability of science as a precondition to dialogue:

> Instead of posturing resistance, we need to forge alliances. To this end, it would help enormously if we were willing to make arguments that did not place constructivism in opposition to reliability. Many scientists feel they

have little stake in defending science's universality, but most believe they have very definite and specific stakes in defending the idea that science can produce reliable knowledge.

(Hayles 1996b: 234)

In *After the Science Wars*, physics professors Ashman and Baringer (2001) claim to present 'an exciting new collection that seeks to move the debate from its stalemate to a more fruitful dialogue'. However, they exclude strong constructivist arguments from a debate which they acknowledge to be not over the existence but the extent of social constructivism (2001: 6), make what I would consider (especially in the light of the evolutionary psychology debates) to be premature claims about the current establishment of 'peace talks', and offer science fiction as the true synthesis between the two cultures of science and literature, thereby glossing over the problem of epistemology. The volume illustrates that dialogue amounts to more than an interesting – and for Baringer 'fun' – juxtaposition of (mildly) conflicting ideas. In another predominantly US anthology, Reid and Traweek (2000) offer a more convincing collaboration which places cultural studies at the heart of interdisciplinary approaches to science and technology which hold the promise of allowing new objects of study to emerge and asking new questions 'not only of practitioners of science, technology, and medicine but also of those researchers who claim to study them' (Reid and Traweek 2000: 7). Locating cultural studies in the post-cold-war era of globalisation, the editors argue that although such approaches 'may be faced with the legacy of spent positivist narratives, they should not be burdened with supplying the linear narrative of a new period or successor paradigm' (5). Cultural studies remains vulnerable to tales of moral and intellectual decline, to the lamentations of modernity (distinct disciplines, conventional narratives of subjectivity, detachment and epistemological mastery) in its quest for intellectual and pedagogical experimentation and reflexivity. For Reid and Traweek (2000), cultural studies is nothing more or less than an implementation of this goal. For them it is meaningless to present a new synthesis of methods, theories, topics and materials involved in science studies, and pointless to deny the centrifugal forces at play (10). Collini (1998), in a similar way, rejects the synthesis or rather, subsuming of one specialism within another which would seem to be the aim of some of those engaged in the current science wars (Gross and Levitt 1998 [1994]; Barlow *et al.* 1992). With reference to Snow's notorious citation of the Second Law of Thermodynamics, or more specifically, of a widespread ignorance of it within the humanities, Collini (1998: lvii) questions 'whether it is most fruitful to think of a common culture so purely in terms of a shared body of *information*'. What is wanted, he suggests, is not to force physicists to read Dickens or literary theorists to mug up on theorems: 'Rather, we need to encourage the growth of the intellectual equivalent of bilingualism, a capacity not only to exercise the language of our respective specialisms, but also to attend to, learn from, and eventually contribute to, wider cultural conversations' (lvii). Such bilingualism is evident in the work of academic physicists Alan Sokal and Karen Barad, both of whom focus their contribution to a potential resolution of the science wars on the relationship between epistemology and ontology. Sokal (2001) clearly validates the cultural and sociological studies of science which adhere to a 'noncontroversial' constructivism and an empiricism consonant with 'the' scientific method. For him, this approach is 'sound' and could 'shed useful light on the social conditions under which good science (defined normatively as the search for truths or at least approximate truths about the world) is fostered or hindered' (Sokal 2001: 16). His hostility toward strong constructivism and constructivists, and the premis of his 1996 hoax revolves around the conflation between postmodernism and post-structuralism, epistemology and semiotics. Accusing postmodern theorists such as Lacan of the abuse of maths and physics, Sokal fails to discern the Saussurian basis of a poststructuralist epistemology which elides rather than eliminates the real within the symbolic, nature within culture. The problem of ontology or the strong 'claim that the nature of the external world plays no role in constraining the course and outcome of a scientific controversy' (Sokal 2001: 20) therefore lies within, is internal not external to epistemology. Sokal calls for a clarification of the relationship within science studies between ontology ('What objects exist in the world? What statements about these objects are true?'), epistemology ('How can human beings obtain knowledge of truths about the world? How can they assess the reliability of that knowledge?'), sociology ('To what extent are the truths known (or knowable) by humans in any given society influenced (or determined) by social, economic, political, cultural, and ideological factors?'), individual ethics ('What types of research ought a scientist (or technologist) to undertake (or refuse to undertake)?') and social ethics ('What types of research ought society to encourage, subsidise, or publicly fund (or, alternatively, discourage, tax or forbid)?') (21). This is an interesting point which is not invalidated by his contestable definitions or his attempt to excise 'science studies' epistemological conceits' (25) in preference for an ontological and sociological emphasis. In her attempt to clarify the relationship between epistemology, ontology and ethics, Barad (2000) insists not on an excision but an imbrication somewhat awkwardly if descriptively presented as an 'epistem-onto-logy' or 'ethico-epistem-onto-logy' (Barad 2000: 225). For her, the notion of a responsible science turns on such a formulation which attempts to re-present the central questions of agency, accountability and objectivity (225) and which incorporates risk. Barad argues that current calls for increased scientific literacy and claims of neutrality are no longer persuasive and that 'public trust in science must be *gained* by making science more accountable and by setting the standards for literacy on the basis of understanding what it means to do responsible science' (230). Her replacement of scientific with 'agential realism' is an attempt to put forward an epistem-ontological framework, inspired by physicist Niels Bohr, which balances the claims of science and the humanities and examines, particularly 'the role of natural, social and cultural factors in scientific knowledge production' (230). Unlike Wilson and Kirby (and indeed Latour and Stengers), Barad seeks to specify the relationship between ontology and epistemology: 'The ontology I propose does not posit some fixed notion of being that is prior to signification (as the classical realist assumes), but neither is being completely inaccessible to language (as in Kantian transcendentalism), nor completely of language (as in linguistic monism)' (Barad 2000: 235). This specification may be partial and cautiously negative – Barad clarifies what her ontology is not, rather than what it is – but it is the central theoretical (if not political) problem at the heart of the science wars and it constitutes a new and necessary reformulation of realism in which nature itself is not faithfully represented but 'our participation *within* nature', our material-discursive intra-action, is. The intra-acting forms of agency proposed by Barad include the human, non-human and cyborgian varieties and they share limited autonomy since, 'according to agential realism, agency is a matter of intra-acting; it is an enactment, not something someone or something has' (2000: 236). In conclusion, my own contribution to the resolution of the science wars is to have shown how debates on evolutionary psychology and especially artificial life are new but not necessary manifestations. More specifically, through a detailed analysis of artificial life in a neo-biological age – of the widespread attempt to humanise HAL – I have sought to become independent of the distinction between nature and culture which forms the 'epistem-onto-logical' ground of the science wars. Instead, I have insisted on the importance of a bioethics of posthuman identity emergent within alife discourse which cyberfeminism might productively contribute to.

Notes

1 Autonomy and artificiality in global networks

1 See Chris Hables Gray (2002) *Cyborg Citizen*. For Hables Gray:

> Cyborgology is a new multidisciplinary field that is concerned with looking at cyborgs and our cyborg society. It includes cyborg anthropologists, medical sociologists, philosophers and historians of science, technology, and medicine and many interdisciplinary scholars from lit crit cult studs (literary criticism/cultural studies) to science fiction writers and science fact journalists.
> (http://www.ugf.edu/CompSci/CGray/cyology.htm)

2 The stories referred to here are elaborated in Chapters 3 and 5 while the importance of science-fiction in relation to AI and ALife is a continuous theme of the book.

3

> If you see a video of *Creatures*, the realistic pace of the on screen life will immediately make you suspect either a massive supercomputer or a trick of time-lapse photography. But there is no trick: these enchanting, irresistible, quasi-conscious little pets live in real time on the screen of your ordinary home computer. Call it a game if you like, but this is the most impressive example of artificial life I have seen.
> (Dawkins in CyberLife 1997)

Other accolades include:

> These are the most advanced versions of artificial life in the entertainment industry, and are quite possibly a good evolutionary head and shoulders above the nearest academic equivalent – which currently has the brain power of a bacterium. Designed for entertainment, the Creatures could teach biologists a thing or two about evolution.
> (*Independent on Sunday*, 21 July 1996, in CyberLife 1997)

4 British mathematician Alan Turing sought to establish the existence of thinking machines by devising the Turing Test:

> Put a machine in one room, he suggested, and a human being in another. Give each a keyboard and a monitor, and connect these to a keyboard and a monitor in a third room. Put a human judge in the third room, and tell him or her that a machine and a human are in the other rooms, but not which is in which. Allow the judge a set amount of time to type questions through the computer to the two other rooms,

and then ask the judge to guess which room houses the human. If a series of judges can do no better than chance at guessing correctly, the machine passes the test.
(Dylan Evans, 'It's the Thought that Counts', *Guardian Weekend* 6 October 2001)

Evans also reports on the annual Loebner contest and the fact that just as no computer has yet passed the Turing Test, so no computer programmer has yet won first prize in the Loebner contest.

5 See Lynne Segal (1999) 'Genes and Gender: The Return to Darwin', in *Why Feminism?*, Cambridge: Polity Press.
6 The concept of life-as-it-could-be clearly has futuristic, science-fictional and political significance. Although predominantly constrained within the parameters of (natural, cultural) life-as-we-know-it (Helmreich 1998a), it remains accessible to dissension, intervention and the construction of (non-Darwinian) alternatives.
7 See Margaret Boden's introduction to *The Philosophy of Artificial Life*: 'Artificial Life (A-Life) uses informational concepts and computer modelling to study life in general, and terrestrial life in particular. It raises many philosophical problems, including the nature of life itself' (Boden 1996b: 1).
8 See AnneMarie Jonson (1999) 'Still Platonic After All These Years: Artificial Life and Form/Matter Dualism', *Australian Feminist Studies*, 14 (29): 47–61.
9 Monica Greco offers a persuasive and eloquent argument about the continued vitality of vitalism through concepts such as emergence: 'On the Vitality of Vitalism', *Bulletin de la Société Americaine de Philosophie de Langue Française* (forthcoming).
10 Like Grand (2000), I designate the discipline of Artificial Life through initial capitals (ALife) and the wider concept and culture of artificial/alien life (which may or may not incorporate the discipline) in lower case.
11 Note the gendered 'nouvelle' as opposed to 'nouveau' AI. Ideas pertaining to the femininity of nouvelle AI/ALife in opposition to the masculinity of classical AI are explored in Chapter 7.
12 The distinction between know-how and know-what is an important one both between ALife and AI and within AI itself.

2 The meaning of life part I: the new biology

1 With reference to sociobiologist E.O. Wilson's book *Consilience* (1998) 'which borrows from EP's [evolutionary psychology's] critique of the social sciences', Steven Rose argues that 'unlike the foundational texts of sociobiology, revisionist sociobiology along with the new EP demands a reply from within the cannon of the social sciences' (Rose and Rose 2000: 8). Evolutionary psychology is then clearly affiliated with revisionist sociobiology and may be deemed to be a facet of it.
2 Although Norbert Wiener worked for the military in the two world wars, he was expressly anti-militarist in outlook, and it is debatable how much the military way of thinking influenced his work and vice versa (for further exposition see Chris Hables Gray's (1997) work on war technology). For Lily Kay:

> From the vantage point of history Wiener's contributions to the cognitive armamentarium of the cold war were more effective than his protests. Naively, if passionately, committed to pacifist ideals and intellectual esthetics, Wiener missed the deeper significance of military pervasiveness: its impact on the world of the mind.
> (Kay 2000: 91)

3 In the context of ALife, it is also of interest that Darwin's model of natural selection is already based on breeding, or processes of artificial selection. Moreover, both

processes offer almost endless possibilities for life: 'Slow though the process of [natural] selection may be, if feeble man can do much by his powers of artificial selection, I can see no limit to the amount of change... which may be effected in the long course of time by nature's power of selection' (Darwin 1985 [1859]: 153).
4 The advent of interactive technologies, such as computer games, and of digital culture more broadly disrupts such polarities as producer/consumer, activity/passivity which have preoccupied both media and cultural studies (see Chapter 4 for the case study on CyberLife's *Creatures*).
5 These ideas, which stimulate a revision of humanist concepts of agency and autonomy in relation to the (post)human organism, are derived from work within theoretical biology on autopoiesis and are further explored in Chapter 7.
6 The word autopoiesis was coined by Humberto R. Maturana and Francisco J. Varela in 1971. *Poiesis*, meaning creation or production (Maturana and Varela 1980 [1971]: xvii), autopoiesis refers to the central self-producing characterisation of all living organisms. Maturana and Varela actually place great emphasis on the unity and autonomy of the organism: 'we wanted a word that would by itself convey the central feature of the organisation of the living, which is autonomy' (1980 [1971]: xvii).
7 Mary Jacobus (1986), 'Is There a Woman in This Text?', in *Reading Woman: Essays in Feminist Criticism*, New York: Columbia University Press.
8 Ashworth (1996: 3) writes that: 'Wilson's assertive tone was too reminiscent of Spencer's and many reacted, not to the new insights that Wilson was popularising, but to the previous attempts of Spencer and the social biologists of the 1930s'.

3 Artificial Life

1 Steve Grand confirms the extent of the influence of science fiction, especially Kubrick's *2001. A Space Odyssey* and its central character the computer HAL 9000 on the related projects of AI and ALife. Helmreich goes on to suggest that within ALife, as within science fiction, 'cyberspace is figured as the new outer space, the cosmos that the various space programs have failed to deliver us to' (1998a: 96). This new outer space yields the new alien life-forms which the space programs have also failed to deliver to the palpable disappointment of key founding figures in ALife such as Langton and Ray. The slippage between artificial and alien life is deliberately and often cynically probed in this project in order to reveal the conscious and unconscious fantasies at play in a technoscientific field which, though relatively unconventional, is always already in retreat from the non-traditional epistemologies it presents.
2 Boden (1996a) explores the controversy surrounding the informational definition of life within ALife in *The Philosophy of Artificial Life*. Adam (1998) analyses informationalism from a feminist perspective in *Artificial Knowing. Gender and the Thinking Machine*, and there are critiques of informationalism as a form of reductionism within debates on artificial life (Emmeche, Helmreich, Hayles) and the new biology (Varela, Kay, Lewontin).
3 Dawkins stresses the high level of programming required 'to simulate an emerging arms race between predators and prey, embedded in a complete, counterfeit ecosystem' (1991: 62), but see later computer games such as *Creatures*, and programs such as Axtell and Epstein's *Growing Artificial Societies*.
4 For Langton, 'Artificial Life' was coined as a deliberate oxymoron to challenge received ideas that life is the sole province of the pure category of nature (Helmreich 1998a: 21). It is also clearly a response, from within biology to the limitations placed on the discipline by the sole availability of carbon-based life on Earth: 'There is nothing in its charter that restricts biology to the study of carbon-based life; it is simply that this is the only kind of life that has been available to study' (1996 [1989]: 39). And again, 'Since it is quite unlikely that organisms based on different physical chemistries will

present themselves to us for study in the foreseeable future, our only alternative is to try to synthesise alternative life forms ourselves – *Artificial Life*: life made by man rather than by nature' (39).
5 Ray states that the original meaning of algorithm referred to an 'Arabic system of counting using numerals instead of a counting-frame, derived from the name of a ninth century Arab mathematician whose system superceded in Europe the earlier method of the "abacus" or counting-frame' (Ray 1996).
6 The analytic method in biology, by exposing the component parts of living beings, has made available a 'broad picture of the mechanics of life on Earth'. But 'there is more to life than mechanics – there is also dynamics. Life depends critically on principles of dynamical self-organisation that have remained largely untouched by traditional analytic methods' (Langton 1996 [1989]: 40). Self-organising dynamics are non-linear and depend on the interaction between parts which can only be captured by the synthetic method. Synthesis entails the combination of separate elements in order to form a coherent whole.
7 Helmreich's is the first sustained published project which is undertaken from an anthropological perspective and which very usefully outlines but cannot developed other key perspectives including that of feminism.
8 Margaret Boden is part of the ALife and cognitive science research programme at the University of Sussex. Risan and Helmreich undertook fieldwork emplacements at the Universities of Sussex and Santa Fe respectively. This project offers a different methodological perspective which engages more with the so-called 'science wars' or conflict between science and the humanities. In particular it effects a reflexive engagement between feminism and biotechnology which examines the basis and potential of dialogue.
9 Boden refers to 1960s psychotherapist Rollo May, who complained of the dehumanising effects of behaviourist psychology and 'the mechanistic implications of the natural sciences in general' (1996a: 96).
10 In an ICA conference, both ALife engineer Steve Grand and musician Brian Eno declared that CAs and particularly, John Conway's Game of Life had 'changed my life' (*What is Life? How Can We Build a Soul?* Tuesday 14 November 2000).
11 These dissenters are mostly consigned to footnote 106, page 186.
12 Helmreich does not suggest that a queer epistemology would or should be concerned primarily with sex, or that it would necessarily exclude reproduction except within a naturalised heterosexual framework (1998a: 220).
13 Aristotle in the *Generation of Animals* argues that in procreation, males provide form and females provide matter (Helmreich 1998a: 115).
14 In re-thinking the relation between matter and discourse in *Bodies that Matter*, Butler regards matter 'not as a site or surface, but as a process of materialisation that stabilises over time to produce the effect of boundary, fixity, and surface we call matter' (1993: 9).
15 The concept of 'rememory' is derived from Toni Morrison's *Beloved* and Hayles uses it in the sense of: 'putting back together parts that have lost touch with one another and reaching out toward a complexity too unruly to fit into disembodied ones and zeros' (1999a: 13).

4 CyberLife's *Creatures*

1 The analysis offered here implicitly rejects interdisciplinary models of new media effects, especially as they have been applied to video and computer games. These models are derived from behavioural psychology (Griffiths 1991) and communications research (Solomon 1990) and have been debated and interrogated within media and cultural studies over a number of years (Morley 1995; Barker and Petley 1997).

2 A clear case in point here is the James Bulger case in which the murder of 2-year-old James by two 10-year-olds – Robert Thompson and Jon Venables – was predicated in the effects debate not simply on their age but on socio-economic factors of class within 'dysfunctional' families (Kember 1997; Blackman and Walkerdine 2001). Against this form of othering, Blake Morrison approaches the case 'as if' the children concerned were him/his (Morrison 1997; Kember 1998).
 3 Characteristic of what Haraway (1991a) terms the 'god-trick' of disembodied knowledge.
 4 There is a spin-off game, *Sims*, involving human figures.
 5 This is interesting due to the fact that all creatures in *Tierra* are either 'mothers' or 'daughters' and so reproduction is asexual.
 6 There have, however, been subsequent artists' impressions and a promotional video which represents Ray's creatures as animal-like figures (Hayles 1999a).
 7 See Eugene Provenzo (1991) *Video Kids: Making Sense of Nintendo*, and findings of CANT (Children and New Technologies) project in Kember (1997) 'Children and Computer Games'. This was a pilot study of the relation between children and computer games which aimed to critically explore the concept of addiction and to question the resumption of a simplistic media effects model in the debate on children and computer games. The project took place in 1995 (supported by a grant from Goldsmiths College Research Fund) and was undertaken in collaboration with Valerie Walkerdine. A group of ten boys and girls aged 10 or 11 were interviewed, as were their parents, and the children were observed playing the games. Socio-economic backgrounds were varied. The project found that the children's favourite games were platformers and beat'em'ups: *Sonic*, *Streetfighter*, *Super Mario Brothers* and *Mortal Kombat*. Children and parents defined addiction in three ways: not being able to stop playing/playing too much; total immersion in the game and acting out elements or sequences from the games (notably the violent sequences). Although these definitions did not adequately describe the activities of any of the children studied, both the children and their parents believed that others were addicted. For boys, these others were often younger boys. For girls, the others were boys. Perhaps most significantly, for middle class parents, the others were working class children and families.
 8 See Steven Rose on methodological reductionism (Chapter 2). Rose argues that reductionist methodology simplifies and facilitates the generation of seemingly linear chains of cause and effect. It has provided 'unrivalled insights' into the mechanisms of the universe 'because it often seems to work, at least for relatively simple systems' (1997: 78).
 9 A simple text parser is attached to the creatures. Nouns are passed to the attention directory lobe of the brain and verbs are passed to the episodic memory and action selection lobes.
10 *Creatures* refers to both the original and second version unless either is specified.
11 The concept of technoscience is a theoretical and strategic attempt to undermine this division which CyberLife reinforces with the splitting of the company into CyberLife Technology and CyberLife Research Ltd in 1999. For Donna Haraway: 'Technoscience extravagantly exceeds the distinction between science and technology as well as those between nature and society, subjects and objects, and the natural and the artifactual that structured the imaginary time called modernity' (1997: 3).
12
> The metaphor of the *toy* rather than *game* is intended to highlight a different style of interaction: A game is usually played in one (extended) session, until an 'end condition' or 'goal state' is reached . . ., in contrast, use of a toy does not imply a score or an aim to achieve some end condition, and interaction with a toy is a more creative, ongoing, open-ended experience.
>
> (Cliff and Grand 1999: 82)

13 This is the narrative of the original *Creatures* 'toy'.
14 Grand points out that in his original mythology the Shee evoked the folk memory of an ancient and unwarlike Neolithic race in Ireland. The colonial narrative was later introduced by Toby Simpson and the oriental artwork supplied by art director Mark Rafter.
15 Feminist debates on the monitoring, regulation and control of pregnancy and childbirth are well established. Treichler and Stabile point out that visual technologies isolate the foetus and eliminate the mother's body from view. The birth process is then figured as an interaction between the doctor and the foetus (Treichler and Cartwright 1992; Stabile 1998).
16 Grand adds:

> It's worth remembering that these more direct means of meddling with creatures were added in C2, based on feedback from the *users*. Originally there were no means of direct genetic control, and I only let the users have my own gene-building tools when it became clear that this was what they intended to do, with or without my help.

17 This is clear from the survey of user groups and CyberLife's estimate of the user base (taken from their own survey) was 40 per cent female and 60 per cent male. There were no apparent patterns of gendered use, partly due to the prescriptive narrative framework and the non-negotiable rules of the game. The responses for and against norn abuse were not clearly gendered.
18 Aspects of identity such as race, gender and sexuality are, however, significant in the re-figuration of ALife politics and epistemology (see Chapters 3 and 7).
19 HAL is the intelligent computer in Kubrick's film *2001*. See Grand's (1999) 'The Year 2001 Bug: Whatever Happened to HAL?' and Chapters 1, 5 and 7 here.
20 Saunders (2000) gives an account of the distinction between routine, innovative and creative design: 'Creative design goes beyond innovative design by requiring the extension of the state space [the space of possible designs] with the addition of new knowledge'. Several operators can introduce new knowledge into a design process and these include: combination, mutation, analogy and emergence. Emergence is of particular interest to Saunders in his account of CyberLife.
21 Note that artificial pilots lie at the astro-military origins of cyborg technology. In as far as ALife is not completely synonymous with the cyborgian technologies of AI/cybernetics, then similar figures may now be characterised as being post-cyborgian.
22 CRL is currently not funded directly, but supported by media publicity and various forms of publication.
23 Natalie Jeremijenko introduced me to the concept of real ALife in her paper 'Cyberfeminist Design: a review of the sensors used to trigger interaction with explosives' (ESRC Seminar Series, Equal Opportunities On-Line: The Impact of Gender Relations on the Design and Use of Information and Communication Technologies, Seminar 3, Cyberfeminism: Issues in Theory and Design) given at the University of Surrey, UK, 16 May 2000. Jeremijenko's real ALife project is 'One Tree', a cloned tree planted in various locations around the San Francisco Bay area and engineered to embody differences in environment and community. A companion CD-ROM is concerned with e-clones and incorporates a CO_2 feeder from the computer's immediate environment which influences the development of the trees.
24 Grand (1997b) discusses the distinction between matter/stuff and stable configurations/persistent phenomena. The obvious and startling basis of the distinction is that matter flows from place to place and 'momentarily' comes together to form a living being.
25 A position popularised in 2000 by Susan Greenfield in her BBC2 television series on the brain. See also Greenfield's 'How Might the Brain Generate Consciousness?' in

S. Rose (ed.) (1999a) *From Brains to Consciousness? Essays on the New Sciences of the Mind*.

5 Network identities

1 One current manifestation of the Turing Test – which to an extent defies the professional sense of its failure as a measure of intelligence – is the annual Loebner contest established in 1990. To date, 'nobody has won the gold medal [and prize of $100,000], which will be awarded when a computer program finally fools the judges into thinking that it is a human being, but a bronze medal and a cash prize of $2,000 is awarded annually to the contestant who comes the closest' (Paul Allen (2001) 'It's the Thought that Counts', *Guardian Weekend*, 6 October).
2 The autumn 2000 floods in the UK produced widespread chaos in a railway network which was already at breaking point due to years of government underfunding followed by privatisation. This rather mechanical sense of breakdown supplanted predictions of an electronic apocalypse and constituted the real millennium bug in the UK.
3 See 'Cog in the Media'(http://www.ai.mit.edu) and 'Research on Cog' (http://www.ai.mit.edu/projects/cog).
4 Human development 'is determined by a continuous interaction between inherited biological predispositions and encounters with the environment' (Burns and Dobson 1984: 403). Cog might be said to aspire to what Piaget terms the sensorimotor stage of development in which 'the young child develops a knowledge of the permanency of objects or the ability to represent internally stimuli that are not immediately present in the environment' (Burns and Dobson 1984: 445).
5 Cog is not funded directly. See Freedman (1994).
6 In psychoanalytic theory (object relations), toys and pets are transitional objects which assist child development by providing a means to work through the boundaries between self and other (see particularly Winnicott 1971).
7 The wargame is entitled *Conflict Zone* (www.conflictzone-thegame.com) and it uses MASA's generic Direct Intelligence Adaptation platform 'to impart the characters with autonomy and adaptation to the player' (Emmanuel Chiva, interview 2001). Emmanuel Chiva of MASA also confirmed that the platform is used in professional training and military simulations. The MASA website (www.animaths.com) states that 'models of each user are automatically constructed, giving a profile of their strengths and weaknesses. These models, which are updated every time the user engages with the system, are used to generate exercises and training plans specially tailored to the user. By adapting to the individual's needs and abilities, our tools make each session a personal experience'. Military simulations model scenarios from actual training exercises and the objectives given to the trainee are the same as those given in 'real life "war games", with the opposing forces commanded by the computer' (Interview 2001). Chiva adds the claim that 'this is to our knowledge the first time an alife product is used in a professional military simulation product'. Space agency projects at MASA are in the field of autonomous robotics (see www.animaths.com).
8 At Microsoft, Microsoft Agent 2.0 is a free software service which enables developers to include interactive animated characters in their applications or webpages: 'characters can be used to extend and enhance conventional interfaces as assistants, guides, avatars or computer opponents' (Microsoft 1998). The aim here is to humanise the relation between users and computers through the provision of believable agents which also form the basis of advanced research projects on intelligent personal agents.
9 See Lars Risan (1996) and Stefan Helmreich's (1998a) ethnographic studies of ALife research at the University of Sussex and the Santa Fe Institute respectively.
10 See Nicholas Gessler (1999) *Artificial Culture: Experiments in Synthetic Anthropology*.

11 The term 'culture' is derived from 'cultura' meaning to cultivate or tend. Williams states that 'culture in all its early uses was a noun of process: the tending *of* something, basically crops or animals' (Williams 1983: 87). Significantly, 'from the eC16 [early sixteenth century] the tending of natural growth was extended to a process of human development, and this, alongside the original meaning in husbandry, was the main sense until lC18 [late eighteenth century] and eC19 [early nineteenth century]' (87).

6 The meaning of life part 2: genomics

1 Sarah Franklin (2000) employs the concept of a genetic imaginary as that which combines images and imagination in the impacted fields of genomic research, institutions, policies and fiction. The imaginary in these fields incorporates the past, present and future and both f/phantasies and fears: 'What are the fantasies and fears catalysed within the new genetics as a domain of millennial cultural practice? What are the forms of recognition, identification and imagination brought into being in response to new genetic technology? What are the pasts, futures and presents of genomic temporality?' (Franklin 2000: 191). My argument is that the genomic imaginary mobilises ancient and modern religious and litarary myths of the overreacher and the doppelganger as well as those (put forward by Haraway) of racial purity and natural kinds.
2 The discourses of myth and science, fantasy and reality, separated in the projects of modernity are, from a postmodern feminist perspective, wholly enveloped. This argument is made persuasively within the subdiscipline of feminist science and technology studies (Jacobus *et al.* 1990; Haraway 1991a; Penley and Ross 1991; Treichler and Cartwright 1992).
3 Chaos and complexity theory are frequently and wrongly conflated. With reference to Jack Cohen and Ian Stewart's (1994) *The Collapse of Chaos*, James Meek points out that complexity, not chaos best describes the effects of the UK rail and fuel crises of autumn 2000:

> Chaos theory . . . involved unpredictable results emerging from minute changes in the data fed into a calculation. It was all about simple systems obeying simple rules – as the weather, for all its unpredictability, does.
>
> Complexity produces unpredictable results from the interaction of a whole host of actions which, by themselves, seem simple. The fuel crisis, says Stewart, was a classic example – a protest outside a few oil refineries could shut down an entire country with astonishing swiftness.
>
> (Meek 2001a: 3)

4 Instant DNA tests, according to *New Scientist*, may not be far away at all. Andy Coghlan reports on the arrival of a computer add-on that can identify DNA traces in blood samples: 'The unit built by Molecular Sensing of Melksham in Wiltshire fits into a disc drive bay. The company says it can be operated by anyone and gives an answer in 8 to 15 minutes' (*New Scientist*, 25 November 2000: 26).
5 See *The Sixth Day* (2000) ('God created man in His own image, and behold, it was very good. And the evening and the morning were the sixth day [Genesis 1. 27, 31]')which features a similar service ('RePet') and explores, through the action movie genre, some of the ethical and ontological ambiguities of cloning humans from an experiential viewpoint.
6 That society seeks to protect and preserve the soul of *man* is underscored by the gendering of the Faustus myth, which explores an attempt to usurp the role of God (Himself).
7 In October 2001, Advanced Cell Technology (ACT) of Worcester, Massachusetts

created cloned human embryos, further stimulating public debate about human cloning despite the fact that the embryos grew to no more than six cells. Although not the first researchers to have claimed such a feat, ACT were the first to publish the results in a scientific paper (online in *Journal of Regenerative Medicine*). The paper was released just as the US government was due to bring in legislation to ban all forms of human cloning. ACT's PR offensive was based on the claim that its research is not about cloning people but the possibility of treating disease and old age. Currently, the UK is separated from the rest of Europe and the US by its receptivity to the potential benefits of therapeutic cloning (*New Scientist*, 1 December 2001).
8 See Melanie Klein (1988) *Envy and Gratitude and Other Works 1946–1963*, London: Virago.
9 Kac articulates this position in an earlier piece which expresses his desire to open up the possibilities of multidisciplinary public and professional dialogue:

> Working with multiple media to create hybrids from the conventional operations of existing communications systems, I hope to engage participants in situations involving biological elements, telerobotics, interspecies interaction, light, language, distant places, times zones, video conferences and the exchange and transformation of information via networks. Often relying on the contingency, indeterminacy and the intervention of the participant, I wish to encourage dialogical interaction and to confront complex issues concerning identity, agency, responsibility and the very possibility of communication.
>
> (Kac 1999: 90)

7 Evolving feminism in Alife environments

1 Holquist argues that Bakhtin's concept of language 'has as its enabling *a priori* an almost Manichean sense of opposition and struggle at the heart of existence, a ceaseless battle between centrifugal forces that seek to keep things apart, and centripetal forces that strive to make things cohere' (Holquist 1981: xviii).
2 I discuss the symbolic potential of the geometry of chaos theory in two articles (Kember 1991, 1996). The strange attractor

> is a computer-generated model of the changes that take place in a complex natural system (such as the weather) over time. The changes are visualised through the movement of a point on the screen. The point on a strange attactor never becomes fixed, never reaches equilibrium but forms a pattern of loops and spirals which follow two wing-like trajectories. The loops and spirals are infinitely deep, creating layers which do not meet but which remain inside finite space. The strange attractor describes a new non-euclidean geometry of infinite length in finite space, which has been called fractional or fractal geometry. It is fractional in as far as it is not so much a geometry of objects but of the boundary between competing forces of order and chaos in natural systems.
>
> (Kember 1996: 266–267)

See also James Gleick's (1987) popular *Chaos. Making a New Science* (London: Viking Penguin).
3 For example, Paul Gross and Norman Levitt (1998 [1994]) in *Higher Superstition. The Academic Left and its Quarrels with Science*. The science wars are discussed in Chapter 8.
4 Interdisciplinarity in this context tends to refer to the conjunction of AI, ALife, cognitive psychology, computational anthropology and so on, or to the conjunction of feminism, sociology and cultural studies. In other words, it signals fields in which, on the whole, the key epistemological and ontological claims reinforce rather than challenge each other.

5 In advertising: the American Express Blue Card ('Its Evolved') and genetic references to cars with breeding and heritage; a recent preoccupation with evolutionary psychology on BBC Radio 4 (such as *In our Time*, 2 November 2000) and in film, *Evolution*, *Final Fantasy*, *AI*, *eXistenZ*, *The Matrix*, *The 6th Day*.

6 Bell and Gardiner (1998) argue that: 'All sociocultural phenomena, according to Bakhtin, are constituted through the ongoing, dialogical relationship between individuals and groups, involving a multiplicity of different languages, discourses, and symbolising practices'. Moreover, 'in prioritising the relation over the isolated, self-sufficient monad, his ideas dovetail neatly with present attempts to supersede what is often called 'subject-centred reason'' (1998: 4).

7 The information paradigm, for Maturana and Varela (1980 [1971]) is non-phenomenological; it belongs to the domain of observations and descriptions but has no ontological basis: 'notions such as coding and transmission of information do not enter in the realisation of a concrete autopoietic system because they do not refer to actual processes in it' (1980 [1971]: 90).

8 For Niklas Luhmann, autopoiesis represents a paradigm shift in both sociology and epistemology where '"openness" to the environment takes place through the self-productive "closure" of self-observation and self-description' (Lury 1998: 140).

9 Maturana and Varela reach no consensus on the full implications of autopoiesis in the social realm, but clearly present it as an alternative to evolution. The Darwinian notion of evolution 'with its emphasis on the species, natural selection and fitness' has sociological significance 'because it seemed to offer an explanation of the social phenomenology in a competitive society, as well as a scientific justification for the subordination of the destiny of the individuals to the transcendental values supposedly embodied in notions such as mankind, the state, or society' (Maturana and Varela 1980 [1971]: 117). The subordination of the individual to the species, society or state and the variety of discriminations sanctioned in reference to natural selection does not, in fact, have any biological basis since 'biological phenomenology is determined by the phenomenology of the individuals' (118). Whatever the social implications of autopoiesis are, 'biologically the individuals are not dispensable' and Maturana and Varela demonstrate the inseparability of the biological and social realms as adeptly as they do those of the organism and its environment. The interesting question then concerns not the autopoietic status of human societies (whether or not human societies 'as systems of coupled human beings' are also biological systems) but the status of their autopoiesis as a bioethics of posthuman systems.

10 The conditions for posthumanism may be described as being primarily bioethical and, I have argued, dialogic. Bell and Gardiner (1998) elucidate the subjective, social and ethical aspects of Bakhtin's work and its 'potential for the development of a new humanist outlook that is not centred in the monolithic, self-contained subject but on the boundary between self and other' (6).

Bibliography

Adam, A. (1998) *Artificial Knowing. Gender and the Thinking Machine*, London and New York: Routledge.
Adam, B. and Van Loon, J. (2000) 'Introduction: Repositioning Risk; The Challenge for Social Theory', in B. Adam, U. Beck and J. Van Loon (eds) *The Risk Society and Beyond. Critical Issues for Social Theory*, London: Sage.
Agre, P.E. (1997) *Computation and Human Experience*, Cambridge: Cambridge University Press.
Alderman, J. (1999) 'Tamagotchi, Schmamagotchi: Here Come the Norns', *Wired*, News, http://www.wired.com.
Aldiss, B. (2001) 'Like Human, Like Machine', *New Scientist*, 15 September: 40–43.
Aleksander, I. (1999) 'A Neurocomputational View of Consciousness', in S. Rose (ed.) *From Brains to Consciousness? Essays on the New Sciences of the Mind*, London: Penguin Books.
Anees, M.A. (1998) 'Re-defining the Human. Triumphs and Tribulations of *Homo Xeroxiens*', http://www.iol.ie.
Argyros, A. (1992) 'Narrative and Chaos', *New Literary History*, 23(3): 659–673.
Ashman, K.M. and Baringer, P.S. (eds) (2001) *After the Science Wars*, London and New York: Routledge.
Ashworth, J. (1996) 'An "ism" for our times', in O. Curry and H. Cronin (eds) *Demos Quarterly. Matters of Life and Death: The World View from Evolutionary Psychology*, issue 10, London: Demos.
Badmington, N. (ed.) (2000) *Posthumanism*, New York and Basingstoke: Palgrave.
Balsamo, A. (1994) 'Feminism for the Incurably Informed', in M. Dery (ed.) *Flame Wars. The Discourse of Cyberculture*, Durham, NC and London: Duke University Press.
Barad, K. (2000) 'Reconceiving Scientific Literacy as Agential Literacy. Or, Learning How to Intra-act Responsibly with the World', in R. Reid and S. Traweek (eds) *Doing Science+Culture. How Cultural and Interdisciplinary Studies are Changing the Way We Look at Science and Medicine*, New York and London: Routledge.
Barker, M. and Petley, J. (eds) (1997) *Ill Effects: the Media/Violence Debate*, London and New York: Routledge.
Barkow, J.H. (1992) 'Beneath New Culture is Old Psychology: Gossip and Social Stratification', in J.H. Barkow, L. Cosmides and J. Tooby (eds) *The Adapted Mind. Evolutionary Psychology and the Generation of Culture*, New York and Oxford: Oxford University Press.

Barkow, J.H., Cosmides, L. and Tooby, J. (eds) (1992) *The Adapted Mind. Evolutionary Psychology and the Generation of Culture*, New York and Oxford: Oxford University Press.
Barthes, R. (1980) *Camera Lucida*, London: Flamingo.
Bateson, P. and Martin, P. (1999) *Design for a Life. How Behaviour Develops*, London: Jonathan Cape.
Baudrillard, J. (1983) *Simulations*, New York: Semiotext(e).
Beardsley, T. (1999) 'Here's Looking at You. A Disarming Robot Starts to Act up', *Scientific American*, http://www.sciam.com.
Beckett, S. (1964 [1958]) *Endgame*, London: Faber and Faber.
Bedau, M. (1996) 'The Nature of Life', in M.A. Boden (ed.) *The Philosophy of Artficial Life*, Oxford: Oxford University Press.
Bedau, M. (2001) 'Artificial Life VII: Looking Backward, Looking Forward (Editor's Introduction to the Special Issue)', *Artificial Life*, 6: 261–264.
Bedau, M., McCaskill, J.S., Packard, N.H., Rasmussen, S., Adami, C., Green, D.G., Ikegami, T., Kaneko, K. and Ray, T.S. (2001) 'Open Problems in Artificial Life', *Artificial Life*, 6: 363–376.
Bell, M.M. and Gardiner, M. (eds) (1998) 'Bakhtin and the Human Sciences: A Brief Introduction', *Bakhtin and the Human Sciences*, London: Sage.
Bender, G. and Druckrey, T. (eds) (1994) *Culture on the Brink. Ideologies of Technology*, Seattle, WA: Bay Press.
Bentley, P.J. (ed.) (1999) *Evolutionary Design by Computers*, San Francisco, CA: Morgan Kauffman.
Bentley, P.J. (2001) *Digital Biology. How Nature is Transforming our Technology*, London: Headline.
Best, S. (1991) 'Chaos and Entropy. Metaphors in Postmodern Science and Social Theory', *Science as Culture*, 2(11), 188–227.
Birke, L. (1994) *Feminism, Animals and Science. The Naming of the Shrew*, Buckingham: Open University Press.
Blackman, L. and Walkerdine, V. (2001) *Mass Hysteria. Critical Psychology and Media Studies*, Basingstoke and New York: Palgrave.
Blackmore, S. (1999) *The Meme Machine*, Oxford: Oxford University Press.
Boden, M.A. (ed.) (1996a) *The Philosophy of Artificial Life*, Oxford: Oxford University Press.
Boden, M.A. (1996b) 'Autonomy and Artificiality', *The Philosophy of Artificial Life*, Oxford: Oxford University Press.
Boden, M.A. (2000) 'Autopoiesis and Life', *Cognitive Science Quarterly*, 1: 117–145.
Bollas, C. (1987) *The Shadow of the Object. Psychoanalysis of the Unthought Known*, London: Free Association Books.
Bonabeau, E.W. and Theraluz, G. (1997) 'Why Do We Need Artificial Life?', in C. Langton (ed.) *Artificial Life. An Overview*, Cambridge, MA and London: MIT Press.
Boutin, P. (2000) 'At Home with the Androids', *Wired*, Archive, 8.09 – September 2000, http://www.wired.com.
Braidotti, R. (1994) *Nomadic Subjects. Embodiment and Sexual Difference in Contemporary Feminist Theory*, New York: Columbia University Press.
Braidotti, R. (1996) 'Cyberfeminism with a Difference', *New Formations. Technoscience*, 29: 9–26.

Bremer, M. (1992) *SimLife. The Genetic Playground*, User's Manual, Orinda, CA: Maxis Inc.

Bremer, M. and Ellis, C. (1993) *SimCity. The Original City Simulator*, User's Manual, Orinda, CA: Maxis Inc.

Brook, J. and Boal, I.A. (eds) (1995) *Resisting the Virtual Life. The Culture and Politics of Information*, San Francisco, CA: City Lights.

Brooks, M. (2000) 'Global Brain', *New Scientist*, 24 June, 22–7.

Brooks, R.A. and Flynn, A. (1989) 'Fast, Cheap and Out of Control: A Robot Invasion of the Solar System', *Journal of the British Interplanetary System*, 42 (10): 478–485.

Brooks, R.A. and Maes, P. (eds) (1994) *Artificial Life IV. Proceedings of the Fourth International Workshop on the Synthesis and Simulation of Living Systems*, Cambridge, MA and London: MIT Press.

Brown, A. (1999) *The Darwin Wars. The Scientific Battle for the Soul of Man*, London: Simon and Schuster.

Brown, K. (2000) 'Close Cousins', *New Scientist*, 4 November: 42–45.

Brown, P. (2000) 'Stepping Stones in the Mist', http://www.paul-brown.com.

Brown Blackwell, A. (1976 [1875]) *The Sexes throughout Nature*, Westport, CT: Hyperion Press.

Browne, K. (1998) *Divided Labours. An Evolutionary View of Women at Work*, London: Weidenfeld and Nicolson.

Buckingham, D. (1997) 'Electronic Child Abuse? Rethinking the Media's Effects on Children', in M. Barker and J. Petley (eds) *Ill Effects: The Media/Violence Debate*, London and New York: Routledge.

Buckingham, D. (2000) *After the Death of Childhood: Growing up in the Age of Electronic Media*, Cambridge: Polity Press.

Burns, R.B. and Dobson, C.B. (1984) *Introductory Psychology*, Lancaster: MTP Press.

Butler, J. (1989) *Gender Trouble. Feminism and the Subversion of Identity*, London and New York: Routledge.

Butler, J. (1993) *Bodies that Matter. On the Discursive Limits of 'Sex'*, London and New York: Routledge.

Calder, J. (ed.) (1979) *Robert Louis Stevenson. Dr Jekyll and Mr Hyde and Other Stories*, London: Penguin Books.

Card, Orson Scott (1985 [1977]) *Ender's Game*, London: Orbit.

Card, Orson Scott (1992 [1986]) *Speaker for the Dead*, London: Legend Books.

Card, Orson Scott (1992 [1991]) *Xenocide*, London: Legend Books.

Castells, M. (2000) *The Rise of the Network Society*, 2nd edn, Oxford: Blackwell.

Cilliers, P. (1998) *Complexity and Postmodernism. Understanding Complex Systems*, London and New York: Routledge.

Clark, A. and Meek, J. (2000) 'Drug Firms Laying Caims to our Genes', *Guardian*, 18 February.

Clarke, Arthur C. (2000 [1968]) *2001. A Space Odyssey*, Special Edition, London: Orbit.

Cliff, D. and Grand, S. (1999) 'The *Creatures* Global Digital Ecosystem', *Artificial Life*, 5(1): 77–93.

Coghlan, A. and Boyce, N. (2000) 'The End of the Beginning', *New Scientist*, 1 July: 4–5.

Cohen, J. and Stewart, I. (1994) *The Collapse of Choas. Discovering Simplicity in a Complex World*, London and New York: Penguin Books.

Collins, S. (1998) 'Introduction', in C.P. Snow, *The Two Cultures*, Cambridge: Cambridge University Press.
Cosmides, L., Tooby, J. and Barkow, J.H. (1992) 'Introduction: Evolutionary Psychology and Conceptual Integration', in J.H. Barkow, L. Cosmides and J. Tooby (eds) *The Adapted Mind. Evolutionary Psychology and the Generation of Culture*, New York and Oxford: Oxford University Press.
Curry, O. and Cronin, H. (eds) (1996) *Demos Quarterly. Matters of Life and Death: The World View from Evolutionary Psychology*, issue 10, London: Demos.
CyberLife (1997) *Creatures*, ECAL 1997 flyer.
CyberLife Research (2000) *The CyberLife Mission: Welcome to Eden*, http://www.cyberlife-research.com/company/mission/index.htm.
Damer, B. (1998) 'Why is Life Trying to Enter Digital Space?', Conference Lecture Notes for Digital Biota II, *Digital Biota II. The Second Annual Conference on Cyberbiology*, http://www.biota.org.
Damer, B., Marcelo, K. and Revi, F. (1999) 'Nerve Garden: A Public Terrarium in Cyberspace', http://www.biota.org.
Dano, M. (2000) 'Media Blitz Works for Thornhill', *Daily Lobo*, University of New Mexico, http://www.dailylobo.unm.edu/news/departments/biology/thornhill/2-4-00thornhill.html.
Darwin, C. (1874) *The Descent of Man, and Selection in Relation to Sex*, 2nd edn. London: John Murray.
Darwin, C. (1901 [1871]) *The Descent of Man, and Selection in Relation to Sex*, London: John Murray.
Darwin, C. (1985 [1859]) *The Origin of Species*, London: Penguin Books.
Davenport, C.B. (1911) *Heredity in Relation to Eugenics*, New York: Henry Holt.
Davidmann, M. (1996) 'Creating, Patenting and Marketing of New Forms of Life', http://www.solbaram.org.
Davidson, C. (1998) 'Agents from Albia', *New Scientist*, 158(2133): 38–44.
Davis, E. (1999) *Techgnosis. Myth, Magic and Mysticism in the Age of Information*, London: Serpent's Tail.
Davis, E. (2000) 'Congratulations, It's a Bot!', *Wired*, Archive, 8.09 – September 2000, http://www.wired.com.
Dawkins, R. (1986) 'Sociobiology: The New Storm in a Teacup', in S. Rose and L. Appignanesi (eds) *Science and Beyond*, Oxford: Basil Blackwell in association with the ICA.
Dawkins, R. (1989 [1976]) *The Selfish Gene*, Oxford: Oxford University Press.
Dawkins, R. (1991) *The Blind Watchmaker*, London: Penguin Books.
Dawkins, R. (1999) 'Foreword', in S. Blackmore, *The Meme Machine*, Oxford: Oxford University Press.
De Landa, M. (1994) 'Virtual Environments and Synthetic Reason', in M. Dery (ed.) *Flame Wars. The Discourse of Cyberculture*, Durham, NC and London: Duke University Press.
Deleuze, G. and Guattari, F. (1988) *A Thousand Plateaus: Capitalism and Schizophrenia*, London: Athlone.
Dennett, D.C. (1995a) *Darwin's Dangerous Idea. Evolution and the Meanings of Life*, London: Allen Lane.
Dennett, D.C. (1995b) *Artificial Life. The Tufts Symposium*, CD-ROM, Oxford: Oxford University Press.

Dennett, D.C. (1997) 'Artificial Life as Philosophy', in C. Langton (ed.) *Artificial Life. An Overview*, Cambridge, MA and London: MIT Press.
Derrida, J. (1978) *Writing and Difference*, London: Routledge and Kegan Paul.
Dery, M. (ed.) (1994) *Flame Wars. The Discourse of Cyberculture*, Durham, NC and London: Duke University Press.
Doyle, R. (1997) *On Beyond Living. Rhetorical Transformations of the Life Sciences*, Stanford, CA: Stanford University Press.
Dreyfus, H. (1991) *Being-in-the-World. A Commentary on Heidegger's* Being and Time, *Division 1*, Cambridge, MA: MIT Press.
Dreyfus, H. (1999) *What Computers Still Can't Do*, Cambridge, MA and London, England: MIT Press.
Dyson, G. (1997) *Darwin among the Machines*, London: Penguin Books.
Easlea, B. (1983) *Fathering the Unthinkable: Masculinity, Scientists and the Nuclear Arms Race*, London: Pluto Press.
ECAL '97, with CyberLife Technology, The Arts Council of England, Brighton Media Centre, Media Communications (1997) *LikeLIFE. A collection of installations, robots, creatures and artworks inspired by living things*, exhibition catalogue, Brighton Media Centre, Brighton.
Edge (1999) 'CyberLife', *Edge*, 75 (September): 40–43.
Edwards, P.N. (1996) *The Closed World. Computers and the Politics of Discourse in Cold War America*, Cambridge, MA and London, England: MIT Press.
Emmeche, C. (1994) *The Garden in the Machine. The Emerging Science of Artificial Life*, Princeton, NJ: Princeton University Press.
Epstein, Joshua M. and Axtell, Robert (1996) *Growing Artificial Societies. Social Science from the Bottom Up*, Washington, DC: Brookings Institution Press and Cambridge, MA: MIT Press.
Fausto-Sterling, A. (2000) 'Beyond Difference: Feminism and Evolutionary Psychology', in H. Rose and S. Rose (eds) *Alas, Poor Darwin. Arguments against Evolutionary Psychology*, London: Jonathan Cape.
Featherstone, M. and Burrows, R. (eds) (1995) *Cyberspace, Cyberbodies, Cyberpunk. Cultures of Technological Embodiment*, London: Sage.
Fernbach, D. (1981) *The Spiral Path. A Gay Contribution to Human Survival*, Boston, MA: Alyson and London: Gay Men's Press.
Foucault, M. (1979) *The History of Sexuality, Vol. 1. An Introduction*, London: Allen Lane.
Foucault, M. (1981) 'Truth and Power', in C. Gordon (ed.) *Power/Knowledge. Selected Interviews and Other Writings, 1972–1977*, Brighton: Harvester.
Foucault, M. (1987) *Discipline and Punish. The Birth of the Prison*, London: Penguin Books.
Foucault, M. (1997) *The Order of Things. An Archaeology of the Human Sciences*, London: Routledge.
Franklin, S. (1993) 'Postmodern Procreation. Representing Reproductive Practice', *Science as Culture*, 3(4/17): 522–561.
Franklin, S. (1997) *Embodied Progress. A Cultural Account of Assisted Conception*, London and New York: Routledge.
Franklin, S. (2000) 'Life Itself. Global Nature and the Genetic Imaginary', in S. Franklin, C. Lury and J. Stacey (eds) *Global Nature, Global Culture*, London: Sage.

Franklin, S., Lury, C. and Stacey, J. (2000) *Global Nature, Global Culture*, London: Sage.
Fraser, M. (2001) 'The Nature of Prozac', *History of the Human Sciences*, 14(3): 57–85.
Freedman, D.H. (1994) 'Bringing Up RoboBaby', *Wired*, Archive, 2.12 – December 1994, http://www.wired.com.
Freud, S. (1919) 'The Uncanny', in J. Strachey (ed.) (1953–1973) *The Standard Edition of the Complete Psychological Works of Sigmund Freud*, 24 vols, London: Hogarth Press.
Gessler, N. (1994) 'Artificial Culture', in Rodney A. Brooks and Pattie Maes (eds) *Artificial Life IV. Proceedings of the Fourth International Workshop on the Synthesis and Simulation of Living Systems*, Cambridge, MA and London: MIT Press.
Gessler, N. (1999) *Artificial Culture: Experiments in Synthetic Anthropology*, http://www.sscnet.ucla.edu/anthro/gessler/synopsis.htm.
Gibson, W. (1986) *Neuromancer*, London: Grafton.
Gilbert, W. (1992) 'A Vision of the Grail', in D.J. Kevles and L. Hood (eds) *The Code of Codes. Scientific and Social Issues in the Human Genome Project*, Cambridge, MA and London, England: Harvard University Press.
Gould, S.J. (2000) 'More Things in Heaven and Earth', in H. Rose and S. Rose (eds) *Alas, Poor Darwin. Arguments against Evolutionary Psychology*, London: Jonathan Cape.
Gowaty, P.A. (ed.) (1997) *Feminism and Evolutionary Biology*, New York and London: Chapman and Hall.
Graham-Rowe, D. (1998) 'Meet Kismet', *New Scientist*, 22 August.
Graham-Rowe, D. (1999) 'Booting up Baby', *New Scientist*, 22 May.
Graham-Rowe, D., Crystall, B. and Lawton, G. (2001) 'Jobs for the Bots', *New Scientist*, 10 February.
Grand, S. (1997a) 'Creatures: An Exercise in Creation', *IEEE Intelligent Systems and their Applications*, IEEE Computer Society Publications, July/August.
Grand, S. (1997b) 'Three Observations that Changed my Life', *IEEE Intelligent Systems and their Applications*, IEEE Computer Society Publications, November/December.
Grand, S. (1998a) 'Battling with GA-Joe', *IEEE Intelligent Systems and their Applications*, IEEE Computer Society Publications, March/April.
Grand, S. (1998b) 'Curiosity Created the Cat', *IEEE Intelligent Systems and their Applications*, IEEE Computer Society Publications, May/June.
Grand, S. (1998c) 'Of Mountains and Molehills', *IEEE Intelligent Systems and their Applications*, IEEE Computer Society Publications, November/December.
Grand, S. (1999a) 'The Year 2001 Bug: Whatever Happened to HAL?', *IEEE Intelligent Systems and their Applications*, IEEE Computer Society Publications, January/February.
Grand, S. (1999b) 'Where Newton Went Wrong', *Guardian Online*, 1 October.
Grand, S. (2000) *Creation. Life and How to Make It*, London: Weidenfeld and Nicolson.
Gray, R. (1997) '"In the Belly of the Monster": Feminism, Developmental Systems, and Evolutionary Explanations', in P.A. Gowaty (ed.) *Feminism and Evolutionary Biology*, New York and London: Chapman and Hall.
Greek, C.R. and Greek, J.S. (2000) 'Xeno Phobia', *Guardian*, 20 September.

Griffiths, M. (1991) 'A Comparative Analysis of Video Games and Fruit Machines', *Journal of Adolescence*, 14: 53–73.
Gross, P.R. and Levitt, N. (1998 [1994]) *Higher Superstition. The Academic Left and its Quarrels with Science*, Baltimore, MD and London: Johns Hopkins University Press.
Grosz, E. (1994) *Volatile Bodies. Toward a Corporeal Feminism*, Bloomington, IN: Indiana University Press.
Hables Gray, C. (1997) *Postmodern War*, New York and London: Guilford.
Hables Gray, C. (2002) *Cyborg Citizen*, New York and London: Routledge.
Hables Gray, C., Figueroa-Sarriera, H.J. and Mentor, S. (eds) (1995) *The Cyborg Handbook*, New York and London: Routledge.
Haldane, J.B.S. (1963) 'Biological Possibilities for the Human Species in the Next Ten Thousand Years', in G. Wolstenholme (ed.) *Man and his Future*, CIBA Foundation Volume, London: J&A Churchill.
Hall, S. (1981) 'Cultural Studies: Two Paradigms', in T. Bennett, G. Martin, C. Mercer and J. Woollacott (eds) *Culture, Ideology and Social Process*, London: Open University Press.
Hansrudi, N., Pandzic, I.S., Capin, T.K., Thalmann, N.M. and Thalmann, D. (1997) 'Playing Games through the Virtual Life Network', in C. Langton and K. Shimohara (eds) *Artificial Life V. Proceedings of the Fifth International Workshop on the Synthesis and Simulation of Living Systems*, Cambridge, MA: MIT Press.
Haraway, Donna J. (1991a) *Simians, Cyborgs and Women*, London: Free Association Books.
Haraway, Donna J. (1991b) 'Cyborgs at Large: An Interview with Donna Harraway', in C. Penley and A. Ross (eds) *Technoculture*, Minneapolis, MN: University of Minnesota Press.
Haraway, Donna J. (1997) Modest_Witness@Second_Millennium.FemaleMan©_Meets_OncoMouseModest_Witness@Second_Millennium.FemaleMan©_Meets_OncoMouse™, London: Routledge.
Haraway, Donna J. (2000) *How Like a Leaf. An Interview with Thyrza Nichols Goodeve*, New York and London: Routledge.
Harding, S. (1986) *The Science Question in Feminism*, Ithaca, NY: Cornell University Press.
Harding, S. (1992) *Whose Science? Whose Knowledge? Thinking from Women's Lives*, Ithaca, NY: Cornell University Press.
Harding, S. and Hintikka, M.B. (eds) (1983) *Discovering Reality. Feminist Perspectives on Epistemology, Metaphysics, Methodology and Philosophy of Science*, Dordrecht: Reidel.
Harding, S. and O'Barr, J.F. (eds) (1987) *Sex and Scientific Inquiry*, Chicago: University of Chicago Press.
Harris, J. (1998) *Clones, Genes, and Immortality. Ethics and the Genetic Revolution*, Oxford and New York: Oxford University Press.
Hart, R. (1996) 'The Flight from Reason: *Higher Superstition* and the Refutation of Science Studies', in A. Ross (ed.) *Science Wars*, Durham, NC and London: Duke University Press.
Harvey, I., Husbands, P., Cliff, D., Thompson, A. and Jakobi, N. (1996) 'Evolutionary Robotics: The Sussex Approach', unpublished research paper produced in the School of Cognitive and Computer Sciences, University of Sussex, 10 May.

Hassan, I. (1980) *The Right Promethean Fire. Imagination, Science, and Cultural Change*, Urbana, IL and London: University of Illinois Press.

Hayles, N. Katherine (1994) 'Narratives of Evolution and the Evolution of Narratives', in John L. Casti and Anders Karlqvist (eds) *Cooperation and Conflict in General Evolutionary Processes*, New York: John Wiley.

Hayles, N. Katherine (1995) 'Simulated Nature and Natural Simulations: Rethinking the Relation between the Beholder and the World', in William Cronin (ed.) *Uncommon Ground. Toward Reinventing Nature*, New York and London: W.W. Norton.

Hayles, N. Katherine (1996a) 'Narratives of Artificial Life', in G. Robertson, M. Mash, L. Tickner, J. Bird, B. Curtis and T. Putnam (eds) *FutureNatural. Nature, Science, Culture*, London and New York: Routledge.

Hayles, N. Katherine (1996b) 'Consolidating the Canon', in A. Ross (ed.) *Science Wars*, Durham, NC and London: Duke University Press.

Hayles, N. Katherine (1999a) *How We Became Posthuman. Virtual Bodies in Cybernetics, Literature and Informatics*, Chicago and London: University of Chicago Press.

Hayles, N. Katherine (1999b) 'Simulating Narratives: What Virtual Creatures Can Teach Us', *Critical Inquiry* 26: 1–26.

Heidegger, M. (1961 [1927]) *Being and Time*, New York: Harper and Row.

Helmreich, S. (1997) 'The Spiritual in Artificial Life: Recombining Science and Religion in a Computational Culture Medium', *Science as Culture*, 6(28): 363–395.

Helmreich, S. (1998a) *Silicon Second Nature. Culturing Artificial Life in a Digital World*, Berkeley, CA: University of California Press.

Helmreich, S. (1998b) 'Replicating Reproduction in Artificial Life: Or, the Essence of Life in the Age of Virtual Electronic Reproduction', in Sarah Franklin and Helena Ragone (eds) *Reproducing Reproduction. Kinship, Power, and Technological Innovation*, Philadelphia, PA: University of Pennsylvania Press.

Helmreich, S. (1998c) 'Recombination, Rationality, Reductionism and Romantic Reactions: Culture, Computers, and the Genetic Algorithm', *Social Studies of Science*, 28(1): 39–71.

Henriques, J., Holloway, W., Urwin, C., Venn, C. and Walkerdine, W. (1998) *Changing the Subject. Psychology, Social Regulation and Subjectivity*, London and New York: Routledge.

Hershman Leeson, L. (ed.) (1996) *Clicking In. Hot Links to a Digital Culture*, Seattle, WA: Bay Press.

Herz, J.C. (1997) *Joystick Nation: How Videogames Gobbled our Money, Won our Hearts and Rewired our Minds*, London: Abacus.

Hogg, J. (1983 [1824]) *The Private Memoirs and Confessions of a Justified Sinner*, London: Penguin Books.

Holquist, M. (ed.) (1981) *The Dialogic Imagination. Four Essays by M.M. Bakhtin*, Austin, TX: University of Texas Press.

Human Fertilisation and Embryology Authority and Human Genetics Advisory Commission (HFEA and HGAC) (1998) 'Cloning Issues in Reproduction, Science and Medicine', http://www.hfea.gov.uk.

ICA (2000) 'What Is Life? How Can We Build A Soul? Steve Grand in Conversation', ICA Flyer, 14: November.

Jacob, F. (1982) *The Possible and the Actual*, Seattle, WA and London: University of Washington Press.
Jacob, F. (1993) *The Logic of Life. A History of Heredity*, Princeton, NJ: Princeton University Press.
Jacobus, M. (1986) 'Is There a Woman in This Text?', in *Reading Woman: Essays in Feminist Criticism*, New York: Columbia University Press.
Jacobus, M., Fox Keller, E. and Shuttleworth, S. (eds) (1990) *Body/Politics. Women and the Discourses of Science*, London: Routledge.
Jameson, F. (1984) 'Post-modernism, or the Cultural Logic of Late Capitalism', *New Left Review*, 146: 53–92.
Jones, S. (1999) 'Introduction', in James D. Watson, *The Double Helix*, London: Penguin Books.
Kac, E. (1999) 'Art at the Biological Frontier', in R. Ascott (ed.) *Reframing Consciousness*, Exeter, Devon and Portland, OR: Intellect.
Kawaguchi, Y. (1997) 'The Art of the GROWTH Algorithm with Cells', in C. Langton and K. Shimohara (eds) *Artificial Life V. Proceedings of the Fifth International Workshop on the Synthesis and Simulation of Living Systems*, Cambridge, MA: MIT Press.
Kay, L. (2000) *Who Wrote the Book of Life? A History of the Genetic Code*, Stanford, CA: Stanford University Press.
Keepers, G.A. (1990) 'Pathological Preoccupation with Video Games', *Journal of the American Academy of Child and Adolescent Psychiatry*, 29(1): 49–50.
Keller, E.F. (1983) *A Feeling for the Organism*, New York: Freeman.
Keller, E.F. (1992) *Secrets of Life. Secrets of Death. Essays on Language, Gender and Science*, New York and London: Routledge.
Keller, E.F. (1995) *Refiguring Life: Metaphors of Twentieth-Century Biology*, New York: Columbia University Press.
Keller, E.F. (2000) *The Century of the Gene*, Cambridge, MA and London, England: Harvard University Press.
Kelly, K. (1994) *Out of Control. The New Biology of Machines*, London: Fourth Estate.
Kember, S. (1991) 'Medical Diagnostic Imaging: The Geometry of Chaos', *New Formations*, 15: 55–66.
Kember, S. (1996) 'Feminist Figuration and the Question of Origin', in G. Robertson, M. Mash, L. Tickner, J. Bird, B. Curtis and T. Putnam (eds) *FutureNatural. Nature, Science, Culture*, London and New York: Routledge.
Kember, S. (1997) 'Children and Computer Games', paper presented at Institute of Education, London, June.
Kember, S. (1998) *Virtual Anxiety. Photography, New Technologies and Subjectivity*, Manchester: Manchester University Press.
Kennedy, B. (2000) 'Cyberfeminisms. Introduction', in D. Bell and B.M. Kennedy (eds) *The Cybercultures Reader*, London: Routledge.
Kerin, J. (1999) 'The Matter at Hand: Butler, Ontology and the Natural Sciences', *Australian Feminist Studies*, 14(29): 91–104.
Kevles, D.J. (1992) 'Out of Eugenics: The Historical Politics of the Human Genome', in D.J. Kevles and L. Hood (eds) *The Code of Codes. Scientific and Social Issues in the Human Genome Project*, Cambridge, MA and London, England: Harvard University Press.
Kevles, D.J. and Hood, L. (eds) (1992) *The Code of Codes. Scientific and Social Issues in*

the Human Genome Project, Cambridge, MA and London, England: Harvard University Press.
Kiely, R. (1972) *The Romantic Novel in England*, Cambridge, MA: Harvard University Press.
Kirby, V. (1997) *Telling Flesh. The Substance of the Corporeal*, New York and London: Routledge.
Kinder, M. (1991) *Playing with Power in Movies, Television and Video Games: from Muppet Babies to Teenage Mutant Ninja Turtles*, Berkeley, CA: University of California Press.
Kolata, G. (1997) *Clone. The Road to Dolly and the Path Ahead*, London: Penguin Books.
Kuchinskas, S. (1999) 'Virtual Plants, Insects Twine through Net', *Wired*, News, http://www.wired.com.
Langton, C. (ed.) (1989) *Artificial Life. The Proceedings of an Interdisciplinary Workshop on the Synthesis and Simulation of Living Systems*, Redwood City, CA: Addison-Wesley.
Langton, C. (1996 [1989]) 'Artificial Life', in M.A. Boden (ed.) *The Philosophy of Artificial Life*, Oxford: Oxford University Press.
Langton, C. (ed.) (1997) *Artificial Life. An Overview*, Cambridge, MA and London, England: MIT Press.
Langton, C. and Shimohara, K. (eds) (1997) *Artificial Life V. Proceedings of the Fifth International Workshop on the Synthesis and Simulation of Living Systems*, Cambridge, MA: MIT Press.
Latour, B. (1987) *Science in Action: How to Follow Scientists and Engineers through Society*, Cambridge, MA: Harvard University Press.
Latour, B. (1993) *We Have Never Been Modern*, New York: Harvester Wheatsheaf.
Latour, B. (1997) 'Foreword', in I. Stengers *Power and Invention. Situating Science*, Minneapolis, MN and London: University of Minnesota Press.
Latour, B. and Woolgar, S. (1979) *Laboratory Life: The Social Construction of Scientific Facts*, London: Sage.
Leonard, A. (1997) *Bots. The Origin of New Species*, San Francisco, CA: HardWired.
Levy, S. (1992) *Artificial Life. The Quest for a New Creation*, London: Jonathan Cape.
Lewontin, R.C. (1993) *The Doctrine of DNA. Biology as Ideology*, London: Penguin Books.
Lewontin, R.C. (1994) 'The Dream of the Human Genome', in G. Bender and T. Druckrey (eds) *Culture on the Brink. Ideologies of Technology*, Seattle, WA: Bay Press.
Lewontin, R. (2000) *It Ain't Necessarily So. The Dream of the Human Genome and Other Illusions*, London: Granta.
Lury, C. (1998) *Prosthetic Culture. Photography, Meaning and Identity*, London: Routledge.
Lury, C. (1999) 'Marking Time with Nike: The Illusion of the Durable', *Public Culture*, 11(3): 499–526.
Lyman, R. (2001) 'E.T.'s Scarier Half-brother', *Observer Review*, 1 July.
Lyotard, J.F. (1984) *The Postmodern Condition: A Report on Knowledge*, Manchester: Manchester University Press.
Macgregor Wise, J. (1998) 'Intelligent Agency', *Cultural Studies. Cultural Studies of Science and Technology*, 12(3): 410–428.

Maes, P. (1996) 'Artificial Life Meets Entertainment: Lifelike Autonomous Agents', in Lynn Hershman Leeson (ed.) *Clicking In. Hot Links to a Digital Culture*, Seattle, WA: Bay Press.

Maes, P. (1997) 'Modelling Adaptive Autonomous Agents', in C. Langton (ed.) *Artificial Life. An Overview*, Cambridge, MA and London, England: MIT Press.

Marchessault, J. (1996) 'The Secret of Life: Informatics and the Popular Discourse of the Life Code', *New Formations. Technoscience*, 29: 120–130.

Martin, L.H., Gutman, H. and Hutton, P.H. (1988) *Technologies of the Self. A Seminar with Michel Foucault*, London: Tavistock.

Maturana, H.R. and Varela, F.J. (1980 [1971]) *Autopoiesis and Cognition. The Realization of the Living*, Dordrecht and London: Reidel.

Maxis Inc. (1995, 1997) *SimEarth: The Living Planet*, CD-ROM, Orinda, CA: Maxis Inc.

Maynard Smith, J. (1998) *Shaping Life. Genes, Embryos and Evolution*, London: Weidenfeld and Nicolson.

Meek, J. (2000) 'Scientists Revel in a Day of Glory', *Guardian*, 27 June.

Meek, J. (2001a) 'Things Fall Apart', *Guardian*, 1 March.

Meek, J. (2001b) 'Team Prepares to Clone Human Being', *Guardian*, 10 March.

Meek, J. (2001c) 'ANDi, First GM Primate. Will Humans be Next?', *Guardian*, 12 January.

Merleau-Ponty, M. (1962 [1945]) *Phenomenology of Perception*, New York: Humanities Press.

Microsoft (1998) 'Press Pass – Microsoft Agent 2.0 Adds a More Human Touch to Computing', http://www.microsoft.com/PressPass/press/1998/Oct98/Agent2pr.asp.

Midgley, M. (1999) 'One World, but a Big One', in S. Rose (ed.) *From Brains to Consciousness? Essays on the New Sciences of the Mind*, London: Penguin Books.

Miller, L. (1995) 'Women and Children First: Gender and the Settling of the Electronic Frontier', in J. Brook and I.A. Boal (eds) *Resisting the Virtual Life. The Culture and Politics of Information*, San Francisco, CA: City Lights Books.

Mitchell, M. and Forrest, S. (1997) 'Genetic Algorithms and Artificial Life', in C. Langton (ed.) *Artificial Life. An Overview*, Cambridge, MA and London, England: MIT Press.

Moran, F., Moreno, A., Minch, E. and Montero, F. (1997) 'Further Steps Towards a Realistic Description of the Essence of Life', in C. Langton and K. Shimohara (eds) *Artificial Life V. Proceedings of the Fifth International Workshop on the Synthesis and Simulation of Living Systems*, Cambridge, MA: MIT Press.

Morley, D. (1995) 'Theories of Consumption in Media Studies', in D. Miller (ed.) *Acknowledging Consumption*, London: Routledge.

Morrison, B. (1997) *As If*, London: Granta.

Murray, I. (1980) 'Introduction', *Oscar Wilde. Plays, Prose Writings and Poems*, London and Toronto: Dent.

Nelkin, D. (1992) 'The Social Power of Genetic Information', in D.J. Kevles and L. Hood (eds) *The Code of Codes. Scientific and Social Issues in the Human Genome Project*, Cambridge, MA and London, England: Harvard University Press.

Nelkin, D. (2000) 'Less Selfish than Sacred? Genes and the Religious Impulse in Evolutionary Psychology', in H. Rose and S. Rose (eds) *Alas, Poor Darwin. Arguments against Evolutionary Psychology*, London: Jonathan Cape.

Nelson, G.L. (1999) 'Sonomorphs: An Application of Genetic Algorithms to the Growth and Development of Musical Organisms', http://www.talbert3.con.oberlin.edu.

Noble, D. (1997) 'The Immortal Mind: Artificial Intelligence', in *The Religion of Technology. The Divinity of Man and the Spirit of Invention*, New York: Alfred A. Knopf.

Panelas, T. (1983) 'Consumption of Leisure and the Social Construction of the Peer Group', *Youth and Society*, 15(1): 51–65.

Pattee, H.H. (1996) 'Simulations, Realizations, and Theories of Life', in M.A. Boden (ed.) *The Philosophy of Artificial Life*, Oxford: Oxford University Press.

Penley, C. (1997) *NASA/Trek: Popular Science and Sex in America*, London and New York: Verso.

Penley, C. and Ross, A. (1991) *Technoculture*, Minneapolis, MN: University of Minnesota Press.

Penley, C., Lyon, E., and Spigel, L. (1990) *Close Encounters: Film, Feminism and Science Fiction*, Minneapolis, MN and Oxford: University of Minnesota Press.

Penny, S. (1994) 'Virtual Reality as the Completion of the Enlightenment Project', in G. Bender and T. Druckrey (eds) *Culture on the Brink. Ideologies of Technology*, Seattle, WA: Bay Press.

Penny, S. (1996) 'The Darwin Machine: Artificial Life and Interactive Art', *New Formations. Technoscience*, 29: 59–69.

Penny, S., Smith, J., Sengers, P., Bernhardt, A. and Schulte, J. (2000) 'Traces: Embodied Immersive Interaction with Semi-Autonomous Avatars', http://www-2.cs.cmu.edu/~jeffrey/research/traces/convergence/traces.pdf. Also in *Convergence: Journal of Research into New Media Technologies*, 7(2).

Penrose, R. (1967) 'The Influence of the English Tradition in Human Genetics', in J.F. Crow and J.V. Neel (eds) *Proceedings of the Third International Congress of Human Genetics*, Baltimore, MD: Johns Hopkins University Press.

Penrose, R. (1999) 'Can a Computer Understand?', in S. Rose (ed.) *From Brains to Consciousness? Essays on the New Sciences of the Mind*, London: Penguin Books.

Picard, R.W. (2000) 'Does HAL Cry Digital Tears? Emotion and Computers', in D.G. Stork (ed.) *HAL's Lecacy. 2001's Computer as Dream and Reality*, Cambridge, MA and London, England: MIT Press.

Plant, S. (1995) 'The Future Looms: Weaving, Women and Cybernetics', in M. Featherstone and R. Burrows (eds) *Cyberspace, Cyberbodies, Cyberpunk. Cultures of Technological Embodiment*, London: Sage.

Plant, S. (1996) 'The Virtual Complexity of Culture', in G. Robertson, M. Mash, L. Tickner, J. Bird, B. Curtis and T. Putnam (eds) *FutureNatural. Nature, Science, Culture*, New York and London: Routledge.

Plant, S. (1997) *Zeros and Ones*, London: Fourth Estate.

Poole, S. (2000) *Trigger Happy: The Inner Life of Video Games*, London: Fourth Estate.

Porter, H. (1999a) 'And How Tall Would you Like your Child Sir?', *Guardian*, 19 April.

Porter, H. (1999b) 'The Joker', *Guardian*, 20 April.

Prigogine, I. and Stengers, I. (1985) *Order Out of Chaos*, London: Flamingo.

Prigogine, I. and Stengers, I. (1997) 'The Reenchantment of the World', in I. Stengers, *Power and Invention. Situating Science*, Minneapolis, MN and London: University of Minnesota Press.

Prophet, J. and Selley, G. (1995) *TechnoSphere*, http://www.technosphere.org.uk.
Provenzo, E.F. (1991) *Video Kids: Making Sense of Nintendo*, Cambridge, MA: Harvard University Press.
Rabinow, P. (1999) *French DNA. Trouble in Purgatory*, Chicago and London: University of Chicago Press.
Radford, T. (2000) 'Pigs Cloned for Organs Down at the "Pharm"', *Guardian*, 15 March.
Radford, T. (2001) 'Door Opens on Deeper Mysteries', *Guardian*, 12 February.
Rawls, J. (1999) *A Theory of Justice*, Oxford: Oxford University Press.
Ray, T.S. (1996) 'An Approach to the Synthesis of Life', in M.A. Boden (ed.) *The Philosophy of Artificial Life*, Oxford: Oxford University Press.
Ray, T.S. (1997) 'An Evolutionary Approach to Synthetic Biology: Zen and the Art of Creating Life', in C. Langton (ed.) *Artificial Life. An Overview*, Cambridge, MA and London, England: MIT Press.
Ray, T. and Virtual Life (1998) 'Documentation for the Tierra Simulator', *Tierra.doc* 17-3-98, Tierra Simulator V4:2.
Reid, R. and Traweek, S. (eds) (2000) *Doing Science+Culture. How Cultural and Interdisciplinary Studies are Changing the Way We Look at Science and Medicine*, New York and London: Routledge.
Ridley, M. (1999) *Genome. The Autobiography of a Species in 23 Chapters*, London: Fourth Estate.
Rifkin, J. (1998) *The Biotech Century. How Genetic Commerce Will Change the World*, London: Phoenix.
Risan, L. (1996) 'Artificial Life: A Technoscience Leaving Modernity?', unpublished thesis, University of Sussex.
Robertson, G., Mash, M., Tickner, L., Bird, J., Curtis, B. and Putnam, T. (eds) (1996) *FutureNatural. Nature, Science, Culture*, London and New York: Routledge.
Rose, H. (1994) *Love, Power and Knowledge. Towards a Feminist Transformation of the Sciences*, Cambridge: Polity Press.
Rose, H. and Rose, S. (eds) (2000) *Alas, Poor Darwin. Arguments against Evolutionary Psychology*, London: Jonathan Cape.
Rose, N. (1999) *Governing the Soul. The Shaping of the Private Self*, London and New York: Free Association Books.
Rose, N. (2001) 'The Politics of Life Itself', *Theory, Culture and Society*, 18(6): 1–30.
Rose, S. (1997) *Lifelines. Biology, Freedom, Determinism*, London: Allen Lane.
Rose, S. (ed.) (1999a) *From Brains to Consciousness? Essays on the New Sciences of the Mind*, London: Penguin Books.
Rose, S. (1999b) 'Brains, Minds and the World', in S. Rose (ed.) *From Brains to Consciousness? Essays on the New Sciences of the Mind*, London: Penguin Books.
Rose, S. and Appignanesi, L. (eds) (1986) *Science and Beyond*, Oxford: Basil Blackwell in association with the ICA.
Rose, S., Lewontin, R.C. and Kamin, L. (1984) *Not in our Genes*, London: Penguin Books.
Ross, A. (1991) *Strange Weather. Culture, Science and Technology in the Age of Limits*, London: Verso.
Ross, A. (1996) *Science Wars*, Durham, NC and London: Duke University Press.
Saunders, R. (2000) 'CyberLife: A Possible Architecture for Complex Design Computing Solutions', http://www.arch.usyd.edu.au/~rob/study/CyberLife.html.

Schiebinger, L. (1999) *Has Feminism Changed Science?*, Cambridge, MA and London, England: Harvard University Press.
Schueller, G. (2001) 'But is it Art?', *New Scientist*, 6 January: 34–37.
Sefton-Greene, J. (ed.) (1998) *Digital Diversions: Youth Culture in the Age of Multimedia*, London: UCL Press.
Segal, L. (1999) 'Genes and Gender: The Return to Darwin', in *Why Feminism?*, Cambridge: Polity Press.
Sengers, P. (1998) 'Anti-Boxology: Agent Design in Cultural Context', unpublished thesis, Computer Science Department and Program in Literary and Cultural Theory, Carnegie Mellon University, Pittsburgh, PA.
Shakespeare, T. and Erickson, M. (2000) 'Different Strokes: Beyond Biological Determinism and Social Constructionism', in H. Rose and S. Rose (eds) *Alas, Poor Darwin. Arguments against Evolutionary Psychology*, London: Jonathan Cape.
Shapin, S. and Schaffer, S. (1985) *Leviathan and the Air-Pump: Hobbes, Boyle, and the Experimental Life*, Princeton, NJ: Princeton University Press.
Shepherd, D. (1993) 'Bakhtin, Carnival and Other Subjects', *Critical Studies*, vol. 3 no. 2 to vol. 4 no. 2, Amsterdam and Atlanta, GA: Rodolpi.
Simpson, T. (1998) *Creatures 2. Strategies and Secrets*, Cambridge: CyberLife and Alameda, CA: Sybex Inc.
Skinner, B.F. (1971) *Beyond Freedom and Dignity*, New York: Alfred Knopf.
Snow, C.P. (1998 [1959]) *The Two Cultures*, Cambridge: Cambridge University Press.
Sokal, A. (1996) 'Transgressing the Boundaries: Towards a Transformative Hermeneutics of Quantum Gravity', *Social Text*, 47/48: 217–252.
Sokal, A. (2001) 'What the *Social Text* Affair does and does not Prove: A Critical Look at "Science Studies"', in K.M. Ashman and P.S. Baringer (eds) *After the Science Wars*, London and New York: Routledge.
Solomon, G. (1990) 'Cognitive Effects with and of Computer Technology', *Communications*, 17(1): 26–44.
Sommerer, C. and Mignonneau, L. (1997) ' "A-Volve" an Evolutionary Artificial Life Environment', in C. Langton and K. Shimohara (eds) *Artificial Life V. Proceedings of the Fifth International Workshop on the Synthesis and Simulation of Living Systems*, Cambridge, MA: MIT Press.
Springer, C. (1996) *Electronic Eros. Bodies and Desire in the Postindustrial Age*, London: Athlone.
Squires, J. (1996) 'Fabulous Feminist Futures and the Lure of Cyberculture', in J. Dovey (ed.) *Fractal Dreams. New Media in Social Context*, London: Lawrence and Wishart.
Stabile, C. (1998) 'Fetal Photography and the Politics of Disappearance', in P.A. Treichler, L. Cartwright and C. Penley (eds) *The Visible Woman. Imaging Technologies, Gender, and Science*, New York and London: New York University Press.
Steane, J.B. (ed.) (1982) *Christopher Marlowe. The Complete Plays*, London: Penguin Books.
Stengers, I. (1997) *Power and Invention. Situating Science*, Minneapolis, MN and London: University of Minnesota Press.
Stengers, I. (2000) *The Invention of Modern Science*, Minneapolis, MN and London: University of Minnesota Press.

Stevenson, R.L. (1979) *Dr Jekyll and Mr Hyde and Other Stories*, London: Penguin Books.
Stone, A.R. (1996) *The War of Desire and Technology at the Close of the Mechanical Age*, Cambridge, MA and London, England: MIT Press.
Stork, D.G. (ed.) (2000) *HAL's Legacy. 2001's Computer As Dream And Reality*, Cambridge, MA and London, England: MIT Press.
Strathern, M. (1992) *After Nature: English Kinship in the Late Twentieth Century*, Cambridge: Cambridge University Press.
Symons, D. (1992) 'On the Use and Misuse of Darwinism in the Study of Human Behaviour', in J.H. Barkow, L. Cosmides and J. Tooby (eds) *The Adapted Mind. Evolutionary Psychology and the Generation of Culture*, New York and Oxford: Oxford University Press.
Tenhaaf, N. (1996) 'Machines for Evolving', *New Formations. Technoscience*, 29: 39–46.
Tenhaaf, N. (1997) 'Aesthetic Issues of ALife', *Fourth European Conference on Artificial Life, ECAL 97*, http://www.cogs.susx.ac.uk/ecal97/present.html.
Terranova, T. (2000) 'Free Labour. Producing Culture for the Digital Economy', *Social Text* 63, 18(2): 33–58.
Thomas, G. and Wyatt, S. (1999) 'Shaping Cyberspace – Interpreting and Transforming the Internet', *Research Policy*, 28: 681–698.
Thornhill, R. (2000) *Home Page*, University of New Mexico, http://biology.unm.edu/~pwatson/rthorn.htm.
Thornhill, R. and Palmer, C.T. (2000) *A Natural History of Rape. Biological Bases of Sexual Coercion*, Cambridge, MA: MIT Press.
Todd, S. and Latham, W. (1992) *Evolutionary Art and Computers*, London: Academic Press.
Tooby, J. and Cosmides, L. (1992) 'The Psychological Foundations of Culture', in J.H. Barkow, L. Cosmides and J. Tooby (eds) *The Adapted Mind. Evolutionary Psychology and the Generation of Culture*, New York and Oxford: Oxford University Press.
Toolis, K. (2000) 'Step by Step to the Human Code', *Guardian*, 6 May.
Tosa, N. and Nakatsu, R. (1997) 'The Esthetics of Artificial Life', in C. Langton and K. Shimohara (eds) *Artificial Life V. Proceedings of the Fifth International Workshop on the Synthesis and Simulation of Living Systems*, Cambridge, MA: MIT Press.
Treichler, P. and Cartwright, L. (eds) (1992) *Camera Obscura. Imaging Technologies, Inscribing Science*, 28 and *Camera Obscura. Imaging Technologies, Inscribing Science* 2, 29.
Treichler, P. Cartwright, L. and Penley, C. (eds) (1998) *The Visible Woman. Imaging Technologies, Gender, and Science*, New York and London: New York University Press.
Turkle, S. (1984) *The Second Self. Computers and the Human Spirit*, New York: Simon and Schuster.
Turkle, S. (1997) *Life on the Screen. Identity in the Age of the Internet*, London: Phoenix.
Turney, J. (1998) *Frankenstein's Footsteps. Science, Genetics and Popular Culture*, New Haven, CT and London: Yale University Press.
Varela, F. (1999) *Ethical Know-How. Action, Wisdom, and Cognition*, Stanford, CA: Stanford University Press.

Varela, F.J. (2001) *Autopoiesis and a Biology of Intentionality*, www.eeng.dcu.ie/~alife/bmcm9401/varela/pdf.

Varela, F. and Bourgine, P. (1992) *Toward a Practice of Autonomous Systems. Proceedings of the First European Conference on Artificial Life*, Cambridge, MA: MIT Press.

Velmans, M. (2000) *Understanding Consciousness*, London and Philadelphia, PA: Routledge.

VNS Matrix (1994) 'All New Gen', in M. Fuller (ed.) *Unnatural. Techno-theory for a Contaminated Culture*, London: Underground.

Wain, J. (1983) 'Introduction', in J. Hogg, *The Private Memoirs and Confessions of a Justified Sinner*, London: Penguin Books.

Watson, J.D. (1999 [1968]) *The Double Helix*, London: Penguin Books.

Wild, L. (2000) 'Would I be Eliminated?', *Guardian*, 29 June.

Wilde, O. (1980) *Plays, Prose Writings and Poems*, London and Toronto: Dent.

Williams, R. (1983) *Keywords*, London: Fontana Press.

Wilson, E.A. (1998) *Neural Geographies. Feminism and the Microstructure of Cognition*, New York and London: Routledge.

Wilson, E.O. (1980 [1975]) *Sociobiology. The New Synthesis*, Cambridge, MA and London: The Belknap Press of Harvard University Press.

Wilson, E.O. (1998) *Consilience: The Unity of Knowledge*, New York: Vintage Books.

Winnicott, D.W. (1971) *Playing and Reality*, London: Tavistock.

Woolley, B. (1992) *Virtual Worlds. A Journey in Hype and Hyperreality*, Oxford: Blackwell.

Wright, B.A. (ed.) (1980) *John Milton. The Complete Poems*, London and Toronto: Dent.

Wright, R. (1996) 'The Dissent of Woman. What Feminists can Learn from Darwinism', in O. Curry and H. Cronin (eds) *Demos Quarterly. Matters of Life and Death: The World View from Evolutionary Psychology*, issue 10, London: Demos.

Index

actor-network theory 178–9, 180–1
Adam, Alison viii, 3, 179, 180–1
Adam, Barbara 161
adaptation: artificial intelligence 110; autonomous agents 66–8; cognition 202; supple 63; universalism 44–5
addiction model 84
Advanced Cell Technology 224–5n7
aesthetics of ALife 130
agency 125–7; atomised 194; autonomy 2, 6; autopoiesis 50; dependency 201; DNA 23; free will 39; genes 15, 38; human 29; Internet 127; Lewontin 22, 23; Maes 66–8; networks 11–12; Rose 22, 23; *see also* autonomous agents
Agent Research Programme 126
aggression 35
Agre, Philip E. 58, 195–6
AI 209
Aldiss, Brian 209
Aleksander, Igor 113–14
algorithm 219–20n5
Alien Resurrection 160, 169
alienation 141–2
ALife 2–4, 63, 181, 183–4; aesthetics 130; artificial intelligence 1, 194; autonomous agents 66–8; autonomy 7, 65–6; autopoiesis 52; Bedau 7–8, 63; bioethics 13; biology 57, 64; complex systems 8–9; connectionism 188; creation 7–8, 54, 55–6, 76; cultural evolutionism 80; cyberfeminism 176–7; CyberLife Technology Limited 105–9; dematerialisation 77; disembodiment 176; embodiment 65; emergence 65; ethics 8, 69; evolution 8; feminism 74, 112–13, 175–6, 183; functionalism 64–5; genetic algorithms 111; Grand 137–8; Helmreich 73–4, 75, 78; Langton 3, 75, 208; matter 112; memetics 50; narratives 77, 78; natural selection 218n3; nature/culture 58; postmodernism 72; queer theory 74; real 222n23; reductionism 135; religious mythology 56; Risan 59, 78; self-organisation 65–6; *Tierra* 91; vitalism 71; witnessing 58; women 73–4; *see also* Artificial Life; artificial life
alife 34, 176–7, 181–2, 184, 206
ALIVE 67
alter-egos 151–2; *see also* doppelgänger
alternative/mirror worlds 84, 87
altruism 35–6
Anderson, Laurie 209–10
ANDi 171–2
Anees, Munawar A. 163, 167
animal protection movement 172
anthropomorphism 59, 121, 175
anti-biologism 176, 191
anti-Cartesian philosophy 58, 195–6; *see also* Descartes, René
anti-humanism 134, 148
anti-instrumentalism 197
antinomianism 150–2
Antinori, Severino 163, 164
AntiNorn 102, 104, 105

244 Index

antiracism 49
apocalypticism 178
applets 95
Aristotle 220n13
Artificial Culture (Gessler) 133, 134–5, 138
artificial intelligence: adaptation 110; aggregate behaviour 106; ALife 1, 194; anthropomorphism 59; artificial life-forms 109–10; cognition 196; cognitive psychology 207–9; consciousness 108–9; CyberLife Research Limited 109–11; emergence 110; epistemology 8–9; functionalism 64; gender 180–1; Grand 108–9, 114; knowledge 9–10; liberal humanism 204–5; schizophrenia 194; Turing Test 109
Artificial Life ix, 58–9, 62, 127
artificial life: autopoiesis 50; chaos theory 128; computer games 50; computer science/biology viii, 72; consciousness 53, 58–9; cyberfeminism vii, viii–ix, 82, 211; Darwin 60–1; discourse ix, 52; evolutionary psychology 50–1, 216; genomics 173–4; God 60–1, 75; Grand 111; Helmreich 207; Kelly 138; Langton 219n4; mirror worlds 91; risks 80–1; sociobiology 3–4, 181; *see also* ALife; alife
Artificial Life 67, 78
artificiality 1, 12, 143–4
Ashman, K. M. 214
Ashworth, John 47, 219n8
astrobiology 63
atomic bomb 27
atomisation 193–4
Australian feminism 183
automata 69–70, 71, 122, 130; *see also* cellular automata
automation 195
autonomous agents 81; adaptation 66–8; aesthetic 130; ALife 66–8; alter-egos 151–2; anthropomorphism 175; artificial life 50; atomisation 193–4; bots 4; Brooks 68–9; commercial applications 8; computer games 83; consciousness 109; Dawkins 67; decentralised 140–1; evolution 8; Helmreich 109; identity 6, 68; intelligence 136; Internet 3; Kelly 67, 140; Levy 69–70; liberal humanism 182; Neo-Biological Age 8; norns 91–5; posthuman subjectivity 144; reproduction 68, 81; robots 67; situated 94; types/kinds 171
autonomy: agency 2, 6; ALife 7, 65–6; autopoiesis 11, 23, 198–204; bots 125; creation 75; determinism 66; genes 15; global networks 12; individual 12, 66; Internet 137; posthumanism 175; robots 118–19, 199; self-organisation 1, 12; *see also* autonomous agents
autopoiesis: agency 50; ALife 52; artificial life 50; autonomy 11, 23, 198–204; bio-aesthetic 129–30; Boden 202–3; embodiment 130; evolutionary biology 202; genetic determinism 129; Grand 199–200; Hayles 203; Helmreich 203; identity 201; Luhmann 226n8; Maturana and Varela 200–1, 202, 219n6, 226n9; robots 202; Rose 50
avatars 205–6
Axtell, Robert 98–9, 138–40
Azarmi, Nader 127

Badmington, N. 207
Bakhtin, Mikhail 6, 176, 198, 225n1, 226n6, 226n10
Barad, Karen 185, 215–16
Barbrook, Richard 101, 141
Baringer, P. S. 214
Barkow, J. H. 41–2, 44, 45–6
Barthes, Roland 9
Baudrillard, Jean 59
Beckett, Samuel 25, 26, 182
Bedau, Mark: ALife 7–8, 63; *Artificial Life* 78; bioethics 80–1; life/mind 79–80; life/not-life 57, 79
behaviour: aggregate 106; culture 45–6; evolutionary biology 190–1; innate 192; life 3; norns 93–4
behaviourism 66, 220n9
Bentley, Peter 129

bilingualism, academic 214–15
binary oppositions 25, 198
bio-aesthetics 129–30
biochemistry 94–5
bioculture 5–6, 7, 138
bioethics vii–viii, 8, 13, 80–1, 167–8
Biogenesis 128
bio-logic 141
Biological Age 5, 135
biologism 184
biology ix, 14; active/passive dichotomy 22–3; ALife 57, 64; computer science viii, 1, 6–7, 72; culture 48–9; decoding 26; determinism 39, 141, 149, 155, 192, 204; essentialism 175–6, 178, 184; Haraway 186–7; hardware/software 145–6; as ideology 20, 21, 40; information theory 16–17; inter-planetary 60; Lewontin 20–1; life-forms 105; life/not-life 57, 79; machine analogy 21; Rose 21; subjectivity 22; technology 184
Biomorph (Dawkins) 54–6, 60, 66, 128, 129
bioports 145–6
biosocial model of disability 46
biosphere 80
Biota.org 132
biotechnology 62, 178, 188–90
biotechnology companies 29–30, 169
birds' flocking behaviour 62–3
birth: computer science 75; feminism 222n15; norns 96; *see also* creation; reproduction
Blackman, L. 83, 204
Blackmore, Susan 38–9
Blackwell, Antoinette Brown 49
Blade Runner 160
Blair, Tony 154
blue-sky project 109, 121
Boden, Margaret 50, 62, 64–6, 200–3, 220n8
body: brain 22; experience 65; genes 22; Haraway 185; machine analogy 21, 145; mind 73, 113, 183, 195–6; posthumanism 143; virtual reality 183
Bohr, Niels 215
Bollas, Christopher 23

Bollen, Johan 136
Bomb 130
Bonabeau, E. W. 65
bots 4, 121, 124–7
Bourgine, P. 78
Boutin, P. 123
Boyce, N. 154
Boyle, Robert 58
The Boys from Brazil 160
Bradshaw, Peter 209
Braidotti, Rosi 146, 177, 178, 181
brain 22, 222n25; *see also* global brain
Breazeal, Cynthia 121
Bremer, Michael 88–9, 89
Brighton Media Centre, *LikeLIFE* exhibition 127
Brooks, Michael 5, 136, 137
Brooks, Rodney A. 68–9, 119, 120, 201
Brown, Andrew 40–1
Brown, Katherine 155
Brown, Paul 128
Browne, Kingsley 43, 47
BT: Information Agents project 127; Intelligent Personal Assistant 126; ZEUS 126, 127
BT exaCT 126
BTG 171
Bulger, James 220–1n2
Bush, George W. 208
Butler, Judith 76, 184–5, 186

Calvin, Jean 150
Calvinism 150–2, 168
capitalism: individualism 20; labour 101, 142; social good 153
Card, Orson Scott: *Ender's Game* 53; *Xenocide* 2, 53
Castells, Manuel: alienation 141–2; biological determinism 141; Information Age 5, 135; network society 6, 140–1; spirituality 6, 142
CAVE 205–6
Celera Genomics 152, 153
cells 61, 202
cellular automata 71, 72, 81, 106, 111, 134
Centre for Computational Neuroscience and Robotics 123, 128

chaos/order 118
chaos theory 128, 189, 224n3, 225n2
chatterbots 117, 125
children/computer games 66, 221n7
Christianity 142–3, 147, 174; *see also* Calvinism
Chung, Caleb 122
Cilliers, Paul 8–9, 148
city life/nature 90
The City of Lost Children 160
Clarke, Arthur C. 1, 116
Cliff, Dave 91, 93, 94, 98–9, 102
Clinton, Bill 153–4
Clonaid® 163
Clonapet 163
cloning 62; commercial applications 163; doppelgängers 152; ethics 160; God 167–8; humans 166; identity 149, 164, 168; Lewontin 166, 168; patenting 171; pigs 171; self-replication 160; soul 160; spirituality 160; therapeutic 165, 224–5n7; transgenesis 146, 152, 165, 169; wetware 168–9; *see also* Dolly, the sheep
COBs (creature objects) 104
cockroach robot 119
coding/decoding 16
co-evolution 83, 101
Cog 120–1, 130, 181, 223n4
Coghlan, Andy 154, 224n4
cognition 196, 202
cognitive psychology 207–9
cold war vii, 12, 16, 208, 212
Collini, S. 211–12, 214
communication 16, 84, 98
complex systems 8–9, 71, 88, 123
complexity: diffraction 177; emergence 190; Lewontin 148–9; networks 148; reductionism 189–90; robots 120–1
computational anthropology 4, 133
computer aided design 129
computer games 221n12; artificial life 50; autonomous agents 83; control 85; creation of life 89; mutations 102; psychology 83–4, 220n1; users 84, 93, 99–105, 221n7; violence 91; *see also individual games*

computer science: artificial life viii, 72; biology viii, 1, 6–7, 72; birth 75; CPU 60; Darwinism 47; reality models 71–2; simulation/synthesis 53, 59, 63–4
conference proceedings 78, 79
Conflict Zone 223n7
Confucianism 11
connectionism: ALife 188; alife 181; embodiment 111–12; metaphor 182; naturalism 184; neural networks 65–6
consciousness 113–14; artificial intelligence 108–9; artificial life 53, 58–9; autonomous agents 109; brain 222n25; norns 109; posthumanism 143
constructivism 57–8, 59, 185, 201, 213–14
consumer involvement 20, 84, 156
Contact Consortium 132
Conway, John: The Game of Life 70–1, 81, 93
corporeality 183; *see also* embodiment
Cosmides, L. 41–2, 45–6
Council of Europe 167
La Cour des Miracles 130
CPU (central processing unit) 60
creation: ALife 7–8, 54, 55–6, 76; Artificial Life 62; autonomy 75; computer games 89; Demiurge 143; DNA 17; evolution 81; God 55–6; Grand 5; narratives 75
creative engineering 59
creativity/emergence 106
Creature Labs 104, 105
Creatures (Grand) 127; Albia 95–8; applets 95; commercial success 104; CyberLife Technology Limited 105; Dawkins 1, 217n3; evolution 97; game playing 96–8; genetic engineering 98, 99; Grendels 95, 102; newsgroups 99, 102–3; norns 91–5, 96–7, 98, 102–4, 109, 199; open sourcing 100, 101; and *Tierra* compared 99; user involvement 84, 93, 99–105; webrings 104; websites 99, 104
Creatures 2 (Grand) 102, 104

Creatures 2 (Karakotsios) 88–9
Creatures 3 (Grand) 104
Creatures Adventures 104
Creatures Development Network 104
Crick, Francis 15, 25
critical psychology 203–4
Cronenberg, David 145
Cronin, Helena 47, 48
cultural evolutionism 4–5, 80, 135
cultural studies 23, 51–2, 214
culturalism/structuralism 23, 134
culture: behaviour 45–6; biology 48–9; evolution 4–5, 133; feminism 48; identity 135; memesis 4, 133; nature 7, 58, 170, 176, 182, 183, 187, 190–1, 216; science 100; sociobiology 37–8; Williams 223n11
Curry, Oliver 47, 48
cyberfeminism: ALife 176–7; artificial life vii, viii–ix, 82, 211; cultural studies 51–2; Plant 182; posthumanism ix, 7
Cyberbiology Conference 132
CyberLife ALife Institute 105
CyberLife Applied Research 105
CyberLife Research 2000 2
CyberLife Research Limited 83, 221n11; artificial intelligence 109–11; blue-sky project 109; CyberLife Technology Limited 109–15; embodiment 110–11; situatedness 111
CyberLife Technology Limited 83, 221n11; ALife 105–9; Creature Labs 104, 105; *Creatures* 105; CyberLife Research Limited 109–15; Darwinism 106; DERA 107–8; neural networks 109; Origin 106; space exploration 108–9; virtual life 106–7
cybernetics 1, 16, 111, 150
cyberpunk 177, 178
cyberspace 53, 76, 177, 219n1
cyborgs vii, 217n1; Haraway vii–viii, 134, 170, 178–9, 185, 187, 208

Damer, Bruce 132–3
Darwin, Charles: artificial life 60–1; *The Descent of Man* 42; evolutionary theory 22, 54; gender differences 42–3; God 81, 86; men/women 42–3; natural selection 20, 34, 44, 54; survival of fittest 54
Darwinism: alife 34; autonomous individual 204; Brown 40–1; computer science 47; CyberLife Technology Limited 106; Dawkins 40–1; Dennett 40–1; feminism 49; Lewontin 20, 40; rape 47; Rose 17, 40, 41–2; sexism 49; Wilson 36
Darwinism Today 47
Darwin@LSE 47
Davenport, Charles 32
Davidmann, Manfred 146, 147
Davidson, C. 107, 108
Davis, Erik 122–3, 142–3
Dawkins, Richard 88; anti-humanism 148; autonomous agents 67; *Biomorph* 54–6, 60, 66, 128, 129; *The Blind Watchmaker* 55, 56; on *Creatures* 1, 217n3; Darwinism 40–1; determinism 39–40; DNA 17; emergence 56; Hayles on 18–19; memes 37–8, 50; memetics 5, 50; reductionism 21–2; *The Selfish Gene* 18, 37; sociobiology 37, 41; 'Sociobiology: The New Storm in a Teacup' 39; ultra-Darwinism 39
decoding 26
dematerialisation 71–2, 77
Demers, Louis-Phillipe 130
Demiurge 143
Demos Quarterly 47–8
Dennett, Daniel 4, 38, 40–1, 44, 67–8
dependency/agency 201
DERA (Defence Evaluation Research Agency) 107–8
Derrida, Jacques 207
Descartes, René 58, 73, 183
determinism: autonomy 66; biology 39, 141, 149, 155, 192, 204; Dawkins 39–40; evolutionary psychology 40; genetic 29, 39, 74, 148, 149–50, 152, 166; genomic imaginary 154; Lewontin 39; predisposition 192; reductionism 148; Rose 39; sociobiology 39–40; ultra-Darwinism 55

dialogue: Bakhtin 6, 198, 226n6, 226n10; communication 84; feminism 12–13, 176, 177; posthumanism 226n10
diffraction 170, 176, 177
Digital Space Corporation 132
disability 46, 156
discourse: artificial life ix, 52; matter 187–8, 220n14
disease 29
disembodiment: ALife 176; epistemology 77; gender 143; Haraway 28–9; queered 179; subject 143; transcendence 132, 146; universalism 143
Distributed Knowledge Systems 136
DNA 15, 16, 17; agency 23; Gilbert 27–9; identity 148, 165; Lewontin 171; personhood 147; polysemy 30–1; race 154–5; reproduction 19; RNA 26; sequencing 28–9; tests 224n4
Dolly, the sheep 146, 163, 165, 166, 171
doppelgänger 152, 155, 168–9
Doyle, Richard 23–5, 24, 31, 76–7
Draves, Scott 130
Dreyfus, Hubert 9, 11, 58
Driessens, Erwin 130
dualism of mind/body 73, 113, 183, 195–6
Dyson, George 137
dystopia 130, 136, 141, 178

Easlea, Brian 27
ECAL 128
Edwards, Paul 207–9
Eliza, chatterbot 125
Ellis, Cliff 89, 90
embodiment 65, 77, 110–12, 130, 180, 182
embryos 28–9, 157, 165, 224–5n7
emergence: ALife 65; alife 181; artificial intelligence 110; bots 125; complexity 190; consciousness 113; creativity 106; Dawkins 56; genetic algorithms 57; Hayles 3; Saunders 222n20; spirituality 81; vitalism 5, 71, 138, 218n9
Emmeche, Claus 71–2, 123, 181

emotion 10, 114
Enlightenment 25, 197
environment of evolutionary adaptation (EEA) 45
epistemology: artificial intelligence 8–9; disembodied 77; feminist 73, 74, 82, 212–13; ontology 216; queer 74, 220n12; scientific 45–6
Epstein, Joshua M. 98–9, 138–40
Erickson, Mark 46
essentialism 175–6, 178, 184
ethics 10–12, 215; ALife 8, 69; cloning 160; genetic engineering 159; genetics 161; Langton 69; life 146; sadism 102–3; Varela 12, 148
ethnicity 143
ethnographer 134
eugenics 29, 31–3, 156
European Conference of Artificial Life 124
Evans, Dylan 217n4
everything, theory of 51
evil/good 151, 163, 174
evolution: ALife 8; artificial 111, 128; autonomous agents 8; control 85; creation 81; *Creatures* 97; cultural 4–5, 133; Darwin 22, 54; feminism 35; Grand 111, 137; intelligence 87; Internet 8; Kelly 4, 68; Latour and Strum's Nine Questions 49–50; narrative 147; natural selection 33–4; norns 92; planetary 88
evolutionary biology 14–16, 44–5, 190–1, 192, 202
evolutionary psychology 41–2; artificial life 50–1, 216; Blackmore 38; determinism 40; Fausto-Sterling 49–50; genotype/phenotype 46; Gould 49; information processing 43–4; natural selection 42; as religion 51; sociobiology 218n1; Wright 48–9
evolutionary theory 22, 48, 54, 101, 137, 201–2
Evolved Octopod 124, 127
eXistenZ 145–6
experience 65, 67
Expressivator 196–7

Fall myth 160–1
Faust (Goethe) 81; *see also* Marlowe, Christopher
Fausto-Sterling, Anne 49–50
feeblemindedness 32
FemaleMan 170, 187
feminism: ALife 74, 112–13, 175–6, 183; alife 176–7, 182; anti-biologism 176; Australia 183; culture 48; Darwinism 49; dialogue 12–13, 176, 177; epistemology 73, 74, 82, 212–13; evolution 35; evolutionary biology 190–1; form/matter discourse 82; globalisation 169–70; Helmreich 182; liberal humanism 143; nature/culture 7; pregnancy/childbirth 222n15; science 75–6, 184; sexual selection 43; Wright 48; *see also* cyberfeminism
feminist epistemology 73, 74, 82, 212–13
fertility clinics 164
fetishism 158–9
figuration 170
flocking behaviour, birds 62–3
Flynn, A. 119
form/matter 81–2, 112–13, 176, 185, 202, 220n13
Forrest, S. 57
Foucault, Michel 12, 19, 24, 75
Frankenheimer, John 172–3
Frankenstein (Shelley) 69, 75, 81, 160, 168
Franklin, Rosalind 15, 26
Franklin, Sarah 7, 146, 147, 169–70, 224n1
Fraser, Mariam 155
free will 39, 152
Freedman, D. H. 120
Freud, Sigmund 122
functionalism 64–5
fundamentalist religion 141
Furby 122

Gaia theory 63, 86, 87, 106
Galton, Francis 32, 33
The Game of Life (Conway) 70–1, 81, 93
GasNets 123
GATT 146
Gattaca 157–8, 160

gender: artificial intelligence 180–1; biosocial model 46; computer game users 100; disembodiment 143; Haraway 169; ideology 42; role blurring 145; sex 184–5
gender differences 42–3, 191
gender selection 156
genes 14–15; agency 15, 38; autonomy 15; body 22; homosexuality 33; information 3, 25; molecular biology 21; norns 95; patenting 153; selfishness 18, 19; *SimLife* 89; soul 168
genetic algorithms: ALife 111; complex systems 123; emergence 57; Gaia 106; mutations 3–4; robotics 123–4
genetic engineering 62, 89, 98–9, 102, 159
genetic imaginary 224n1
genetic screening 155, 156, 157–8
genetically modified organisms 147
genetics: determinism 29, 39, 74, 129, 148–50, 152, 166; ethics 161; identity 155, 167; information 3, 25; Lewontin 156–7; parental choice 156; patriarchy 49; self-modification 155; social class 158
GeneWatch 172
Genghis 119, 127
genomic imaginary 154, 168
genomics 149–50; artificial life 173–4; imagination 224n1; posthumanism 174; risk discourse 161; self-regulating market 156; unity of life 154–5; *see also* human genome; Human Genome Project
genotype 46, 171
Gessler, Nicholas 4–5, 133, 134–5, 138
Gibson, William 53
gift economy 101, 141–2
Gilbert, Walter 27–9, 30
glider pattern 70–1, 76–7
global brain 5, 136–7, 208–9
global networks vii, 12
globalisation 5, 169–70, 208
gnosticism 143
God: artificial life 60–1, 75; cloning 167–8; creation 55–6; Darwin 81, 86; Milton 151; Nine Laws of 138; Satan 151, 153

god-trick 75, 159
gods 5, 75, 134–5
Goethe, J. W. von 81, 161
good/evil 151, 163, 174
Goodeve, Thyrza Nichols 188
gothic novel 151–2, 168–9
Gould, Stephen Jay: evolutionary biology 44–5, 50; evolutionary psychology 49; sociobiology 41
Gowaty, Patricia Adair 190–1
Graham-Rowe, D. 121, 123
Grand, Steve: ALife 137–8; artificial intelligence 108–9, 114; artificial life 111; autopoiesis 199–200; Biological Age 5, 135; creation 5; *Creation, Life and How to Make It* 118; digital naturalism 102; evolution 111, 137; HAL 117; Lucy 85, 114–15, 198–9; norns 93, 94, 103–4, 109, 199; robotics 6; science fiction 219n1; *SimLife* 91; 'Three Observations that Changed my Life' 112–13; Turing Test 2; Unmanned Air Vehicles 108; *see also Creatures*
Gray, Chris Hables vii, 217n1
Gray, Russell 191–3
Greek, C. R. 171
Greek, J. S. 171
Grendels 95, 102
Griffiths, Mark 84
Gross, Paul 211–12, 213
Guardian 124, 152, 209
Gulf War 53

Habermas, Jürgen 10
hacking 100, 105, 177
Haig, David 67, 68
HAL 9000 1–2, 10, 116–18
Haldane, J. B. S. 32
Hall, S. 134
Haraway, Donna: actor-network theory 178–9; animal protection movement 172; biology 186–7; body 185; cyborgs vii–viii, 134, 170, 178–9, 185, 187, 208; diffraction 7, 176; disembodiment 28–9; fetishism 158–9; figuration 170; gender 169; god-trick 75, 159; Gross and Levitt 213; *How Like a Leaf* 178; matter/discourse 187–8; Modest Witness 179–80; *Modest_Witness* 178, 179; nature-culture 182; panoptic vision 86; *Primate Visions* 52; primatology 187; refigurations 170, 179, 182; representations of science 51; sanctity of life 169; seeds 146; situated knowledge 9, 179, 187, 213; transgenesis 190
Harding, Sandra 180, 213
Harris, John 146, 149, 155, 156, 168, 172
Hart, Roger 213
Harvey, I. 123
Hasbro 122
Hassan, Ihab 207
Hayles, N. Katherine: autopoiesis 203; constructivism 185, 201; on Dawkins 18–19; dematerialisation of ALife 77; embodiment 77, 182; emergence 3; liberal humanism 6, 143; narrative 77, 78, 122; posthuman identity 6, 143; re-embodiment 143–4; rememory 77, 220n15; science/cultural institutions 213–14; simulated creations 121; subjectivity 22
Hegel, G. W. F. 73
Heidegger, Martin 58
Helmreich, Stefan: ALife 73–4, 75, 78; artificial life 207; autonomous agents 109; autopoiesis 203; cyberspace 219n1; embodiment 180; feminist intervention 182; science fiction 54; Sugarscape 140; terratorial metaphors 76
Henriques, J. 83
heredity 31–2, 33, 35, 192
heteroglossia 198
Heylighen, Francis 5, 136
Hogg, James: *Confessions of a Justified Sinner* 150–1, 168
Holland, John 57
Holquist, Michael 198, 225n1
homosexuality 33, 74, 165; *see also* queer theory
Hood, L. 27–8
HUGO (Human Genome Organisation) 30

human characteristics: agency 29; human nature 35, 42; identity 37; self-organisation 77; universalism 35–6, 42
Human Fertilisation and Embryology Authority 157, 164–5, 166
Human Genetics Advisory Commission 164–5, 166
human genome 16, 17, 152
Human Genome Diversity Project 29, 159
Human Genome Project 159; Kevles and Hood 27–8; Lewontin 29–31; Sulston 152
humanism 144, 158, 207
humanoids 120
Hurry, Mark 130
Husbands, Phil 123
Huxley, Aldous 159
hybrids 126, 146–7, 173, 225n9

ideality 64, 183
identity: autonomous agents 6, 68; autopoiesis 201; cloning 149, 164, 168; culture 135; DNA 148, 165; genetic 155, 167; human characteristics 37; meaning 141, 142; network vii, ix, 11–12; persistence 112–13, 222n24; posthumanism 6, 116, 122, 143; self 12; twins 165–6
imagination 114, 224n1
immediate coping mechanism 11
immortality 168
individualism 20, 35
individuals 12, 66, 204
information 6, 16–17; bio-logic 141; communication 16; genetics 3, 25; global networks vii, 12; life 3, 14; literalisation 16–17; materialism 142, 143; molecular biology 15–16; political economy analogy 119–20
Information Age 5, 6, 135, 143
Information Agents project 127
information processing 9, 43–4
information technology 140–1
informationalism 141, 202, 219n2
inheritance, laws of 31
insect-like robots 119, 127
Insuraclone® 163

Integrated Causal Model 45–6
intellectual property 155–6
intelligence 10, 87, 118, 136
Intelligent Personal Assistant 126
interdependence 101
interdisciplinarity 195, 225n4
International Associated Research Centre for Human Reproduction 164
Internet: agency 127; autonomous agency 3; autonomy 137; culture/identity 135; dystopia 136; evolution 8; global brain 5, 136; globalisation 5; self 141; self-organisation 137; technoscience 178; utopianism 101; virtual reality 101
intuitiveness 11, 43
irony 59
IVF treatment 157

Jacob, François 33–4
Jacobus, Mary 27
Jakobi, Nick 124, 127
jellyfish gene 172
Jeremijenko, Natalie 222n23
Johnson, Norman 136–7
Jones, Steve 26, 31
Julia, chatterbot 125
Jurassic Park 160
justice theory 10–11

Kac, Eduardo 173, 225n9
Kant, Immanuel 10
Karakotsios, Ken 88
Kay, Lily 16–17, 26, 31, 218n2
Keepers, G. A. 83
Keller, Evelyn Fox: atomic bomb 27; constructivism 185; language 23; machine/organism 19–20; molecular biology 15–16, 25–6, 186
Kelly, Kevin: artificial life 138; autonomous agents 67, 140; evolution 4, 68; natural selection 138; Neo-Biological Age 5, 135, 138; 'Network Culture' 5, 135–6; robots 118
Kember, Sarah 71, 75, 84, 169, 179, 225n2
Kerin, Jacinta 183, 185, 186, 187
Kevles, D. J. 27–8, 30, 32–3

kinship 36, 178
Kirby, Vicky 183, 185, 187
Kismet 121, 130
Klein, Melanie 168
knobots 67
knowledge: artificial intelligence 9–10; enaction 11; Foucault 24; language 17–18; power 157; situated 9, 179, 187, 213
Kolata, Gina 160
Kubrick, Stanley 209, 219n1

labour 43, 101, 142
Lacan, Jacques 215
Langton, Christopher 88; ALife 3, 75, 208; artificial life 219n4; 'Artificial Life' 55, 61–3; conference papers 78; ethics 69; form/matter 112; hang-gliding crash 76; life 62, 65; self-organisation 66
language 17–18, 23; *see also* metaphor
Latham, William 127, 128, 129
Latour, Bruno 49–50, 59, 180, 185–6, 189
Leavis, F. R. 211
Lenat, Douglas 9–10
Lenin, Vladimir Ilyich 149
Leonard, Andrew 4, 117, 124, 127, 142
Levin, Ira 160
Levitt, Norman 211–12, 213
Levy, Steven 69–70
Lewontin, Richard: agency 22, 23; biology 20–1; cloning 166, 168; complexity 148–9; Darwinism 20, 40; determinism 39; DNA 171; genetics 156–7; Human Genome Project 29–31; reductionism 20; sociobiology 34–5, 41
liberal humanism 6, 143, 182, 204–5
life 8; behaviour 3; *Biomorph* 66; children's understanding 66; co-evolution 83; dematerialisation 71–2; ethics 146; information 3, 14; Langton 62, 65; meaning 14; mind 79–80; networks 25; not-life 56–7, 64, 79; sanctity of 8, 80, 169; secret of 25–6; self-organisation 220n6; self-replication 111; *Tierra* 61, 66; unity of 154–5; virtual 60–1, 106–7; *see also* life forms
Life 2.0 International Competition 130
life forms 88, 103, 105
LikeLIFE exhibition 127
literary texts 77
Los Alamos National Laboratory 136
Lovelock, James 86
Lucy robot 85, 114–15, 198–9
Luhmann, Niklas 226n8
Lury, Celia 169–70, 201

McClintock, Barbara 186
Macgregor Wise, J. 129
machine analogy 19–20, 21, 145
machine/witness 58
Maes, Pattie 3–4, 66–8, 125, 197
mailbots 124
Malthus, Thomas 20
market forces 140–1, 156, 208
Marlowe, Christopher: *Dr Faustus* 161–2
MASA 124, 126, 223n7
Matel 122
materialism 112–13, 142, 143; *see also* matter
matter: ALife 112; discourse 187–8, 220n14; form 81–2, 112–13, 176, 185, 202, 220n13; ideality 183
Maturana, Humberto R.: autopoiesis 23, 129–30, 199, 200–1, 202, 219n6, 226n9; evolutionary theory 201–2
Mauldin, Michael 125
Maxis Inc. 86–8
May, Rollo 220n9
Mayer, Sue 172
meaning: complex systems 9; identity 141, 142; life 14
media 83–4, 220n1
Meek, James 154, 164, 224n3
memes 37–8, 50
memesis 4, 133
memetics 5, 38–9, 50, 80
memory 112–13
Mendel, Gregor 31
Mengele, Josef 32
metaphenomenon 115
metaphor 19, 27, 50, 76–7, 182

Microsoft Agent 2.0 223n8
Midgley, Mary 115
Milton, John 151
mind: body 73, 113, 183, 195–6; life 79–80
Ministry of Defence 107
Miracle Moves Baby 122
mirror worlds 84, 87, 91, 109
Mitchell, M. 57
Mixotricha paradoxa 188
mobots 122
Modest Witness 170, 179–80, 187
modulator mechanisms 112
molecular biology 23, 24–5; eugenics 32–3; genes 21; information 15–16; Keller 15–16, 25–6, 186; Monod 15; Rose 17
Monod, Jacques 15
Motorola 107
multinational corporations 146–7
MUSE (multi-user simulation environment) 125
mutations 3–4, 68, 92, 102
My Dream Baby 122
My Real Baby 121–2
mythology 81, 147, 161, 174

narcissism 109
Narcissus myth 160–1, 163
narratives: ALife 77, 78; creation 75; evolution 147; Hayles 77, 78, 122
NASA 119
National Bioethics Advisory Committee 149, 166
National Museum of Photography, Film and Television 131
natural selection: ALife 218n3; altruism 36; Darwin 20, 34, 44, 54; evolution 33–4; evolutionary psychology 42; heritable differences 192; human nature 35; Kelly 138; kinship 36; *Tierra* 85–6
naturalism: connectionism 184; constructivism 57–8, 59; digital 81, 98, 99, 102
nature: city life 90; culture 7, 58, 170, 176, 182, 183, 187, 190–1, 216; first/second 73; gothic novel 152; nurture 152, 192–3; reductionist approach 20; women 25
Nature 154, 159, 171
NCR 107
Nelkin, Dorothy 51, 157
Nelson, G. L. 129
Neo-Biological Age 5, 8, 135, 138
neo-Darwinism 139
Nerve Garden 132
Net: *see* Internet
Netscape 101
Network Culture, Kelly's 5, 135–6
network society 6, 140–1
networks ix, 11–12, 25, 148; *see also* neural networks
Neumann, John von 69–70, 71, 81
neural networks: biochemistry 94–5; complex systems 8, 9; connectionism 65–6; consciousness 114; CyberLife Technology Limited 109; ethics 11–12; object recognition 10; self-replication 116
New Scientist 5, 123, 136, 154, 167, 224n4
Niccol, Andrew 157–8
norns: autonomous agents 91–5; behaviour 93–4; birth 96; communication 98; consciousness 109; evolution 92; genes 95; Grand 93, 94, 103–4, 109, 199; mutations 92; nutrition 94–5; reproduction 96–7; torture 102–4
nouvelle AI: *see* ALife
nurture/nature 152, 192–3

objectivity/subjectivity 21, 58
OncoMouse 170, 179, 187
ontology 185, 216
open sourcing 100, 101
Oppenheimer, (Julius) Robert 27
oppositional science studies 180
optimisation, robotics 124
order/chaos 118
Origin 106
ownership of software 100

Palmer, Craig 47, 51
panoptic vision 75, 86, 155

parental choice 156
patenting 146–7, 153, 171
patriarchy 49, 75–6
Pattee, H. H. 63–4
penetration 145
Pengi 196
Penny, Simon 205–6
Penrose, Lionel 32
Penrose, Roger 113
performativity of language 23
persistence of identity 112–13, 222n24
personhood 147
personoids 5, 134
pharming 146, 170–1, 172
phenomenology 10–11, 201–2
phenotype 46, 156
philosophy 57, 58, 114
photography 9
Picard, Rosalind 10
Plant, Sadie 176, 177–8, 182
Plato 32, 64, 176
pleiotropy 149
pluralism 149
politics 48, 186
polysemy 30–1, 198
Polyworld (Yaeger) 99
Porter, Henry 152–3
posthumanism: alife 206; autonomous agents 144; autonomy 175; body 143; consciousness 143; cyberfeminism ix, 7; cyborgs vii; dialogue 226n10; ethics 12; genomics 174; Hayles 6, 143; humanism 207; identity 6, 116, 122, 143; Information Age 143; subjectivity 12, 143, 144
post-menopausal conception 164
postmodernism 72, 206, 212–13, 215
post-structuralism 207, 215
power/knowledge 157
PPL Therapeutics PLC 165, 171
predestination 150–2
predetermination, genetic 158
predisposition 192
pregnancy/childbirth: atomic bomb metaphor 27; feminism 222n15
preimplantation genetic diaagnosis 157
Prigogine, Ilya 189
Primate Research Centre 171–2

primatology research 49, 52, 184, 187
Principia Cybernetica Web 136
procreation 165; *see also* reproduction
Prometheus myth 160–1, 163
Prophet, Jane 4, 130, 131
protection: of animals 172; of humanity 167
Psycho/Cyber 127–8
psychoanalytic approach 23, 223n6
psychology 83–4, 220n1
purgatory 148

Quastler, Henry 16–17
queer theory 74, 179, 182, 220n12

Rabinow, Paul 147–8, 158, 161
race 32, 143, 154–5
racism 29, 49
Raelian movement 163–4
RAM 60, 85
rape 47, 49, 51
Rawls, John 10–11
Ray, Thomas: algorithm 219–20n5; 'An Approach to the Synthesis of Life' 60; digital naturalism 98; irony 59; *NetTierra* 99; *see also Tierra*
realism 71–2, 73; *see also* virtual reality
reductionism: ALife 135; complexity 189–90; computational 65; Dawkins 21–2; determinism 148; informationalism 219n2; Lewontin 20; methodological 221n8; nature 20; Rose 21, 22
re-embodiment 143–4
refigurations 170, 179, 182
Reid, R. 214
relativism 42
religion 51, 56, 141; *see also* Christianity; Confucianism
rememory 77, 220n15
replicators 19
reproduction: asexual 55; autonomous agents 68, 81; DNA 19; homosexual couples 165; norns 96–7; post-menopausal 164; self-reproduction 69–70; sex 74; *TechnoSphere* 131; women 25
Reynolds, Craig 62–3, 93

rhizomatic approach 24
Ridley, Matt 149–50, 155–6, 161
Rifkin, Jeremy 161, 168
rights vii, 103, 167
Riot Girls 177
Risan, Lars 57, 58–9, 78
risk: artificial life 80–1; Butler 186; genomics 161; Latour 186, 189; Stengers 188–90
RNA 15, 16, 26
Robosapiens 123
robotics 6, 123–4, 180–1
robots: anthropomorphism 121; autonomous agents 67; autonomy 118–19, 199; autopoiesis 202; complexity 120–1; domestic 123; dystopia 130; humanoids 120; situated 118; toy industry 121–2
Rogers, Carl 125
Rose, H. 41–2, 51
Rose, N. 83, 155
Rose, Steven: agency 22, 23; autopoiesis 50; biology 21; consciousness 113; Darwinism 17, 40, 41–2; determinism 39; evolutionary biology 15; methodological reductionism 221n8; molecular biology 17; reductionism 21, 22; religion 51
Roslin Institute 165
Ross, Andrew 100, 177, 193, 212

sadism 102–3
same sex preferences: *see* homosexuality
sanctity of life 8, 80, 169
Sanger Centre 152, 153
Santa Fe Institute 76, 77, 138, 140
Satan 151, 153
Saunders, Robert 106, 109, 222n20
Schiebinger, Londa 184
schizophrenia 194
Schueller, G. 173
Sci-Art competition, Wellcome Institute 193
science: cold war 12, 16; culture 100; epistemology 45–6; feminism 75–6, 184; institutions 213–14; literary texts 77; representation 51; science fiction 53–4

Science 31, 154, 171
science fiction 2, 53–4, 177, 219n1
science wars viii, 6–7, 195, 211–12
scientism 34
Seed, Richard 163
seeds, hybrids 146–7
Segal, Lynne viii, 47, 50, 82
segmentation 56
self 12, 141, 155, 163; *see also* identity
self-modification 155
self-organisation: adaptive agent systems 127; ALife 65–6; autonomy 1, 12; cells 202; human beings 77; informational paradigm 141; intelligence 118; Internet 137; Langton 66; life 220n6
self-replication 111, 116, 160
self-reproduction 69–70
Selley, Gordon 4, 130
Sengers, Phoebe 144, 193–5, 196–7, 204–5
servometer 114
Severe Combined Immune Deficiency 173
sex 42, 74, 184–5
sexism 26, 49, 156
sexual morality 43
sexual selection 42, 43
Shakespeare, Tom 46
Shannon, Claude 16, 195
Shelley, Mary 160, 168
Shimohara, K. 78
SIGGRAPH conference 132
SimCity (Bremer and Ellis) 84, 89–91
SimEarth 84, 86–8
SimLife 84, 88–9, 91, 98
Simpson, T. 93, 94, 95, 96
Sims, Karl 121, 127, 128
simulacra 72
simulation 121; biological 58–9; mirror worlds 84; synthesis 53, 59, 60, 63–4
situatedness 111; knowledge 9, 179, 187, 213; robots 118
Skinner, B. F. 66
Smith, Jeffrey 205–6
Snow, C. P. 211–12, 214
social class 42, 158, 220–1n2
social determinism 149
social good 153

social ills 29, 32, 33
social sciences 138–40
society, artificial 138–43
sociobiology: artificial life 3–4, 181; critiqued 176; culture 37–8; Dawkins 37, 41; determinism 39–40; Epstein and Axtell 139; evolutionary psychology 218n1; Gould 41; Lewontin 34–5, 41; prediction 37; Sugarscape 139–40; Wilson 36–7, 41, 48, 219n8
software ownership 100; *see also* intellectual property
Sokal, Alan 212–13, 215
soul 160, 166, 168
Soviet Union 208
space exploration 108–9
species miscegenation 170
species-self 150
Spencer, Herbert 48
Spielberg, Steven 209
spirituality: Castells 6, 142; cloning 160; emergence 81; Rabinow 147–8, 161
Spivak, Gayatri 181
splitting 168
Springer, C. 146, 177
Stacey, J. 169–70
Standard Social Science Model 45–6
standpoint politics 186
Steane, J. B. 161, 162
Stelarc 127–8
Stengers, Isabelle 176, 185, 188–9
Stent, Gunther 159–60
Stevenson, Robert Louis: *Dr Jekyll and Mr Hyde* 168
Stone, Allucquere Roseanne 183
strange attractor 225n2
structuralism 23, 134
subjectivity: artificial 143–4; biology 22; disembodiment 143; Foucault 12, 19; Hayles 22; liberal humanism 143; nomadic 178; objectivity 21, 58; posthumanism 12, 143, 144; re-embodiment 143–4; transformation 12
Sugarscape 138–43
Sulston, Sir John 152, 153, 154
survival of fittest 54
Sussex University 59, 123, 128

Symbiotic Intelligence Project 136
synthesis/simulation 53, 59, 60, 63–4

taboo 170
techgnosis 143
technology 55, 177–8, 184
technoscience 178
TechnoSphere 130–2
Tenhaaf, Nell 127, 129, 202
Terminator, The 209
Terminator 2 209
Terranova, T. 101, 142
terrorism 12, 208
Theraulaz, G. 65
Thornhill, Randy 47, 51
Tickle 130
Tierra (Ray) 84; algorithm creatures 91; ALife 91; and *Creatures* compared 99; Hayles on 77; Helmreich on 76; life 61, 66; natural selection 85–6
Time 70
Todd, S. 128
Tooby, J. 41–2, 45–6
Toolis, Kevin 153
torture 102–4
Tortured Norns website 104–5
town planning 90
toy industry 121–2, 221n12
toys/pets 223n6
Traces: Embodied Immersive Interaction with Semi-Autonomous Avatars 205–6
transcendence 132, 146, 177
transCendenZ 145
transgenesis 62; and cloning 146, 152, 165, 169; Haraway 190; Harris 172; hybrids 173
transgenic organisms 147
Traweek, S. 214
tribe formation 140
Tufts University symposium 67
Turing, Alan 2
Turing Test 217n4, 222–3n1; artificial intelligence 109; bots 125; Grand 2; HAL 1, 117
Turkle, Sherry 67, 68, 128, 181–2
twins' identity 165–6
2001. A Space Odyssey 1–2, 209, 219n1

ultra-Darwinism 39, 55, 88–9
understanding 113–14
UNESCO 167
United Systems Military 169
universalism: adaptation 44–5; disembodiment 143; human characteristics 35–6, 42; relativism 42
Unmanned Air Vehicles 107–8
utopianism 101

Van Loon, Joost 161
Varela, Francisco J.: autopoiesis 23, 129–30, 199, 200–1, 202, 219n6, 226n9; conference papers 78; ethics 12, 148; evolutionary theory 201–2; phenomenology 10–11
variability 45–6
Vaucanson, Jacques de 122
vehicles 19
Velmans, Max 113
Venter, Craig 152, 153–4
Verstappen, Maria 130
Videodrome 145
violence: computer games 91; media 83, 220–1n2
virtual life 60–1, 106–7
virtual reality 53, 58–9, 101, 183
vitalism 5, 69, 71, 138, 218n9
VNS Matrix 177
Vorn, Bill 130
vulnerability 83–4

Wain, J. 150–1
Walkerdine, V. 83, 204

Watson, James 15, 17, 25, 26, 30
Wellcome Institute 193
Wells, H. G.: *The Island of Dr Moreau* 160, 172–3
wetware 146, 168–9
White, Norm 127, 128
Wiener, Norbert 16, 144, 218n2
Wild, Leah 157
Wilde, Oscar: *The Picture of Dorian Gray* 160, 162–3
Williams, Raymond 135, 223n11
Wilmut, Ian 163
Wilson, E. O.: *Consilience* 218n1; Darwinism 36; new synthesis 140; sociobiology 36–7, 41, 48, 219n8
Wilson, Elizabeth A. 183–5, 187, 188
Wired 123, 135
witnessing 58
women: ALife 73–4; madonna/whore dichotomy 49; nature 25; reproduction 25; *see also* cyberfeminism; feminism
Woolley, B. 53, 177
World Wide Web 124, 167–8
Wright, B. A. 151
Wright, Richard 48–9
Wright, Robert 45

xenotransplantation 62, 169, 171

Yaeger 99

ZEUS 126, 127